Materials for Photovoltaics

MATERIALS RESEARCH SOCIETY
SYMPOSIUM PROCEEDINGS VOLUME 836

Materials for Photovoltaics

Symposium held November 29–December 2, 2004, Boston, Massachusetts, U.S.A.

EDITORS:

Michael Durstock
Air Force Research Laboratory/WPAFB
Wright Patterson Air Force Base, Ohio, U.S.A.

Daniel Friedman
National Renewable Energy Laboratory
Golden, Colorado, U.S.A.

Russell Gaudiana
Konarka Technologies
Lowell, Massachusetts, U.S.A.

Angus Rockett
University of Illinois
Urbana, Illinois, U.S.A.

Materials Research Society
Warrendale, Pennsylvania

CAMBRIDGE
UNIVERSITY PRESS

University Printing House, Cambridge CB2 8BS, United Kingdom

One Liberty Plaza, 20th Floor, New York, NY 10006, USA

477 Williamstown Road, Port Melbourne, VIC 3207, Australia

314-321, 3rd Floor, Plot 3, Splendor Forum, Jasola District Centre, New Delhi - 110025, India

79 Anson Road, #06-04/06, Singapore 079906

Cambridge University Press is part of the University of Cambridge.

It furthers the University's mission by disseminating knowledge in the pursuit of education, learning and research at the highest international levels of excellence.

www.cambridge.org
Information on this title: www.cambridge.org/9781558997844

Materials Research Society
506 Keystone Drive, Warrendale, PA 15086
http://www.mrs.org

First published 2005
First paperback edition 2013

Single article reprints from this publication are available through University Microfilms Inc., 300 North Zeeb Road, Ann Arbor, MI 48106

CODEN: MRSPDH

A catalogue record for this publication is available from the British Library

ISBN 978-1-558-99784-4 Hardback
ISBN 978-1-107-40908-8 Paperback

CONTENTS

Materials Research Society Symposium Proceedings.......................................xi

DYE-SENSITIZED SOLAR CELLS

Progress in Producing Large Area Flexible Dye Sensitized Solar Cells.......................................3
Krishna C. Mandal, Michael Choi, Caleb Noblitt, and R. David Rauh

Amphiphilic Dye for Solid-State Dye-Sensitized Solar Cells.......................................11
Lukas Schmidt-Mende, Shaik M. Zakeeruddin, and Michael Grätzel

Dye Sensitized Solar Cells Incorporating Polyelectrolyte Multilayer Composites.......................................17
Geoffrey M. Lowman, Hiroaki Tokuhisa, Jodie L. Lutkenhaus, and Paula T. Hammond

Synthesis, Characterization and Thermal Stability of Highly Crystallized Titania Nanotubes.......................................23
B. Poudel, W.Z. Wang, C. Dames, J.Y. Huang, S. Kunwar, D.Z. Wang, D. Banerjee, G. Chen, and Z.F. Ren

Effect of Anodization Bath Chemistry on Photochemical Water Splitting Using Titania Nanotubes.......................................29
Gopal K. Mor, Oomman K. Varghese, Maggie Paulose, Karthik Shankar, and Craig A. Grimes

Performance of a Solid-State Photoelectrochromic Window.......................................35
Ursa Opara-Krasovec, Anneke Georg, Andreas Georg, and Marko Topic

NANOPARTICLE-HYBRID SOLAR CELLS

Time Evolution of Photoconductivity in TiO_2 Electrodes Fabricated by a Sol Gel Method.......................................43
Zhibin Xie, Victor M. Burlakov, Bernard M. Henry, Kiril R. Kirov, Christopher R.M. Grovenor, Hazel E. Assender, G. Andrew D. Briggs, Mitsuru Kano, and Yusuke Tsukahara

Fabrication of Nanorod Arrays for Organic Solar Cell Applications 49
Susan Huang, Harry Efstathiadis, Pradeep Haldar, Hee-Gyoun Lee, Brian Landi, and Ryne Raffaelle

Quantum Dot-Single Wall Carbon Nanotube Complexes for Polymeric Photovoltaics 55
Brian J. Landi, Stephanie L. Castro, Chris M. Evans, Herbert J. Ruf, Sheila G. Bailey, and Ryne P. Raffaelle

Nanoporous Si – Organic Composite Photovoltaics 61
I.A. Levitsky, W.B. Euler, N. Tokranova, B. Xu, and J. Castracane

POLYMER-BASED DEVICES

High Efficiency P_3HT/PCBM Solar Cell 69
Kanzan Inoue, Ross Ulbricht, Pallavi C. Madakasira, Miaoxin Zhou, Sergey B. Lee, John Ferraris, and Anvar A. Zakhidov

Plasticized Conjugated Polymers: A Possible Route to Higher Voltage Solar Cells 75
Paul Wentzel and Aurelien Du Pasquier

Polymer Solar Cells: Screen-Printing as a Novel Deposition Technique 81
Tom Aernouts, Peter Vanlaeke, Jef Poortmans, and Paul Heremans

SMALL MOLECULE-BASED DEVICES

Star-Shaped Heptamers of Discotic Dyes as New Materials for Photovoltaic Devices 89
S. Holger Eichhorn, Nicholas Fox, and Bryan Bornais

Controlling Organization in Photovoltaic Diodes From Discotic Liquid Crystals via Anode Surface Energy Alteration 93
Johanna P. Schmidtke, Klaus Müllen, and Richard H. Friend

POSTER SESSION

Temperature Effects on Photocurrent Generation in Polymer Hetero-Junction Photovoltaic Devices 101
Mi Yeon Song, Kang-Jin Kim, and Dong Yong Kim

New Application of Electrospun TiO_2 Nanofibers as an Electrode for Dye-Sensitized Solar Cell 107
Mi Yeon Song, Young Rack Ahn, Seong Mu Jo, and Dong Yong Kim

Photocurrent of an Individual ZnO Nanorods Synthesized by Sol-Gel Route on a Pulse Laser Deposited ZnO Film 113
Seung Eon Ahn, Gyu Tae Kim, Jong Soo Lee, Hyunsuk Kim, Sangsig Kim, Chang Hyun Bae, Seung Min Park, and Jeong Sook Ha

Synthesis and Characterization of $Cu_x(In,Ga)_ySe_z$ Nanoparticles by Colloidal Route 119
Ki-Hyun Kim, Young-Gab Chun, Byung-Ok Park, and Kyung-Hoon Yoon

Preparation of $CuInGaSe_2$ Absorber Layer by Nanoparticles-Based Spray Deposition 125
Ki-Hyun Kim, Young-Gab Chun, Byung-Ok Park, and Kyung-Hoon Yoon

Dye Sensitized Solar Cells Using Nanostructured Thin Films of Titanium Dioxide 131
Douglas A. Gish, Gregory K. Kiema, Martin O. Jensen, and Michael J. Brett

Novel Strategies for the Preparation of TiO_2 Nanofibers 137
Kenneth J. Balkus Jr., Chunrong Xiong, and Minedys Macias-Guzman

Bulk Heterojunction Photovoltaic Cells From Polymer Mixtures With Soluble Oxadiazole and Quinoline Polymers as Electron Acceptors 143
Solon Economopoulos, Christos L. Chochos, Giannis K. Govaris, Panagiotis Yiannoulis, Joannis K. Kallitsis, and Vasilis G. Gregoriou

Efficient Bulk Heterojunction Photovoltaic Cells Using Sublimable Transition Metal Complex as Photosensitizers 149
Hei Ling Wong, Kitty Ka Yan Man, Wai Kin Chan, Chung Yin Kwong, and Aleksandra B. Djurišić

Dendritic CuIn Films Grown by Electroless Deposition 155
David W. Lane, Jonathan D. Painter, Keith D. Rogers, Ian Forbes, Robert W. Miles, and Kathleen M. Hynes

Bath pH Dependence on the Structural and Optical Properties of Chemical Bath Deposited CdS Thin Films 161
Udaya S. Ketipearachchi, David W. Lane, Keith D. Rogers, Jonathan D. Painter, and Michael A. Cousins

Absorber Films of Antimony Chalcogenides via Chemical Deposition for Photovoltaic Application 167
M.T.S. Nair, Y. Rodríguez-Lazcano, Y. Peña, S. Messina, J. Campos, and P.K. Nair

$AgSbSe_2$ Thin Films for Photovoltaic Structures Produced Through Reaction of Chemically Deposited Selenium Thin Films With Ag and Sb_2S_3 173
K. Bindu, M.T.S. Nair, and P.K. Nair

Structural Studies of Chloride-Treated RF Sputtered $Cd_{1-x}Mn_xTe$ Films 179
S.L. Wang, S.H. Lee, A. Gupta, and A.D. Compaan

The Band Offsets of Isomeric Boron Carbide Overlayers 185
A.N. Caruso, P. Lunca-Popa, Y.B. Losovyj, A.S. Gunn, and J.I. Brand

Crystal Growth of Photovoltaic Polycrystalline $Si_{1-x}Ge_x$ by Die-Casting Growth 191
H. Hirahara, T. Iida, Y. Sugiyama, T. Baba, Y. Takanashi, and S. Sakuragi

Characterization of Polycrystalline Si Silicon Sheet Grown by Die Casting Combined With the Bridgman Technique 197
K. Saito, T. Iida, D. Akimoto, A. Nose, Y. Takanashi, S. Sakuragi, H. Nanba, G. Sakuragi, and T. Shimazaki

Current Transport Study of Schottky and P-N Junction Solar Cells Using Metal-Induced Growth Poly-Si Thin Films 203
Chunhai Ji, Joon-Dong Kim, and Wayne A. Anderson

III-V SEMICONDUCTORS

* **III-V Multi-Junction Materials and Solar Cells on Engineered SiGe/Si Substrates** 211
Steven A. Ringel, Carrie L. Andre, Matthew Lueck, David Isaacson, Arthur J. Pitera, Eugene A. Fitzgerald, and David M. Wilt

* **Metamorphic GaInP-GaInAs Layers for Photovoltaic Applications** 223
A.W. Bett, C. Baur, F. Dimroth, and J. Schöne

Mid-10^5 cm^{-2} Threading Dislocation Density in Optimized High-Ge Content Relaxed Graded SiGe on Si for III-V Solar on Si 235
David M. Isaacson, Carl L. Dohrman, Arthur J. Pitera, Saurabh Gupta, and Eugene A. Fitzgerald

1MeV Electron Irradiation Effects of GaAs/Si Solar Cells 241
N. Chandrasekaran, T. Soga, Y. Inuzuka, M. Imaizumi, H. Taguchi, and T. Jimbo

II-VI SEMICONDUCTORS AND TRANSPARENT CONDUCTING OXIDES

* **Scanning Tunneling Luminescence of Semiconductors** 249
M.J. Romero

Expanded Experimental Space for Luminescence Studies of Thin Film CdS/CdTe Solar Cells 259
Scott Feldman, Tim Ohno, Victor Kaydanov, and Reuben Collins

Preparation and Characterization of Monolithic HgCdTe/CdTe Tandem Cells 265
S.L. Wang, J. Drayton, V. Parikh, A. Vasko, A. Gupta, and A.D. Compaan

*Invited Paper

SILICON THIN FILMS

Effects of Grain Boundaries in Amorphous/Multicrystalline Silicon Heterojunction Photovoltaic Cells **273**
M. Farrokh Baroughi and S. Sivoththaman

Improved Efficiency in Hydrogenated Amorphous Silicon Solar Cells Irradiated by Excimer Laser **279**
A.A. Damitha T. Adikaari, S. Ravi P. Silva, Michael J. Kearney, and John M. Shannon

Harvesting Betavoltaic and Photovoltaic Energy with Three Dimensional Porous Silicon Diodes **285**
Wei Sun, Nazir P. Kherani, Karl D. Hirschman, Larry L. Gadeken, and Philippe M. Fauchet

Density of States in Tritiated Amorphous Silicon Measured Using CPM **291**
Simone Pisana, Stefan Costea, Tome Kosteski, Nazir P. Kherani, Stefan Zukotynski, and Walter T. Shmayda

Author Index **297**

Subject Index **301**

Materials Research Society Symposium Proceedings

Volume 807— Scientific Basis for Nuclear Waste Management XXVII, V.M. Oversby, L.O. Werme, 2004, ISBN: 1-55899-752-0
Volume 808— Amorphous and Nanocrystalline Silicon Science and Technology—2004, R. Biswas, G. Ganguly, E. Schiff, R. Carius, M. Kondo, 2004, ISBN: 1-55899-758-X
Volume 809— High-Mobility Group-IV Materials and Devices, M. Caymax, E. Kasper, S. Zaima, K. Rim, P.F.P. Fichtner, 2004, ISBN: 1-55899-759-8
Volume 810— Silicon Front-End Junction Formation—Physics and Technology, P. Pichler, A. Claverie, R. Lindsay, M. Orlowski, W. Windl, 2004, ISBN: 1-55899-760-1
Volume 811— Integration of Advanced Micro- and Nanoelectronic Devices—Critical Issues and Solutions, J. Morais, D. Kumar, M. Houssa, R.K. Singh, D. Landheer, R. Ramesh, R. Wallace, S. Guha, H. Koinuma, 2004, ISBN: 1-55899-761-X
Volume 812— Materials, Technology and Reliability for Advanced Interconnects and Low-k Dielectrics—2004, R. Carter, C. Hau-Riege, G. Kloster, T-M. Lu, S. Schulz, 2004, ISBN: 1-55899-762-8
Volume 813— Hydrogen in Semiconductors, N.H. Nickel, M.D. McCluskey, S. Zhang, 2004, ISBN: 1-55899-763-6
Volume 814— Flexible Electronics 2004—Materials and Device Technology, B.R. Chalamala, B.E. Gnade, N. Fruehauf, J. Jang, 2004, ISBN: 1-55899-764-4
Volume 815— Silicon Carbide 2004—Materials, Processing and Devices, M. Dudley, P. Gouma, P.G. Neudeck, T. Kimoto, S.E. Saddow, 2004, ISBN: 1-55899-765-2
Volume 816— Advances in Chemical-Mechanical Polishing, D. Boning, J.W. Bartha, G. Shinn, I. Vos, A. Philipossian, 2004, ISBN: 1-55899-766-0
Volume 817— New Materials for Microphotonics, J.H. Shin, M. Brongersma, F. Priolo, C. Buchal, 2004, ISBN: 1-55899-767-9
Volume 818— Nanoparticles and Nanowire Building Blocks—Synthesis, Processing, Characterization and Theory, O. Glembocki, C. Hunt, C. Murray, G. Galli, 2004, ISBN: 1-55899-768-7
Volume 819— Interfacial Engineering for Optimized Properties III, C.A. Schuh, M. Kumar, V. Randle, C.B. Carter, 2004, ISBN: 1-55899-769-5
Volume 820— Nanoengineered Assemblies and Advanced Micro/Nanosystems, J.T. Borenstein, P. Grodzinski, L.P. Lee, J. Liu, Z. Wang, D. McIlroy, L. Merhari, J.B. Pendry, D.P. Taylor, 2004, ISBN: 1-55899-770-9
Volume 821— Nanoscale Materials and Modeling—Relations Among Processing, Microstructure and Mechanical Properties, P.M. Anderson, T. Foecke, A. Misra, R.E. Rudd, 2004, ISBN: 1-55899-771-7
Volume 822— Nanostructured Materials in Alternative Energy Devices, E.R. Leite, J-M. Tarascon, Y-M. Chiang, E.M. Kelder, 2004, ISBN: 1-55899-772-5
Volume 823— Biological and Bioinspired Materials and Devices, J. Aizenberg, C. Orme, W.J. Landis, R. Wang, 2004, ISBN: 1-55899-773-3
Volume 824— Scientific Basis for Nuclear Waste Management XXVIII, J.M. Hanchar, S. Stroes-Gascoyne, L. Browning, 2004, ISBN: 1-55899-774-1
Volume 825E—Semiconductor Spintronics, B. Beschoten, S. Datta, J. Kikkawa, J. Nitta, T. Schäpers, 2004, ISBN: 1-55899-753-9
Volume 826E—Proteins as Materials, V.P. Conticello, A. Chilkoti, E. Atkins, D.G. Lynn, 2004, ISBN: 1-55899-754-7
Volume 827E—Educating Tomorrow's Materials Scientists and Engineers, K.C. Chen, M.L. Falk, T.R. Finlayson, W.E. Jones Jr., L.J. Martinez-Miranda, 2004, ISBN: 1-55899-755-5
Volume 828— Semiconductor Materials for Sensing, S. Seal, M-I. Baraton, N. Murayama, C. Parrish, 2005, ISBN: 1-55899-776-8
Volume 829— Progress in Compound Semiconductor Materials IV—Electronic and Optoelectronic Applications, G.J. Brown, M.O. Manasreh, C. Gmachl, R.M. Biefeld, K. Unterrainer, 2005, ISBN: 1-55899-777-6
Volume 830— Materials and Processes for Nonvolatile Memories, A. Claverie, D. Tsoukalas, T-J. King, J. Slaughter, 2005, ISBN: 1-55899-778-4
Volume 831— GaN, AlN, InN and Their Alloys, C. Wetzel, B. Gil, M. Kuzuhara, M. Manfra, 2005, ISBN: 1-55899-779-2
Volume 832— Group-IV Semiconductor Nanostructures, L. Tsybeskov, D.J. Lockwood, C. Delerue, M. Ichikawa, 2005, ISBN: 1-55899-780-6

MATERIALS RESEARCH SOCIETY SYMPOSIUM PROCEEDINGS

Volume 833— Materials, Integration and Packaging Issues for High-Frequency Devices II, Y.S. Cho, D. Shiffler, C.A. Randall, H.A.C. Tilmans, T. Tsurumi, 2005, ISBN: 1-55899-781-4
Volume 834— Magneto-Optical Materials for Photonics and Recording, K. Ando, W. Challener, R. Gambino, M. Levy, 2005, ISBN: 1-55899-782-2
Volume 835— Solid-State Ionics—2004, P. Knauth, C. Masquelier, E. Traversa, E.D. Wachsman, 2005, ISBN: 1-55899-783-0
Volume 836— Materials for Photovoltaics, R. Gaudiana, D. Friedman, M. Durstock, A. Rockett, 2005, ISBN: 1-55899-784-9
Volume 837— Materials for Hydrogen Storage—2004, T. Vogt, R. Stumpf, M. Heben, I. Robertson, 2005, ISBN: 1-55899-785-7
Volume 838E—Scanning-Probe and Other Novel Microscopies of Local Phenomena in Nanostructured Materials, S.V. Kalinin, B. Goldberg, L.M. Eng, B.D. Huey, 2005, ISBN: 1-55899-786-5
Volume 839— Electron Microscopy of Molecular and Atom-Scale Mechanical Behavior, Chemistry and Structure, D. Martin, D.A. Muller, E. Stach, P. Midgley, 2005, ISBN: 1-55899-787-3
Volume 840— Neutron and X-Ray Scattering as Probes of Multiscale Phenomena, S.R. Bhatia, P.G. Khalifah, D. Pochan, P. Radaelli, 2005, ISBN: 1-55899-788-1
Volume 841— Fundamentals of Nanoindentation and Nanotribology III, D.F. Bahr, Y-T. Cheng, N. Huber, A.B. Mann, K.J. Wahl, 2005, ISBN: 1-55899-789-X
Volume 842— Integrative and Interdisciplinary Aspects of Intermetallics, M.J. Mills, H. Clemens, C-L. Fu, H. Inui, 2005, ISBN: 1-55899-790-3
Volume 843— Surface Engineering 2004—Fundamentals and Applications, J.E. Krzanowski, S.N. Basu, J. Patscheider, Y. Gogotsi, 2005, ISBN: 1-55899-791-1
Volume 844— Mechanical Properties of Bioinspired and Biological Materials, C. Viney, K. Katti, F-J. Ulm, C. Hellmich, 2005, ISBN: 1-55899-792-X
Volume 845— Nanoscale Materials Science in Biology and Medicine, C.T. Laurencin, E. Botchwey, 2005, ISBN: 1-55899-793-8
Volume 846— Organic and Nanocomposite Optical Materials, A. Cartwright, T.M. Cooper, S. Karna, H. Nakanishi, 2005, ISBN: 1-55899-794-6
Volume 847— Organic/Inorganic Hybrid Materials—2004, C. Sanchez, U. Schubert, R.M. Laine, Y. Chujo, 2005, ISBN: 1-55899-795-4
Volume 848— Solid-State Chemistry of Inorganic Materials V, J. Li, M. Jansen, N. Brese, M. Kanatzidis, 2005, ISBN: 1-55899-796-2
Volume 849— Kinetics-Driven Nanopatterning on Surfaces, E. Wang, E. Chason, H. Huang, G.H. Gilmer, 2005, ISBN: 1-55899-797-0
Volume 850— Ultrafast Lasers for Materials Science, M.J. Kelley, E.W. Kreutz, M. Li, A. Pique, 2005, ISBN: 1-55899-798-9
Volume 851— Materials for Space Applications, M. Chipara, D.L. Edwards, S. Phillips, R. Benson, 2005, ISBN: 1-55899-799-7
Volume 852— Materials Issues in Art and Archaeology VII, P. Vandiver, J. Mass, A. Murray, 2005, ISBN: 1-55899-800-4
Volume 853E—Fabrication and New Applications of Nanomagnetic Structures, J-P. Wang, P.J. Ryan, K. Nielsch, Z. Cheng, 2005, ISBN: 1-55899-805-5
Volume 854E—Stability of Thin Films and Nanostructures, R.P. Vinci, R. Schwaiger, A. Karim, V. Shenoy, 2005, ISBN: 1-55899-806-3
Volume 855E—Mechanically Active Materials, K.J. Van Vliet, R.D. James, P.T. Mather, W.C. Crone, 2005, ISBN: 1-55899-807-1
Volume 856E—Multicomponent Polymer Systems—Phase Behavior, Dynamics and Applications, K.I. Winey, M. Dadmun, C. Leibig, R. Oliver, 2005, ISBN: 1-55899-808-X
Volume 858E—Functional Carbon Nanotubes, D.L. Carroll, B. Weisman, S. Roth, A. Rubio, 2005, ISBN: 1-55899-810-1
Volume 859E—Modeling of Morphological Evolution at Surfaces and Interfaces, J. Evans, C. Orme, M. Asta, Z. Zhang, 2005, ISBN: 1-55899-811-X
Volume 860E—Materials Issues in Solid Freeforming, S. Jayasinghe, L. Settineri, A.R. Bhatti, B-Y. Tay, 2005, ISBN: 1-55899-812-8
Volume 861E—Communicating Materials Science—Education for the 21st Century, S. Baker, F. Goodchild, W. Crone, S. Rosevear, 2005, ISBN: 1-55899-813-6

Prior Materials Research Society Symposium Proceedings available by contacting Materials Research Society

Dye-Sensitized Solar Cells

Mater. Res. Soc. Symp. Proc. Vol. 836 © 2005 Materials Research Society L1.2

Progress in Producing Large Area Flexible Dye Sensitized Solar Cells

Krishna C. Mandal, Michael Choi, Caleb Noblitt and R. David Rauh
EIC Laboratories, Inc., 111 Downey Street, Norwood, MA 02062-2612, USA

ABSTRACT

Dye sensitized nanocrystalline TiO_2 solar cells have been reported with over 11% efficiency and are extremely promising as very low cost and lightweight photovoltaic sources. However, most reports are for cells of low area fabricated on glass, which withstands processing temperatures of ~450°C. In this paper, we describe the fabrication and performance of cells made on flexible ITO-coated polyethylene terephthalate (PET) substrates with 6" x 3" dimensions. To improve the efficiency in the cells, we enhanced the ITO current collection efficiency with metallization fingers. The fingers resulted in a >10 fold increase in short-circuit current under normal solar illumination compared to cells without metallization. Further improvements were realized by passivating the metallization fingers at the metal/polymer electrolyte interface.

INTRODUCTION

Dye sensitized solar cells (DSSC) are usually prepared by first depositing colloidal TiO_2 onto a transparent conductive oxide coated glass substrate. The colloidal layer is then sintered at ~450°C to form a porous pure anatase layer 5-10 μm thick. Next, the layer is exposed to a solution of a sensitizing dye carrying -COO^- functional groups, which are believed to bind to the Ti^{+4} surface sites. The Ru(II) complexes usually employed as sensitizing dyes apparently undergo electron transfer to the Ti(IV) sites on nearly the optical time scale, leaving the oxidized sensitizer, a process which has been argued to protect the dye against slower irreversible photodegrative pathways. These sensitized electrodes may be used as the photoanode in regenerative photoelectrochemical (PEC) cells. The regenerative redox couple is typically I^-/I_3^- in a liquid nonaqueous electrolyte. Grätzel and co-workers were the first to have reported >11% solar conversion efficiencies for small area devices [1].

At its current state of development, at least in the published literature, the DSSC has relied mostly on the use of glass substrate materials. Glass is used because it is optically transparent and it can withstand the high temperature annealing currently required to achieve high photoactivity of the nanoporous TiO_2. However, lightweight polymer substrates are clearly advantageous for portability, adaptability and ease of handling. Furthermore, the use of liquid electrolytes is unacceptable due to sealing difficulties and inhomogeneous electrolyte distribution, especially if a flexible version of the DSSC is to be realized [2].

It is the purpose of this study to evaluate DSSCs in a flexible, large-area cell configuration. To achieve this, we have employed a commercial ITO-coated polymer as the substrate rather than glass. The electrolyte is based on a plasticized polyvinylidene difluoride (PVDF) composition derived from a material used in flexible Li-ion batteries. Cells were nominally 6"x3" (active area somewhat lower), which is much larger than has been reported previously for flexible DSSCs. However, such large areas are essential for most practical applications.

EXPERIMENTAL DETAILS

Baseline cell

The basic DSSC cell design is a trilayer laminate with a polymer gel electrolyte. The supporting substrate was a commercially available polyester (polyethylene terephthalate, PET) coated with ITO provided by CPFilms, Inc. The lowest available sheet resistance is 60 ohms/sq. We obtained this in 5 mil thickness. The maximum temperature without significant distortion is given by the manufacturer as 150°C. Substrates were cut to the desired cell size, in this case 6"x3". The photoanode was spray coated while heating to 95°C with 5-10 microns of a Degussa P25 nanocrystalline TiO_2 photoanode precursor slurry, with the carrier solvent evaporating on contact. The TiO_2 slurry consisted of 36g TiO_2, 165ml deionized water, 2.1ml acetyl acetone, and 0.15g of polyethylene glycol. The PET/ITO/TiO_2 substrates were then annealed at 150°C under flow of O_2 for 8 hours. Following annealing, the coated substrates were transferred to an Ar atmosphere glove box, heated to 95° C to drive off remaining water. They were then placed in a sensitizing dye solution (cis-$(SCN)_2$bis(2,2'-bipyridyl-4,4'-dicarboxylate) ruthenium, or "N3") in anhydrous ethanol (0.045g/100 ml). The films remained in the dye for several hours and were then rinsed several times with ethanol. The dye-sensitized films appeared dark purple color.

The counter electrodes were made by dc magnetron sputter coating PET|ITO substrates with titanium (~40 nm) and then platinum (~60 nm) to provide a uniform adherent metallization.

The electrolyte was a plasticized polymer based on polyvinylidene difluoride, PVDF (Solvay Solef 21216). The PVDF was combined with plasticizer solvents (propylene carbonate, PC, and ethylene carbonate, EC) and redox electrolyte (t-butyl ammonium iodide plus I_2) in the weight ratio 1:1.2:3.0:.95:0.03. The PVDF was ground in a mortar with a small amount of the plasticizer until a smooth slurry was formed and then allowed to stir overnight with the rest of the plasticizer. The semisolid electrolyte was then placed on the Pt counter electrode and heated at 90°C for about 5 minutes following which the TiO_2 was placed on top face down and clamped in place for several hours to set. The two opposite facing electrodes were offset to allow for contacts and edge sealing. Finally the cells were cleaned and sealed with Dow Corning RTV 9-1363 adhesive along the edges to prevent entry of water and oxygen. Edge contacts were reinforced with a wire adhesive mesh. Cell assembly and sealing were conducted in the argon-controlled glove box.

Metallization of baseline cells

To improve current collection on the 6" x 3" solar cells, metallization fingers were deposited onto the PET/ITO substrates before TiO_2 deposition. The metallization fingers were made by sputter-depositing a metal onto the PET/ITO through an aluminum stencil mask. The metallization pattern consists of 7 fingers, each 2.5" long and 0. 1" wide, along with 0.25" wide busbar along the edge of the PET/ITO. The resulting loss of area from depositing the grid is approximately 13%.

Cells were assembled in an identical manner to the baseline cell. Several kinds of metallization grid treatments were compared: 1) a "thick" 100 nm Ir grid with a 50 nm Ti adhesion layer made by dc magnetron sputtering, 2) a "thin" ~1 nm semitransparent Ir grid, 3) a Ir grid with a layer of TiO_2, and 3) the entire substrate and grid structure with a 100 nm layer of sputtered TiO_2. The TiO_2 coated grids, configuration 3), were made by a wet oxidation process in which the Ir grids were first coated with 100 nm Ti in the same pump-down cycle. The substrates were then introduced into a furnace at 150°C under a flow of oxygen bubbled through DI water.

RESULTS AND DISCUSSION

Cell characteristics

The current-voltage (I-V) characteristics of the cells were measured using a potentiostat in the dark and under outdoor sunlight, ~40 mW/cm^2. A compilation of the dark current responses of the cells with different grid treatments is shown in Figure 1. The bare metallization grids, which were deposited onto the ITO to reduce the sheet resistance, caused a marked increase in the dark current. This is a result of the oxidation/reduction reaction of the iodide redox system at the metal/electrolyte interface. Such reactions effectively create a short-circuit in the cell.

Since the metallizations need only create a lower resistance current path for electrons on the ITO, it should be possible to passivate the metal/electrolyte interface by adding an additional resistive layer on the metallization surface (see below). Sputtering Ti onto the Ir grids during the grid deposition process produced grids with a Ti/Ir/Ti structure. The top Ti layer was converted to TiO_2 by a wet oxidation process. Cells made from the resulting substrates showed markedly reduced dark current on polarization between -0.6 to 0.6V. Metallizations were also passivated by reactive dc sputter deposition of TiO_2 over the entire active area, a process which reduced the dark current even more, and below that observed for cells without grids.

The I-V characteristics of the cells were greatly affected by the grid structures. Examples are shown in Figure 2. Cells without grids had very low photocurrents, on the order of 5 mA at short-circuit. Open-circuit photovoltages exceeded 0.6V, however. Introducing the grids increased the short circuit photocurrents approximately 10-fold, but the photovoltage, the point where $I_L+I_D=0$, was reduced to less than 0.4V due to the large dark currents. The most effective treatment was passivation of the grids and the ITO with a thin layer of sputtered TiO_2. This reduced the dark current to the level of the ungridded cells, increasing the photovoltage to >0.5V. Photocurrents for these cells were on the order of 60 mA at 0V.

As a demonstration, we fabricated 16 cells with the sputtered TiO_2 passivation over the photoanode substrate and arranged them into a series-parallel array, shown in Figure 3, by making interconnects between them. The total active area of each cell (without grid) is 5.5"x 2.5"=13.75 in^2 = 88.71 cm^2. The opaque grids cover 12.7% of the ITO, producing a real active area per cell of 77.4 cm^2. Before assembly, and up to one month after their fabrication, cells were tested under solar illumination. Typical output of the as-prepared individual cells is shown in Figure 2. The maximum power point of ~ 8 mW/cm^2 resulted in a single cell efficiency of 0.25 %. However, some of the cells to be used in construction the array gave less than optimal performance, possibly due to leakage of the edge seals. Nevertheless, this was the first demonstration of a complete array of individual, externally connected DSSC cells on flexible polymer substrates.

A test of the array of 16 cells in sunlight at a measured intensity of 25 mW/cm^2 provided a V_{oc} of 2.51-2.56V and an I_{sc} of 0.095-0.12 amps. The V_{oc} represents four cells in series and the I_{sc} four cells in parallel. The values are close to those expected for this run of cells under the given intensity.

Current Collectors in DSSCs

The current collection at the transparent conducting oxide layer is a major source of power loss in the DSSC and other photoelectrochemical cells of this basic design. Typical transparent conducting oxide layers on polymer substrates, for example, are on the order of 100 ohms/sq. The effect can be seen by considering two solar DSSC solar cells, one with an area of 1 cm^2 and the other with an area of 100 cm^2. Consider under identical illumination conditions that both cells are generating a current density of 10 mA/cm^2. Thus, the total current that must be withdrawn form the 1 cm^2 and 100 cm^2 cells is 10 and 1000 mA, respectively. The resistive contribution from the transparent oxide is roughly the same for the two square sizes. Thus, the IR loss is approximately 100 times greater in the larger cell.

A well-established method to reduce the sheet resistance on traditional solar cells is to employ current collection metallization patterns, in the form of strips, grids, etc. These are able to collect current over smaller local areas and direct it to a current collection busbar via a low resistance metallization pathway. The patterns can be designed to occlude a minimum fraction of the active area while maximizing the overall reduction in resistive losses [3].

In the case of the DSSC, these metallization lines need to be disposed in direct electrical contact with the transparent conductive oxide, which is in turn disposed on the inner surface of the front transparent substrate. Figure 4A illustrates this arrangement schematically. It is seen, however, that current (electrons) collected on the grid is available at the grid/electrolyte interface. There, electrons can react with the I_3^-, reducing it to I^-. This in effect creates a short-circuit in the cell and reduces the current flowing through the external circuit. We observe in Figure 1 and 2 that, while the placement of metallization fingers on the ITO does indeed generate a large "dark current" indicative of this short-circuiting, large increases in the short-circuit photocurrent are still realized with this approach.

The cell design in Figure 4B further provides metallizations in which the metal/conductive oxide interface is preserved, but the metal/electrolyte reaction is prevented or suppressed. This could be achieved by several approaches, such as 1) using a metallization in which the electrochemical reduction of the redox species is extremely slow, 2) coating the metallization with an ion exchange material which rejects the redox species, or 3) coating the metallization with an electrical insulator. Figure 4 illustrates this modification with an insulator. Very recently 10 x 10 cm^2 DSSCs have been reported on glass with a 4.3% solar conversion efficiency using Ni grids to reduce the sheet resistance of the FTO (F-doped tin oxide) to 0.28 ohms/sq [4]. These cells employed very low sheet resistance FTO to begin with (4.8 ohms/sq) compared to ~60 ohms/sq that we are able to obtain on PET [3] and a TiO_2 processing temperature of 450°C. They also employed a liquid electrolyte.

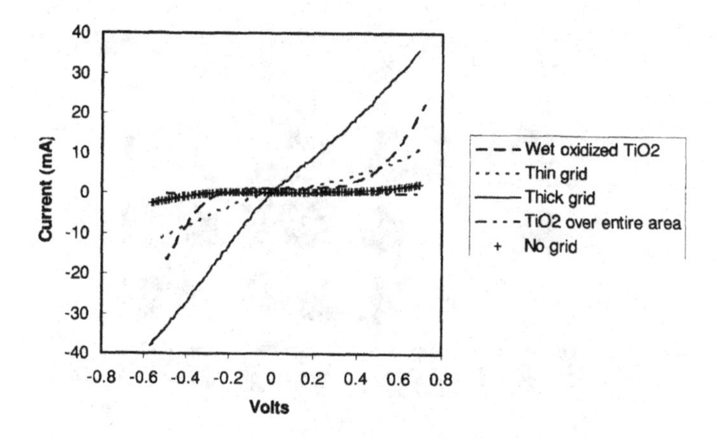

Figure 1. Dark current vs. voltage for 6"x3" cells with and without Ir grids. Thin and thick grids correspond to 1 nm and 100 nm.

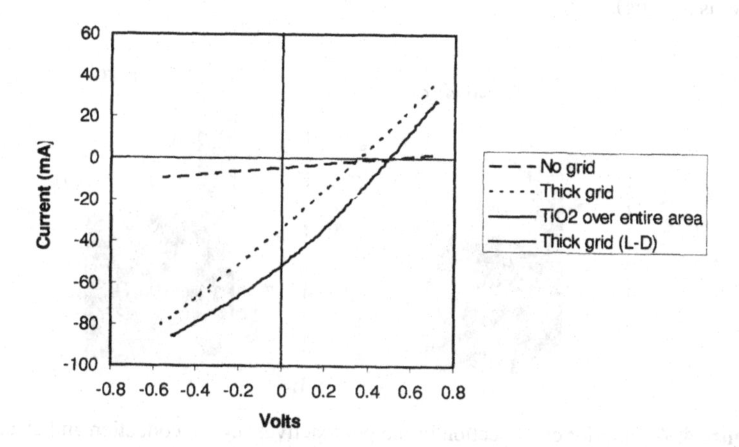

Figure 2. Photocurrent vs. voltage for cells with and without Ir metallization grids. The L-D trace subtracts the large dark current from the total current in the light. Also shown is the output of a cell with a uniform dense layer of TiO_2 sputtered over the entire grid and active area.

Figure 3. DSSC 16-cell panel. Cells are arranged in 4 parallel sets of 4 cells in series. (Ruler is 8" long).

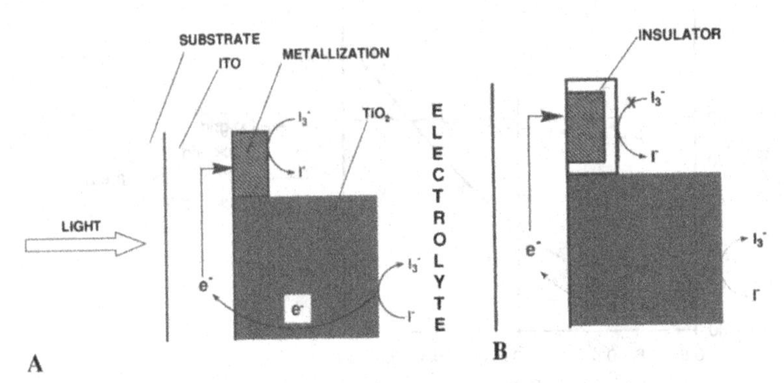

Figure 4. An interior cross section of the photoactive, current collection and electrolyte elements of a DSSC, along with the metallization segments. A. metallization creating a short-circuit at the metal/electrolyte interface. B. Addition of insulation on the metal surface prevents back reaction with redox system.

We employed Ir as the metallization since it is a very conductive noble metal which will not corrode in contact with the electrolyte. However, it is also an excellent electrocatalyst, which results in a large dark current. Recently Ni was used as the metallization grids [4] , which was shown to give lower dark currents than Au or Pt. In our work, cells were generally improved by insulating the Ir metallization, either by a layer of TiO_2 directly on the grids or over the entire active surface. Preliminary results also indicate that the cells with a dense, thin TiO_2 layer over the entire surface had a longer shelf-life. This is to be expected since the conductive oxides available on plastics tend to be of lower density and generally poorer quality than that obtainable on glass using higher processing temperatures.

CONCLUSIONS

There are several areas for improvement of these large area flexible cells before they can be considered as viable candidates for portable power. In particular, we would like even closer spaced and finer conductive grids to reduce the ITO sheet resistance further. These would be produced by photolithographically defining the pattern, then possibly by electroplating of Ni, which has been shown to produce low dark currents in DSSCs. We also seek flexible substrates that can endure higher processing temperatures, such as high T_g plastics or microsheet glass. A related issue is with the quality and high sheet resistance of current commercially available ITO coated PET, which is much inferior to what is available on glass. Indeed, to scale up these cells on flexible substrates will require new developments in these areas. Further improvements in the electrolyte conductivity and effective thickness might be realized in order to optimize the efficiencies of laminated devices. Finally, DSSCs in general have yet to undergo the kind of durability testing that will be required for the wide range of environmental operating conditions.

ACKNOWLEDGMENTS

This work is partially supported by the DOE (DE-FG02-98ER82567), NASA (NAS5-01173 & NNC04CA98C), United States Air Force (AFRL/MLQD, F08630-02-C-0074) and MDA (AFRL/VSSV FA9453-04-M-0283).

REFERENCES

1. B. O'Regan and M. Grätzel, *Nature (London)* **335**, 737 (1991); M. Grätzel, *J. Photochem. Photobiol. A* **164**, 3 (2004).
2. Krishna C. Mandal, Anton Smirnov, D. Peramunage, and R. David Rauh, (Mater. Res. Soc. **737**, Boston, MA, 2003) pp. 739-744.
3. See, for example, A. R. Burgers, *Prog. Photovolt: Res. Appl.* **7**, 457 (1999).
4. K. Okada, H. Matsui, T. Kawashima, T. Ezure and N. Tanabe, *J. Photochem. Photobiol. A* **164**, 193 (2004).

Mater. Res. Soc. Symp. Proc. Vol. 836 © 2005 Materials Research Society L1.4

Amphiphilic Dye for Solid-State Dye-Sensitized Solar Cells

Lukas Schmidt-Mende, Shaik M. Zakeeruddin and Michael Grätzel
Institut des Sciences et Ingénierie Chimiques (ISIC), Laboratoire de Photonique et Interfaces (LPI), Ecole Polytechnique Fédérale de Lausanne (EPFL), CH-1015 Lausanne, Switzerland

ABSTRACT

We report a solid-state dye-sensitized solar cell with a record efficiency of 4% under simulated sunlight (AM1.5global, 100mW/cm^2). This was made possible by using a new amphiphilic dye with hydrophobic spacers in combination with spiro-OMeTAD. We attribute the significant improvement in the device performance to the self-assembly of the dye to form a compact layer on the TiO_2 surface and to the hydrophobic chains working as blocking layer between spiro-OMeTAD and TiO_2 to reduce the back electron transfer. In addition, we studied the influence of nanoporous TiO_2 film thickness on the performance of the device. These results demonstrate the high potential for solid state dye sensitized solar cells to compete with amorphous silicon cells as low cost alternative.

INTRODUCTION

The world energy consumption is raising, which makes the limitation of fossil fuels more and more apparent. Oil prizes are as high as never before. If it does not succeed to find cheap and easy accessible renewable energy sources the energy problem will further increase in the near future. This is certainly one reason, why the market of solar cells is one of the fastest growing markets, although the production of conventional solar cells is expensive and combined with a high energy consumption. Therefore the interests in and the perspectives of cheap alternative solar cells are very high. Dye-sensitized solar cells are a viable alternative for conventional solar cells. Since the first report of a highly efficient dye-sensitized solar cell in 1991[1] many groups are working on this type of cell and could improve the efficiency further[2-4]. Values exceeding 10% efficiency at one sun illumination have been reached[5]. It seems to be advantageous to replace the liquid electrolyte of these cells with a solid holeconductor to avoid any sealing and long-term stability problems. Not only the possibility of low cost production is an important advantage over conventional solar cells, also the versatile design possibilities, such as production of flexible cells and differently colored devices. We report here a solid-state dye-sensitized solar cell that has an efficiency of 4% under simulated sunlight (AM 1.5 global, 100mW/cm^2)[6] with an amphiphilic ruthenium sensitizer.

EXPERIMENT

The structure of our cells is schematically shown in Figure 1. The cells consist of a F-doped SnO_2 conducting glass substrates onto which a compact TiO_2 layer was deposited by spray pyrolysis. This is necessary to avoid direct contact between the holeconductor and the SnO_2, which would short circuit the cell. The holeconductor forms an ohmic contact to the SnO_2 and charges would recombine at this interface. A dense TiO_2 blocking layer impedes this current loss. The nanoporous TiO_2 layers was deposited by doctor-blading a TiO_2 paste containing 19 nm-sized anatase particles. This method allowed us to produce homogenous nanoporous TiO_2 films with controlled thickness by using different spacers for doctorblading the same TiO_2 paste. After sintering the TiO2 layer at 450°C it was sensitized by soaking it in a $3x10^{-4}$ M solution of amphiphilic polypyridyl ruthenium complex, *cis*-RuLL`(SCN)2 (L=4,4`-dicarboxylic acid-2,2`-bipyridine, L`=4,4`-dinonyl-2,2`-bipyridine) (Z907) (Figure 1) in acetonitrile:tert.-butanol (1:1 vol%) at room temperature. This dye contains two carboxylate groups to anchor to the TiO_2 surface and hydrophobic spacer providing an insulating barrier to avoid charge carrier recombination. It has already successfully been employed in polymer gel-electrolyte cells and showed a remarkably long-term stability[7-9].

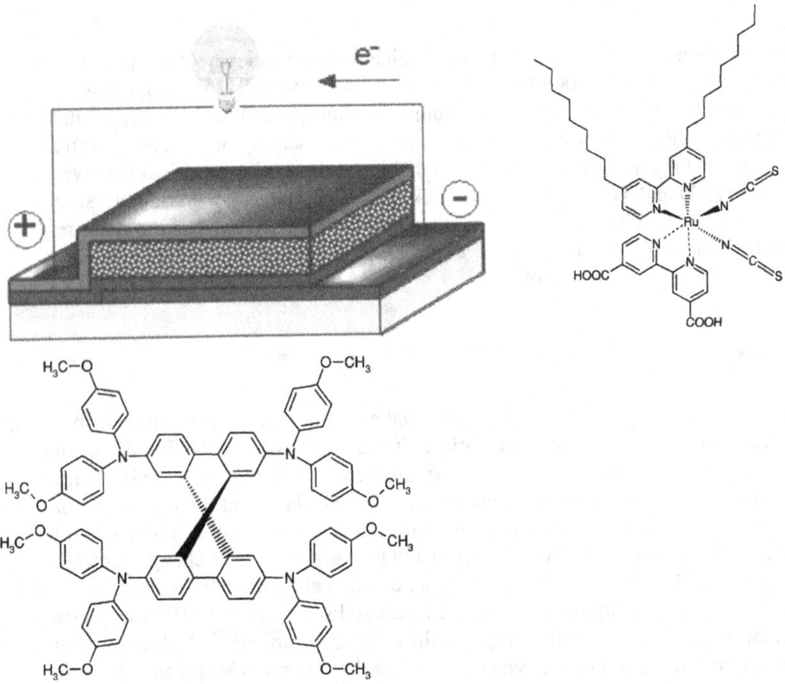

Figure 1. The device structure consists of a glass substrate coated with F-doped SnO_2. On top of this a dense layer of TiO_2 is deposited by spray pyrolysis and a nanoporous TiO_2 layer be doctor-blading. The nanoporous layer is soaked in dye and filled with the holeconductor. A gold electrode is evaporated on top as counter electrode. The molecular structure of the Z-907 (right) and the spiro-OMeTAD (below).

2,2',7,7'-tetrakis-(N,N-di-p-methoxyphenyl-amine)9,9'-spirobifluorene (spiro-OMeTAD) (Figure 1) is used as an organic holeconductor. We dissolved the spiro-OMeTAD in chlorobenzene (0.17 M) and added $Li[CF_3SO_2]_2N$ (13 mM), tert.-butylpyridine (0.13 M) and $N(PhBr)_3SbCl_6$ (0.3 mM) and then deposited it onto the dye sensitized substrate and let it penetrate into the pores of the TiO_2 layer for one minute prior to spincoating. The device was fabricated by evaporating a 30nm gold electrode on top. The cell design allowed us to have 4 cells on each substrate with a size of around 0.16 cm^2.

RESULTS AND DISCUSSION

The molecular extinction coefficient of the Z907 dye is lower than the previously used ruthenium dye N719 dye (13900 l mol^{-1} cm^{-1})[10]. Nevertheless we could reach much higher efficiencies with Z907 dye despite the lower molecular extinction coefficient. In Table I we correlated the photovoltaic performance of these two dyes under identical conditions.

Table I: Comparison of N719 and Z907 devices.

Dye	Jsc (mA/cm^2)	Voc (mV)	Fill factor (%)	η (%)
N719	4.40	796	68.8	2.4
Z907	7.36	882	63.5	4.1

The devices reported here had a dipping time of 16hrs in the dye-solution. The nanoporous TiO_2 had a film thickness of ~2μm and the device size is 0.148 cm^2 and 0.188 cm^2 for the N719 and the Z907 device, respectively. For comparison, both devices were made under identical conditions at the same time by using the same hole-conductor solution. This way we can exclude a batch to batch variation as a reason for the significant difference in performance. The immense difference in performance is reproducible. Also the device performance of our N719 device is very similar to values reported earlier[11].

Figure 2 shows the current-voltage characteristic of Z907 device at different light intensities.

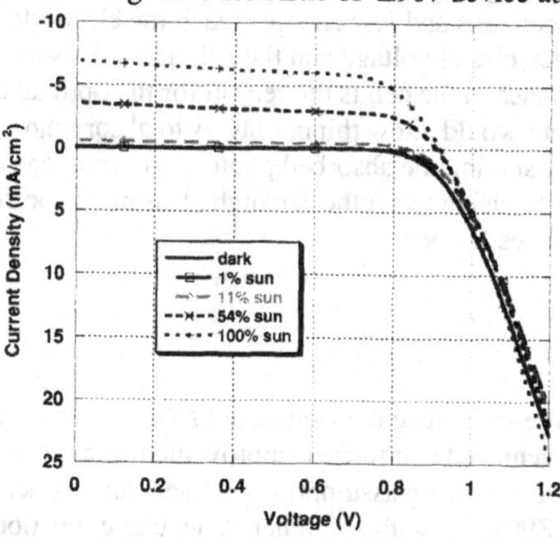

Figure 2: Current-voltage characteristic of a Z907 device under different illumination intensities.

The device performance is not linear with the incident light intensity and the efficiencies under different illumination intensities are shown in Table II.

Table II: Device efficiency under different illumination intensities.

Z907 device	1% Sun	11% Sun	54% Sun	100% Sun
η (in %)	1.3	2.7	3.9	4.1

This increase in efficiency with higher illumination makes it difficult to get the incoming photon to current conversion efficiency (IPCE). The IPCE curve for the shown device is very low, which can be explained by the low light intensity of the single wavelength, which we have, when we use the monochromator to scan the wavelength. This makes it difficult to investigate the absorbed photon to current conversion efficiency under one sun conditions at different wavelengths. We assume that this value has to be very high for the Z907 devices, otherwise an efficiency of 4% could not be reached.

We attribute the high efficiencies of these devices is mainly due to the structure of the dye, that self-assembles and forms a compact layer by anchoring through the carboxylic groups to the TiO_2 surface. Its hydrophobic chains might act as blocking layer between holeconductor and TiO_2, at the same time allow a very good wetting of the dye covered surface with the holeconductor. This enables an efficient electron injection from the dye into the conduction band of the TiO_2 and also a good hole transfer from the dye to the hole-conductor, regenerating the dye's original ground state.

In the optimized devices the TiO_2 film thickness is found to be around 2μm. If the nanoporous TiO_2 layer is considerably thinner, the devices performance goes down, because of a smaller current density. This can be explained by the fact, that already in 2μm thick layers only ~60% of the incoming light is absorbed at the peak wavelength of the Z907. In thinner layers the amount of absorbed light decreases further. Unfortunately it is not possible to increase the film thickness significantly because it leads to an increased series resistance in the cell, which is caused by the relative low charge carrier mobility of the holeconductor. Therefore the charge recombination in the cell increases and few charges reach the electrode. The current density decreases as well as the open circuit voltage and the fill factor. This trade off between light harvesting and series resistance in the cell is the reason for the optimal thickness. Dyes with higher absorption coefficient would allow thinner layers to absorb more light and therefore lead to more efficient devices, assuming the absorbed photon to current conversion efficiency stays the same. Presently we are focusing to synthesize high absorption coefficient amphiphilic dyes to enhance the device efficiencies further.

CONCLUSIONS

This work shows that the solid-state dye-sensitized solar cells have a great potential. An efficiency of 4% could be achieved and further improvement is expected. Our studies of the device thickness support the following assumptions: Molecular engineering of a dye with an amphiphilic structure like Z907, but with an higher molecular extinction coefficient could lead to significant higher efficiency in solid-state dye-sensitized solar cells. With this prospective these cells seem to have a high potential as cheap alternative to conventional photovoltaic devices.

ACKNOWLEDGMENTS

The authors wish to acknowledge the European Commission for partial funding (MOLYCELL project, contract SES6-CT-2003-502783). LSM thanks the German Research Foundation (DFG) for funding (Emmy-Noether Stipendium).

REFERENCES

[1] B. O'Regan, M. Grätzel, *Nature* **1991**, *353*, 737.

[2] K. Tennakone, P. K. M. Bandaranayake, P. V. V. Jayaweera, A. Konno, G. R. A. Kumara, *Physica E* **2002**, *14*, 190.

[3] K. Tennakone, G. R. R. A. Kumara, I. R. M. Kottegoda, V. P. S. Perera, *Chemical Communications* **1999**, *1*, 15.

[4] Z. S. Wang, C. H. Huang, Y. Y. Huang, Y. J. Hou, P. H. Xie, B. W. Zhang, H. M. Cheng, *Chem Mater* **2001**, *13*, 678.

[5] M. Grätzel, *J. Photochem. and Photobio. C: Photochem Rev* **2003**, *4*, 145.

[6] L. Schmidt-Mende, S. M. Zakeeruddin, M. Grätzel, *Applied Physics Letters* **2005**, 86, 13504.

[7] P. Wang, S. M. Zakeeruddin, I. Exnar, M. Grätzel, *ChemComm* **2002**, *24*, 2972.

[8] P. Wang, S. M. Zakeeruddin, R. Humphry-Baker, J. E. Moser, M. Grätzel, *Adv. Mater.* **2003**, *15*, 2101.

[9] P. Wang, S. M. Zakeeruddin, J. E. Moser, M. K. Nazeeruddin, T. Sekiguchi, M. Grätzel, *Nature Materials* **2003**, *2*, 402.

[10] J. Krüger, R. Plass, L. Cevey, M. Piccirelli, M. Grätzel, U. Bach, *Appl. Phys. Lett.* **2001**, *79*, 2085.

[11] J. Krüger, R. Plass, M. Grätzel, H. J. Matthieu, *Appl. Phys. Lett.* **2002**, *81*, 367.

Mater. Res. Soc. Symp. Proc. Vol. 836 © 2005 Materials Research Society L1.5

Dye Sensitized Solar Cells Incorporating Polyelectrolyte Multilayer Composites

Geoffrey M. Lowman, Hiroaki Tokuhisa, Jodie L. Lutkenhaus, Paula T. Hammond
Department of Chemical Engineering, Massachusetts Institute of Technology
Cambridge, MA 02139, U.S.A.

ABSTRACT

Dye sensitized solar cells (DSSC's) are constructed using TiO_2 electrodes synthesized by aqueous liquid phase deposition in combination with microcontact printing techniques or porous thin film template methods. Layer-by-layer deposition of polyelectrolytes is used to produce an ionic conducting solid-state electrolyte thin film, which is enhanced by post-processing in oligoethylene glycol diacid (OEGDA). The impact of TiO_2 film architecture, as well as the thin film electrolyte, on device performance is discussed.

INTRODUCTION

The emerging need for inexpensive renewable energy sources has stimulated new techniques and materials for the production of efficient photovoltaic devices. One of the most heavily studied of this new class of inexpensive photovoltaic device is the dye-sensitized solar cell (DSSC).[1] Over the last decade, DSSC's based on mesoporous TiO_2 have shown the potential to replace silicon as an inexpensive semiconductor material for solar energy conversion. Typically, high performance is achieved (over 10% efficiency) when an I_3^-/I^- is employed as the redox couple in a liquid electrolyte. The ability to create thin polymer films that can effectively replace liquid electrolytes is one of the most important steps toward the improvement of photovoltaic performance, due to the long-term durability of solid-state electrolytes compared to liquid-based devices. Several methods have been developed for the production of solid-state or quasi-solid-state DSSC's, including traditional solid-state polymer electrolytes[2], gels formed from ionic liquids[3], or hole conducting polymers[4] and small molecules.[5]

Layer-by-layer (LbL) deposition of oppositely charged polyelectrolytes[6] has proven a viable solid-state alternative to liquid electrolytes for photovoltaic applications in a DSSC architecture. Our laboratory has recently demonstrated relatively high room-temperature ionic conductivity in LbL polyelectrolyte thin films.[7] By using weak polyelectrolytes (where charge density can be modified by adjusting deposition pH conditions), such as linear poly(ethyleneimine) (LPEI) and poly(acrylic acid) (PAA), we have been able to create thin films containing greater numbers of mobile ions. In the fully hydrated state these composites show ionic conductivities on the order of 10^{-5} S/cm^2.[7] In order to achieve comparable conductivities at atmospheric humidity, oligoethylene glycol dicarboxylic acid (OEGDA) is introduced via direct adsorption into the fully formed multilayer.

The second critical component of the DSSC photovoltaic investigated in this work is the TiO_2 electrode. Traditionally, this layer is made by repetitive screen-printing of a paste of nanoparticles, which are sintered into a mesoporous matrix. In addition, light scattering particles can be added to increase light harvesting capability. Based on this concept, another promising approach toward improving the TiO_2 electrode is the introduction of micro- or nanoscale patterning of the electrode to improve both active surface area and potential photonic effects by patterning on the same length scale as the wavelength of the input radiation. We have demonstrated previously the ability to pattern TiO_2 on the micro- and nanoscale using a technique similar to soft-lithography or polymer-on-polymer stamping[8] combined with solution synthesis of TiO_2 thin films.[9] Specifically, the technique involves microcontact printing of a polystyrene-*b*-poly(*N*-methyl 4-vinyl pyridinium iodide) copolymer (PS-*b*-PVPQ) onto a negatively charged hydrophilic substrate, followed by the templated deposition of TiO_2. The stamped regions act to resist the deposition of TiO_2 due to the hydrophobic PS groups presented at the solution-substrate interface. We have also begun the production of porous TiO_2 thin films using a template approach. The template consists of a porous LbL film that can be filled with TiO_2 using the same solution phase deposition strategy. Once the TiO_2 film is synthesized, the polymer can be removed by heat treatment. The resulting framework presents a great amount of surface area, as well as the potential for filling of the pores with polymer or gel electrolyte, resulting in increased TiO_2-electrolyte contact area analogous to heterojunction photovoltaics constructed from electron and hole conducting materials.

EXPERIMENTAL DETAILS

TiO_2 electrodes are prepared from a solution process on indium tin oxide (ITO) coated glass substrates (Donelly Applied Films, patterned by DCI, Inc.). The substrates are cleaned by ultrasonication in a series of solvents including, detergent, Milli-Q water, acetone, methanol, and 1,1,1-trichloroethane for 30 minutes each. Before use the ITO substrates are plasma etched in a Harrick PCD 32G plasma cleaner with oxygen bleed for 5 minutes. The liquid phase deposition bath consists of 0.02 M $(NH_4)_2TiF_6$ aqueous solution and a 0.015 M boric acid aqueous solution prepared separately and mixed under a nitrogen environment. When the mixture is heated to 50°C the substrates are suspended vertically into the solution bath. 1.0 M HCl is added to adjust the pH to 2.5 – 3.0 and the reaction is left to run for 5 – 10 hours. The thickness of the resulting film from one reaction cycle is typically 3 – 4 µm. TiO_2 films are subsequently calcined at 400°C for 40 minutes in air. Sensitizing dyes used for solar cell applications included 4,4',4'',4'''-(21*H*-23*H*-porphine-5,10,15,20-tetrayl)tetrakis(benzoic acid) (TPPAc), *cis*-di(thiocyanato)-*N,N*-bis(2,2'-bipyridyldicarboxylate)-ruthenium(II) (N3) and (tri(cyanato)2,2',2''-terpyridyl-4,4',4''-tricarboxylate)ruthenium(II) (black dye).

For cases in which the TiO_2 electrode is patterned, the copolymer is stamped upon a layer of TiO_2 made by the procedure above. A 10 mM solution of PS-*b*-PVPQ/NMP and water solution (1:1; vol) is used as the ink for microcontact stamping using a PDMS stamp with a pattern made from a silicon master made by traditional lithography techniques. The stamp is inked by spincoating a small amount of the ink solution directly onto the stamp surface. After evaporation of the solvent, the stamp is dried in nitrogen, and brought into contact with the TiO_2 surface for 10 – 40 minutes under light heating (40°C). PDMS stamps with dot patterns with 10, 5, 1, and 0.32 µm dimensions are used in this work. The PS-*b*-PVPQ is oriented with the PVPQ block bound to the negatively charged OH groups of the TiO_2 surface, presenting a PS hydrophobic block to resist TiO_2 growth while the liquid phase deposition process is repeated.

Porous titania films are made using a template made from a porous film of $(LPEI/PAA)_n$ treated in pH 1.8 – 2.0 aqueous solution. This pH treatment has been documented previously.[10] The LbL film is deposited on a layer of TiO_2, undergoes the porosity transition, and is then placed in the same liquid phase TiO_2 synthesis bath used to produce the patterned samples noted above.

LbL polyelectrolyte films are assembled as follows. LPEI (25,000 M_w) and PAA (90,000 M_w) (Polysciences, Inc.) are used as the cation and anion. The polyelectrolytes are dissolved in Milli-Q water at a concentration of 20 mM with respect to repeat unit molecular weight, and pH adjusted with dilute HCl or NaOH to pH = 5.0. Films are constructed using a modified Carl Zeiss DS50 programmable slide stainer. After preparing the LbL film, substrates are dried with nitrogen and heated at 120 °C for 30 min. OEGDA, LiI, and I_2 are introduced by immersing the device into an aqueous solution at pH = 3.0 for 20 min. The concentration of OEGDA is 20 wt.-%. Thermally evaporated Au, or platinized conductive glass substrates are used as counter electrodes. Certain devices utilized ionic liquid electrolyte gels containing 1-alkyl-3-methylimidazolium iodide (alkyl: C-3-C-9) and I_2.[11]

Patterned and porous TiO_2 samples are characterized using a JEOL 6320 FEGSEM. UV-Vis characterization is carried out using a Cary 5E UV-Vis-NIR dual beam spectrophotometer. The current – voltage (*I-V*) characteristics of the solar cells are tested using an EG&G 263A potentiastat, controlled with PC software. An Oriel 150 W xenon lamp served as a light source in combination with a heat-absorbing filter, an air mass (AM) 0 filter and AM 1.5 direct filter to simulate direct solar irradiance. Light intensity is measured by an Oriel radiant power energy meter with a thermopile detector.

DISCUSSION

Polymer-on-polymer stamping has been shown to be a viable means of creating micro- and nanoscale templates for the deposition of a range of materials, including inorganics such as TiO_2.[8,12] We have adapted this process to enable patterning of a block copolymer (PS-*b*-PVPQ), where the PVPQ block binds electrostatically to the negatively charged oxide surface, leaving a monolayer of PS on the surface. The result is an alternating hydrophobic/hydrophilic patterned surface on a previously deposited TiO_2 film. TiO_2 deposition on the stamped area is inhibited due to the hydrophobicity of PS.

The patterned TiO_2 electrode gives two distinct advantages over a non-patterned surface: 1) increased surface area for binding of sensitizing dye, and 2) possible optical interaction leading to trapping of light within the TiO_2 matrix and better light harvesting capability. When this TiO_2 electrode is combined with a LbL ionic conducting film consisting of $(LPEI/PAA)_{60}$ treated with OEGDA(LiI, I_2) a working solid-state dye sensitized solar cell shows typical current-voltage behavior with nearly 10x improvement in cell efficiency between a non-patterned surface and a 1 μm patterned TiO_2 when utilizing a porphyrin dye (TPPAc) (previously published).[12]

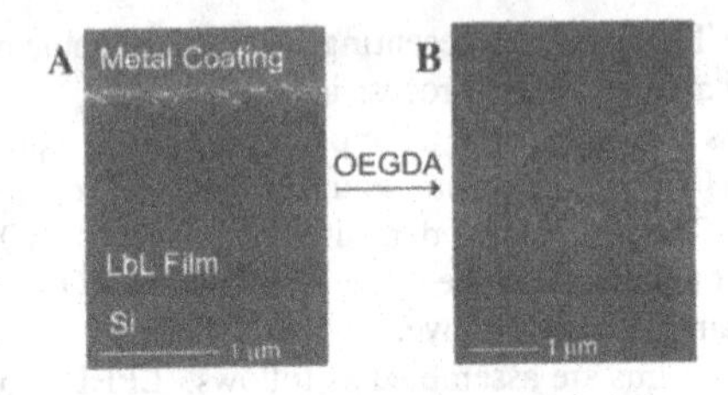

Figure 1. Cross-sectional SEM micrographs of a $(LPEI/PAA)_{25}$ film (a) before and (b) after treatment in OEGDA aqueous solution at pH = 3.0.

One of the primary drawbacks to ionic conducting solid polymer thin films is their low conductivity compared to that of liquid or gel electrolytes. In an attempt to overcome this low conductivity in LbL ionic conducting thin films, we have worked primarily with hydrophilic polyelectrolytes (LPEI) that are amenable to transport of ions along their backbones, similar to traditional PEO solid electrolytes. In addition, the completed multilayer thin films are exposed to a low pH aqueous solution containing OEGDA, which acts to increase the ionic conductivity of the LbL matrix by two primary mechanisms: 1) the OEGDA acts as a humectant which causes plasticization of the LbL matrix by water, and 2) the low pH OEGDA treatment induces a porous transition in the LbL films which causes uptake of OEGDA molecules into the film bulk. The uptake of OEGDA has been confirmed by both FTIR and DSC (results not shown).[13] The porous transition, due to protonation of the PAA acid groups, is shown to be important to increase the ionic conductivity. When changing the pH of the OEGDA immersion solution from pH = 9.0 (no porous transition) to pH = 2.0 (strong porous transition), the ionic conductivity increases from 5×10^{-7} S/cm^2 to 9×10^{-4} S/cm^2. Examination of the OEGDA treated samples using cross-sectional SEM indicated the presence of a more open film structure, although not a completely open network, as is seen in low pH treatment in water. The main structural difference between films treated in OEGDA is the OEGDA molecules remain within the composite film after drying. Figure 1 presents electron micrographs of a (LPEI/PAA) film before and after OEGDA treatment at pH = 3.0. There is clear transition observed after OEGDA treatment, as shown in Figure 1b, where the film thickness is increased 2 to 3-fold.

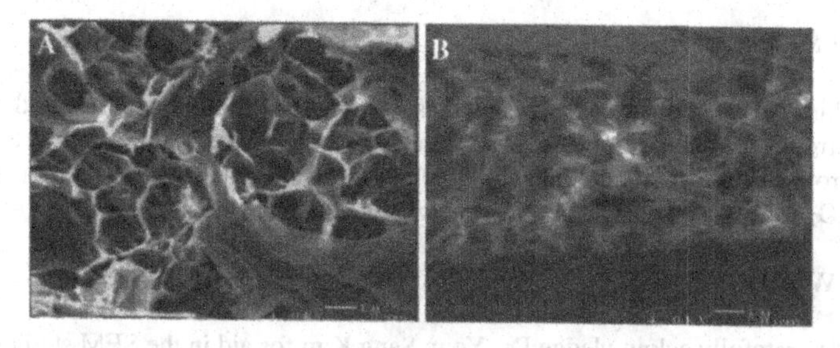

Figure 2. SEM micrograph of the surface (a) and cross-section (b) of a TiO_2 film produced from an LPEI/PAA polyelectrolyte film template.

In an effort to obtain higher surface area TiO_2 films, porous films made from the LPEI/PAA polyelectrolyte pair treated in pH = 1.8 – 2.0 water are used as templates for the deposition of TiO_2 by liquid phase deposition. A similar technique for producing porous TiO_2 films using poly(allylamine hydrochloride) and PAA and TiO_2 deposition from TiF_4 was recently reported.[14] Figure 2 presents scanning electron micrographs of porous TiO_2 films produced from a polyelectrolyte template. Figure 2a shows the highly porous nature of the surface of the TiO_2 film, and Figure 2b shows a cross-sectional view of the film showing the porosity being continuous from the surface of the film to the substrate.

The high surface area of the porous TiO_2 electrode is shown through a higher sensitizing dye (black dye) loading capacity, as presented in the UV-Vis absorption spectra Figure 3a, compared to both patterned and non-patterned TiO_2 surfaces. Figure 3b presents the performance of a photovoltaic device produced using a porous TiO_2 electrode combined with a (PEO/PAA)OEGDA LbL electrolyte layer.

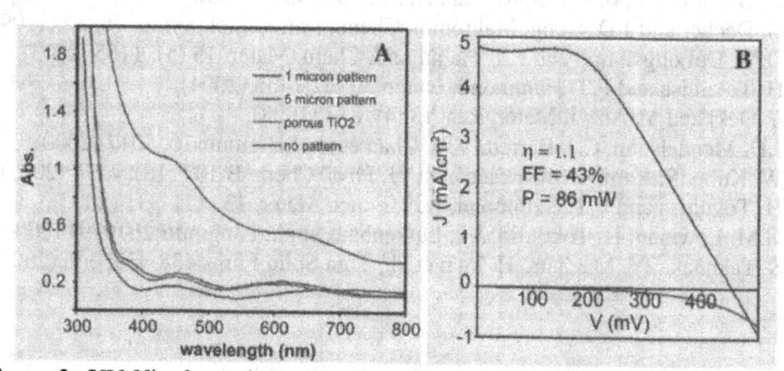

Figure 3. UV-Vis absorption spectra (a) of black dye loading capacity of a various TiO_2 electrodes (patterned, non-patterned, porous), and performance (b) of a device produced using a porous TiO_2 electrode with a (PEO/PAA)OEGDA LbL electrolyte layer, under dark (red) and illuminated (blue) conditions.

CONCLUSIONS

This study demonstrates novel synthesis routes for forming patterned and porous TiO_2 electrodes for photovoltaic devices. When combined with an ionic conducting LbL polyelectrolyte thin film, treated in an oligoethylene glycol diacid, porous-template grown TiO_2 solid-state devices show an energy conversion efficiency of 1.1%.

ACKNOWLEDGMENTS

We gratefully acknowledge Dr. Youn Sang Kim for aid in the SEM studies, Prof. Michael Rubner for helpful discussions, the Office of Naval Research Polymers Program for funding, the use of the facilities of the MIT Center for Materials Science and Engineering, and MRSEC program of the National Science Foundation, and the visiting scientist program of JSR Corporation.

REFERENCES

1. B. O'Regan and M. Grätzel, Nature 353, 737 (1991).
2. K. Tennakone, G.K.R. Senadeera, V.P.S. Perera et al., Chem. Mater. 11 (9), 2474 (1999); A.F. Nogueira, J.R. Durrant, and M.A. DePaoli, Adv. Mater. 13 (11), 826 (2001); E. Stathatos, P. Lianos, U. Lavrencic-Stangar et al., Adv. Mater. 14, 354 (2002).
3. P. Wang, S.M. Zakeeruddin, P. Comte et al., J. Am. Chem. Soc. 125 (5), 1166 (2002); P. Wang, S.M. Zakeeruddin, J.E. Moser et al., Nature Mater. 2 (6), 402 (2003).
4. K.M. Coakley, Y. Liu, M.D. McGehee et al., Adv. Funct. Mater. 13 (4), 301 (2003).
5. U. Bach, D. Lupo, P. Comte et al., Nature 395, 583 (1998).
6. G. Decher and J.D. Hong, Makromol. Chem., Marcromol. Symp. 46, 321 (1991).
7. D.M. DeLongchamp and P.T. Hammond, Chem. Mater. 15 (5), 1165 (2003).
8. H. Tokuhisa and P.T. Hammond, Langmuir 20, 1436 (2004).
9. S. Deki and Y. Aoi, J. Mater. Res. 13 (4), 883 (1998).
10. J.D. Mendelsohn, C.J. Barrett, V.V. Chan et al., Langmuir 16, 5017 (2000).
11. W. Kubo, S. Kambe, S. Nakade et al., J. Phys. Chem. B 107 (18), 4374 (2003).
12. H. Tokuhisa and P.T. Hammond, Adv. Funct. Mater. 13 (11), 831 (2003).
13. G.M. Lowman, H. Tokuhisa, J.L. Lutkenhaus et al., Langmuir 20, 9791 (2004).
14. S. Takenaka, Y. Maehara, H. Imai et al., Thin Solid Films 438-439, 346 (2003).

Mater. Res. Soc. Symp. Proc. Vol. 836 © 2005 Materials Research Society L1.8

Synthesis, Characterization and Thermal Stability of Highly Crystallized Titania Nanotubes

B. Poudel, [1] W. Z. Wang, [1] C. Dames, [2] J. Y. Huang, [1] S. Kunwar, [1] D. Z. Wang, [1] D. Banerjee, [1] G. Chen, [2] and Z. F. Ren [1]
[1] Department of Physics, Boston College, Chestnut Hill, MA 02467
[2] Department of Mechanical Engineering, MIT, Cambridge, MA 02139

ABSTRACT

Gram quantities of titania (TiO_2) nanotubes, with typical outside diameter about 9 nm, wall thickness about 2.5 nm, and length about 600 nm, were synthesized from anatase nano- and micro-powder using the hydrothermal method. The crystallization, structure, and phase stability of the nanotubes at high temperatures were systematically studied. A morphology change from nanotube to nanowire was observed at 650°C. The as-prepared nanotubes were usually contaminated with sodium impurities, other TiO_2-derived phases and were poorly crystallized, but under optimized synthesis conditions the impurity phases was completely removed, resulting in highly crystallized pure nanotubes. The volume filling fraction of the autoclave as well as the concentration of the acid treatment were found to be particularly important for controlling the purity and crystallinity of the resulting nanotubes.

INTRODUCTION

Low-dimensional titania (TiO_2) nanostructures such as nanotubes, nanowires, and nanobelts have attracted much attention in recent years because they have potential applications in electronics, optics, catalysts, sensors, and energy conversion [1-7]. For example, titania nanotubes have shown more than a two-fold increase in short-circuit current density compared to nanoparticles in thin film dye-sensitized solar cells [8]. For all of these applications, purity, crystallinity, and stability at elevated temperatures are particularly important.

Titania nanotubes of various dimensions have been synthesized by different groups using techniques like anodization [9], template method [10], and hydrothermal method [1, 2]. A simple and cost-effective hydrothermal method for the large-scale production of titania nanotubes with small diameters has been introduced by Kasuga *et al.* There is currently much discussion about the formation mechanism and phases of the resulting nanotubes. The formation mechanism has been described in various reports [11-16]. In this context, we have systematically studied the synthesis conditions in order to determine optimal parameters for the growth of pure crystalline anatase nanotubes from the hydrothermal method. By systematically studying the alkali treatment and acid washing steps, we have identified conditions leading to each of the various phases reported previously by different groups, and established the optimal conditions for the formation of pure crystalline titania nanotubes.

EXPERIMENTAL DETAILS

Titania nanotubes were synthesized following the method of Kasuga *et al.* [1].The starting materials were anatase nano-powder or micro-powder, NaOH, and HCl. In the optimized synthesis using micro-powder, 3 g of anatase powder was first mixed with 168 mL of 10 M NaOH, stirred for 1 hour in a beaker, and transferred into a 200 mL autoclave (84% filling fraction). After 30 hours of reaction in the sealed autoclave at 120°C, the product was treated with 1 M HCl solution for at least 2 hours, washed several times with distilled water, and dried in

Sample	Autoclave filling fraction (%)	HCl concentration (M)	Crystallinity	Impurities	
				Sodium	Hydrogen titanate
A	84	0	Good	Present	Present
B	84	0.1	Good	Present	Present
C	84 *	1 *	Excellent	None	None
D	50	0.1	Poor	Present	Present
E	50	1	Poor	Present	None

Table 1. Synthesis conditions, crystallinity, and impurities for various nanotube samples. * represents the optimal condition.

an oven for several hours at 100°C. The synthesis conditions and properties of the resulting nanotubes of all of these samples are summarized in Table 1.

The overall phase of the nanotube samples was analyzed using an x-ray diffractometer (XRD, Bruker-AXS, G8 GAADS) with Cu Kα radiation, and the phase of an individual nanotube or a small region of the nanotubes was determined using selected-area electron diffraction (SAED). The morphology of the samples was examined using scanning electron microscopy (SEM, GEOL-6340F) and transmission electron microscopy (TEM, GEOL-2010F), while the atomic composition was determined using an EDX spectrometer attached to the TEM.

RESULTS AND DISCUSSION

Synthesis of pure crystalline titania nanotubes requires optimization of the temperature

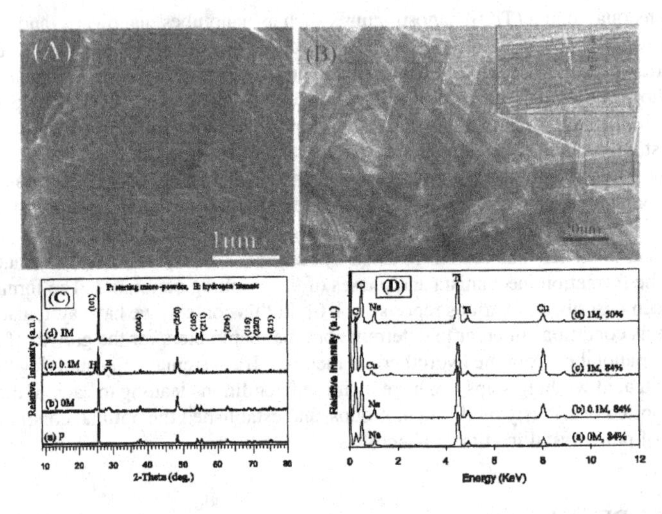

Figure 1. (A) SEM, (B) TEM images, and XRD patterns (C) of the nanotubes synthesized under optimal autoclave conditions, showing good uniformity, crystallinity, and anatase phase. EDX pattern (D) shows that only in the optimal condition sodium impurity is absent.

during the hydrothermal treatment, filling fraction, and the concentration of the acid during the wash step. SEM [Fig. 1(A)] and TEM [Fig. 1(B)] images of a sample from optimal synthesis conditions (sample C) clearly show the abundance of highly crystallized titania nanotubes. The typical length of these nanotubes is ~ 600 nm and the outside diameter is ~ 9 nm. The spacing between layers of the wall is ~0.71 nm [Fig. 1(B) inset], significantly less than that observed by other groups [3, 14]. This may be because the excess H_2O and/or hydrogen titanate leading to larger layer spacing.

For all of the nanotube samples from optimal synthesis conditions, EDX studies show the molar ratio of Ti and O to be close to 1:2, verifying the TiO_2 composition. Figs. 1C(b)-(d) show the XRD patterns of the as-prepared nanotubes from the optimal autoclave conditions treated with various acid concentrations, as compared to the original starting anatase micro-powder [Fig. 1C(a)], confirming the anatase phase of the nanotubes.

To our surprise, the autoclave filling fraction had a strong influence on the crystallinity of the resulting nanotubes, with different optimal filling fractions for micro- and nano-powders. For the micro-powder in the standard hydrothermal protocol at 120°C for 30 hours reaction time, it was necessary to fill the autoclave at least 84% by volume to achieve excellent crystallinity. At lower volume fractions, varying the temperature between 110°C and 135°C did not improve the crystallinity. For nano-powder, it was necessary to fill the autoclave to at least 90% to achieve good crystallinity. Similar requirements of high autoclave filling fraction were seen for experiments using a smaller autoclave vessel (125 mL) or in an argon environment, ruling out the possible effect on the crystallinity caused by the presence of atmospheric gases. One possible explanation for this unusual observation is that the pressure inside the vessels during the autoclave step is strongly dependent on the filling fraction above ~80% because of the thermal expansion of the different volumes of solution: although we have not measured the pressure directly, simple calculations suggest that the pressure inside an 84% filled vessel exceeds that of a 50% filled vessel by about 50 kPa (392 vs. 342 kPa). For the 90% filling fraction required for nano-powder, the calculated change is 136 kPa (478 vs. 342 kPa).

We found that acid treatment was necessary and identified an optimal concentration between 0.5 and 1 M for high yield of pure nanotubes. Concentrations above 2 M destroyed the nanotubes leaving only ~ 100 nm clumps, while concentrations below 0.5 M failed to remove the sodium impurities. However, even at these optimal concentrations the sodium impurities could be removed only when the autoclave step was performed close to the optimal volume faction (≥84% for micro-powder, and ≥90% for nano-powder).

This observation is supported by comparing the EDX spectra in Fig. 1D, which shows that for 84% autoclave filling fraction, sodium remains after washing with water [sample A, Fig. 1A (a)] or treating with 0.1 M acid [sample B, Fig. 1A (b)], but can be removed completely after treating with 1 M acid [sample C, Fig. 1A (c)]. At 50% autoclave filling fraction, however, sodium remains even after treating with 1 M acid [sample E, Fig. 1A (d)]. These results suggest that good crystallinity prior to the acid treatment may cause the sodium atoms to be more weakly bound, and more easily removed by the acid. Thus, we conclude that the nanotube crystallinity is controlled primarily by the autoclave step, while the impurities are controlled primarily by the acid treatment.

An additional outcome of the systematic study outlined above is that we have identified synthesis conditions that consistently result in the other titanium oxide phases that have been previously reported, including pure sodium titanate, pure hydrogen titanate, and mixtures of

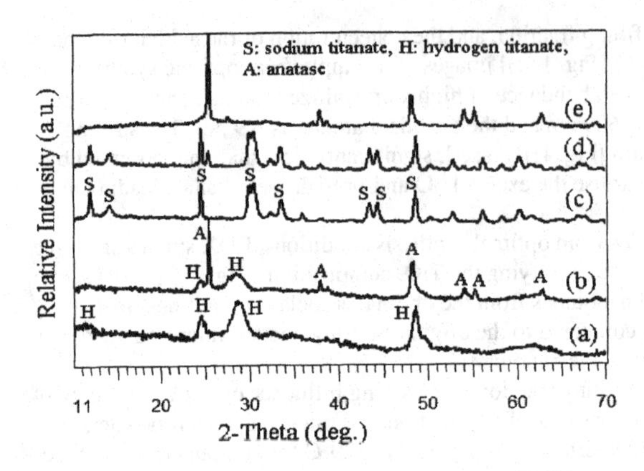

Figure 2. XRD patterns of nanotubes synthesized in different synthesis conditions and annealed at various temperature resulting in various phases.

either of these with titania. The XRD spectra of these phases are shown in Fig. 2 and discussed further below.

Sodium impurities are apparently always present prior to the acid treatment regardless of the autoclave temperature and filling fraction. However, for these impurities to crystallize into pure sodium titanate [Fig. 2(c)], we have found that the autoclave filling fraction should be less than 84%, the acid concentration should be less than 0.5 M, and the annealing temperature should be above 550°C. If the autoclave filling fraction is higher and the acid concentration is lower, the resulting nanotubes will be mixture of hydrogen titanate and anatase at room temperature [Fig. 2(b)], and become a mixture of sodium titanate and anatase after annealing at temperatures above 550°C [Fig. 2(d)]. If the acid concentration is higher, all the sodium will be removed and no sodium titanate phase will appear after the annealing [Fig. 2(e)]. Hydrogen titanate phase [Fig. 2(a)] is observed when the acid concentration is 0.1 M, treatment temperature 110°C, and the autoclave filling fraction less than 80%.

We have also studied the temperature stability of nanotubes prepared at optimal autoclave conditions and treated with 0.1 M and 1 M acid (sample B and C respectively), including the crystallization of the sodium titanate phase, the transformation of anatase to rutile, and an unexpected change from nanotube to nanowire morphology. We observed that both the samples were stable in phase up to 700°C and completely transformed into 870°C. However, the sample treated with 0.1 M developed new phases at 600°C matching well with sodium titanate spectra (XRD not shown here).

Although the original morphology is preserved for nanotubes annealed at temperatures up to 550°C, the SEM image in Fig. 3(a) shows that the nanotubes annealed at 650°C have changed into shorter rods with larger diameters. HRTEM [Fig. 3(c)] shows that these shorter rods are, in fact, nanowires with diameters more than 20 nm. Similar morphology changes are also observed for nanotubes treated with 1 M acid [sample C, Figs. 3(b) and 3(d)]. To better understand the formation, phase, and morphology change during heat treatment, we are performing a systematic study using in-situ high-temperature TEM observations, which will be reported in a forthcoming paper.

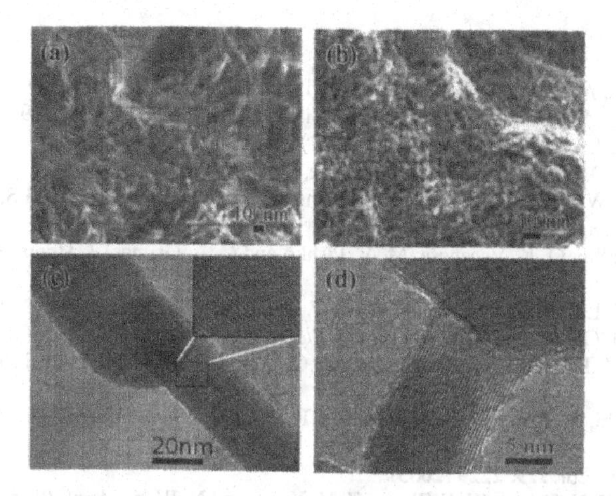

Figure 3. SEM (a) and TEM (c) images of nanotubes synthesized under optimal autoclave condition and treated with 0.1 M HCl (sample B), and annealed at 650°C, (b) SEM and (d) TEM images of similar samples but treated with 1 M HCl (sample C).

CONCLUSIONS

We have identified conditions to synthesize highly crystallized titania nanotubes, with no sodium impurities and no hydrogen titanate phase, from the hydrothermal method starting from either nano- or micro-powders of bulk anatase. We conclude that the nanotube crystallinity is controlled primarily by the autoclave step, while the impurities are controlled primarily by the acid treatment. It is necessary to fill the autoclave more than 84% by volume for micro- and 90% for nano-powder and treat with 0.5 to 1 M HCl to obtain nanotubes with high purity and crystallinity. The anatase phase of the as-prepared nanotubes is stable up to 700°C, but the morphology changes from nanotubes to nanowires above 550°C. If sodium impurities are present, they typically crystallize into sodium titanate above 600°C. Synthesis conditions have also been identified to obtain pure hydrogen titanate and sodium titanate nanotubes, thus clarifying some of the current controversy in the literature about the phases of nanotubes grown with the hydrothermal method.

ACKNOWLEDGEMENTS

The work performed by Boston College is supported by the US Army Research Development and Engineering Command, Natick Soldier Center, under grant DAAD16-03-0052 and NASA. The work performed by MIT is supported by DOE, under grant FG02-02ER45977.

REFERENCES

1. T. Kasuga, M. Hiramatsu, A. Hoson, T. Sekino, and K. Nihara, *Langmuir* **14**, 3160 (1998).
2. T. Kasuga, M. Hiramatsu, A. Hoson, T. Sekino, and K. Nihara, *Adv. Mater.* **11**, 1307 (1999).
3. W. Wang, O. Varghese, M. Paulose, and C. A. Grimes, *J. Mater. Res.* **19**, 417 (2004).
4. W. Wang, C. A. Grimes, O. K. Varghese, and M. Paulose, Unpublished.
5. Y. Lei, L. D. Zhang, G. W. Meng, G. H. Li, X. Y. Zhang, C. H. Liang, W. Chen, and S. X. Wang, *Appl. Phys. Lett.* **78**, 1125 (2001).
6. G. K. Mor, M. A. Carvalho, O. K. Varghese, M. V. Pishko and C. A. Grimes, *J. Mater. Res.* **19**, 628 (2004).
7. Z. R. Tian, J. A. Voigt, J. Liu, B. Mckenzie, and H. Xu, *J. Am. Chem. Soc* **125**, 12384 (2003).
8. M. Adachi, Y. Murata, I. Okada, and S. Yoshikawa, *J. Electrochem. Soc.* **150**, G488 (2003).
9. K. Varghese, D. Gong. M. Paulose, C. A. Grimes, and E. C. Dickey, *J. Mater. Res.* **18**, 156 (2003).
10. J. H. Jung, H. Kobayashi, K. Bommel, S. Shinkai and T. Shimizu, *Chem. Mater.*, **14**, 1445 (2002).
11. X. Sun and Y. Li, Chem. Eur. J. **9**, 2229 (2003).
12. B. D. Yao, Y. F. Chan, X. Y. Zhang, W. F. Zhang, Z. Y. Yang, and N. Wang, *Appl. Phys. Lett.* **82**, 281 (2003).
13. G. Du, Q. Chen, R. C. Che, Z. Y. Yuan, and L. M. Peng, *Appl. Phys. Lett.* **79**, 3702 (2001).
14. Y. Suzuki, S. Yoshikawa, *J. Mater. Res.* **19**, 982 (2004).
15. D. V. Bavykin, V. N. Parmon, A. A. Lapkin and F. C. Walsh, *J. Mater. Chem.* **14,** 3370 (2004).
16. Q. Chen, W. Zhou, G. Du, and L. M. Peng, *Adv. Mater.* **14**, 1208 (2002).

Effect of Anodization Bath Chemistry on Photochemical Water Splitting Using Titania Nanotubes

Gopal K. Mor, Oomman K. Varghese, Maggie Paulose, Karthik Shankar, and Craig A. Grimes*
Department of Electrical Engineering, and Department of Materials Science and Engineering
The Pennsylvania State University, University Park, PA 16802 USA.
*Email: cgrimes@engr.psu.edu

ABSTRACT

In this study highly-ordered titania nanotube arrays of variable wall-thickness and length are used to photocleave water under ultraviolet irradiation. We demonstrate that the wall thickness, and length, of the nanotubes can be controlled via anodization bath composition and temperature. The nanotube length and wall thickness are key parameters influencing the magnitude of the photoanodic response and the overall efficiency of the water-splitting reaction. For 22 nm inner-pore diameter nanotube-arrays 6 μm in length, with 9 nm wall thickness, upon 320-400 nm illumination at an intensity of 100 mW/cm^2, hydrogen gas was generated at the power-time normalized rate of 51 mL/hr•W at an overall conversion efficiency of 12.5%. To the best of our knowledge, this hydrogen generation rate is the highest reported for a titania-based photoelectrochemical cell.

Keywords: Titania, anodization, nanotube, porous, anatase.

INTRODUCTION

The principal impetus towards fabricating nano-dimensional materials lies in the promise of achieving unique properties and superior performance due to their inherent nano-architectures. Our interest has been in fabrication of ordered titania nanotube arrays, by anodization of a starting titanium thick or thin film [1]. Titania nanotubes fabricated by anodization are highly-ordered, high-aspect ratio structures with nanocrystalline walls oriented perpendicular to the substrate. The nanotubes have a well-defined and controllable pore size, wall thickness and tube-length. The nanotube arrays demonstrate unique material properties; for example titania nanotube array based resistive gas sensors exhibit an amazing 1,000,000,000% change in electrical resistance upon exposure to 1000 ppm of hydrogen gas at 23°C [2-4].

The electrolyte composition and applied anodic potential primarily determine the oxide structure resulting from an anodization. Sulfuric acid has been the most widely used electrolyte, for which a non-porous TiO_2 film is formed at low potentials, and porous TiO_2 film can be formed at high potentials due to electrical breakdown of the oxide [5,6]. In fluoride containing electrolytes, the anodization of titanium is accompanied with the chemical dissolution of titanium oxide due to the formation of TiF_6^{2-}. Highly ordered nanotube arrays, in place of porous or non-porous structures, are formed at relatively low potentials, say 10V, as a result of the competition between the electrochemical etching and the chemical dissolution [7]. These arrays were reported to be obtained only in fluoride containing acids [6], or in a mixture of fluoride containing acid and other acids, including sulfuric acid [8], and acetic acid [9]. Since the high rate of chemical dissolution of the oxide in hydrofluoric acid containing electrolytes is the factor directly limiting the nanotube length, our efforts have focused on controlling the dissolution rate by adjusting the pH of the electrolyte through additives and using different fluorine containing salts, such as potassium fluoride (KF) and sodium fluoride (NaF) instead of HF.

We make note that potassium fluoride (KF) is totally dissociated in water, with F^- hydrolyzing with water to form HF: $F^- + H_2O = HF + OH^-$. Therefore a KF electrolyte, without the addition of other acids, will result in an alkaline solution. The ultimate concentration of F^- and HF are determined by the solution pH. Prior to the addition of KF, the electrolyte pH was adjusted by adding sodium hydroxide, sulfuric acid, sodium hydrogen sulfate, citric acid and in one case phosphoric acid. By optimizing the anodization conditions we have achieved a highly ordered nanotube arrays approximately 6.0 μm in length, an order of magnitude higher than the previously reported arrays [9]. Here we detail methods for fabrication of these nanotube arrays, that are substantially longer than any previously reported, detailing the effect of anodization process variables such as electrolyte pH. As the results obtained from sodium fluoride containing electrolyte were identical to that from potassium fluoride, for brevity we confine our discussion to the fabrication of nanotubes using the latter electrolyte.

Table 1. Electrolyte pH and composition, anodization conditions, and size of the resulting nanotubes.

No.	Electrolyte[(1)]				pH[(2)]	V	t	D	L	Q[(3)]
	F^-	SO_4^{2-}	PO_4^{3-}	Cit		(V)	(hr)	(nm)	(μm)	
01	0.1	1.0	-	-	<1	5	1	10±2	-	No NT
02	0.1	1.0	-	-	<1	10	1	40±5	0.28±0.02	NT
03	0.1	1.0	-	-	<1	15	1	80±9	-	NT
04	0.1	1.0	-	-	<1	20	1	100±11	0.48±0.03	NT
05	0.1	1.0	-	-	<1	25	1	110±12	0.56±0.04	NT
06	0.1	1.0	-	-	<1	30	1	-	-	No NT
07	0.1	1.0	-	-	<1	20	6.5	100±11	0.43±0.03	NT
08	0.1	2.0	-	-	<1	20	1	100±11	0.45±0.03	NT
09	0.1	1.0	-	0.2	1.3	10	20	30±5	0.32±0.03	NT
10	0.1	1.0	-	0.2	2.8	10	20	30±5	0.59±0.05	NT
11	0.1	1.0	-	0.2	2.8	15	20	50±5	1.00±0.05	NT
12	0.1	1.0	-	0.2	2.8	25	20	115±10	1.50±0.04	NT
13	0.1	1.0	-	0.2	3.8	10	20	30±5	0.80±0.06	NT
14	0.1	1.0	-	0.2	3.8	10	60	30±5	1.80±0.06	NT
15	0.1	1.0	-	0.2	3.8	10	90	30±5	2.30±0.08	NT
16	0.1	1.0	-	0.2	4.5	10	20	30±5	1.05±0.04	NT
17	0.1	1.0	-	0.2	4.5	25	20	115±5	6.00±0.10	NT
18	0.1	1.0	-	0.2	5.0	10	20	30±5	1.40±0.06	NT
19	0.1	1.0	0.1	0.2	6.4	10	24	-	-	No NT
20	-	2.0	-	-	<1	10	24	-	-	No NT

(1) Electrolyte components are in mol/L. (2) pH < 1 represents a 1.0 or 2.0 mol/L H_2SO_4 medium. (3) Quality Q of resulting nanotubes. NT: nanotubes uniformly across substrate. No NT: no nanotubes or partly developed nanotube/porous structures. (4) Cit: citrate; t: time; D: inner diameter of nanotube; L: length of nanotube. (5) SO_4^{2-} is from addition of H_2SO_4 or $NaHSO_4$; PO_4^{3-} is addition of potassium hydrogenphosphate $K_2HP_3O_4$; Cit denotes citric acid from its salt, $HO(CO_2Na)(CH_2CO_2Na)_2{\bullet}2H_2O$.

EXPERIMENTAL

Titanium foils (99.8% pure), and needed chemicals were purchased from Aldrich. The titanium foils were ultrasonically cleaned in Micro-90 clean solution, followed by distilled water and acetone ultrasonic rinse prior to anodization. The electrolyte compositions are shown in **Table 1**. Prior to KF addition the desired pH was obtained by adding NaOH, sulfuric acid (pH 1-2), sodium hydrogen sulfate, or citric acid (pH 2.5-7.5). The anodization was conducted in a two-electrode electrochemical cell with a platinum foil as cathode at a constant potential. The photocurrent at the TiO_2 electrode, i.e. the working electrode, was measured using a scanning potentiostat, at a scan rate of 20 mV/sec. The illuminated area of the working electrode is 1 cm^2. 1 M KOH purged with N_2 was used as the electrolyte. A 50 W metal hydride lamp (EXFO Lite) was used as the UV light source. The light was passed through broadband optical filters, which allowed only wavelengths between 320 nm and 400 nm to be incident on the TiO_2 photoanode at a measured intensity of 100 mW/cm^2. The gaseous photoproducts were identified and measured by a SRI 8610C gas chromatograph (GC) equipped with a thermal conductivity detector.

RESULTS AND DISCUSSION

The overall reactions for anodic oxidation of titanium can be represented as $2H_2O \rightarrow O_2 + 4e + 4H^+$, and $Ti + O_2 \rightarrow TiO_2$. At the start of the process an oxide layer forms on the surface of titanium which at constant potential brings a decrease in current in the circuit. In the absence of F^- the resistance will increase to a point where the anodic reaction stops. However in the presence of F^- the oxide layer partially dissolves forming pits: $TiO_2 + 6F^- + 4H^+ \rightarrow TiF_6^{2-} + 2H_2O$. This chemical dissolution, the key for the self-organized formation of the nanotube arrays, reduces the thickness of the oxide layer (barrier layer) keeping the electrochemical etching (field assisted oxidation and dissolution) process active. No nanotubes can be formed if the chemical dissolution is too high or too low. The electrochemical etch rate depends on anodization potential as well as concentration of electrolytes. If the electrochemical etch proceeds faster than the chemical dissolution the thickness of the barrier layer increases, which in turn reduces the electrochemical etching process to the rate determined by chemical dissolution. The nanotube length increases until the electrochemical etch rate equals the chemical dissolution rate of the top surface of the nanotubes. After this point is reached the nanotube length will be independent of the anodization duration, as determined for a given electrolyte concentration and anodization potential. The chemical dissolution rate is determined by the F^- concentration and solution pH. With increasing F^- and H^+ concentrations chemical dissolution increases. Previous investigations [10] have shown that only in a certain F^- concentration range can nanotube arrays be achieved; from 0.05 to 0.3 mol/L in acidic solution. The anodic potential at which nanotubes are formed is related to the F^- concentration, with higher potentials requiring higher F^- concentrations.

Electrolyte pH affects both the behavior of the electrochemical etch, and chemical dissolution owing to the hydrolysis of titanium ions. With increasing pH the hydrolysis content increases, which slows the rate of chemical dissolution. As shown in **Table 1,** longer nanotubes can be formed in higher pH solution. For a potential of 25V, with pH increasing from strong acidity (Sample 05, pH<1) to weak acidity (Sample 17, pH 4.5), nanotube length increased from 0.56 μm to 6.0 μm; for 10V, the length increased from 0.28 μm (Sample 2, pH<1) to 1.4 μm (Sample 18, pH 5.0). For a particular pH, the length increases with applied potential (Samples 10-12 and 16-17). When the potential increased from 10V to 25V, the length increased from 0.59 μm to 1.5 μm for pH=2.8 and from 1.05 μm to 6.0 μm for pH=4.5. At a particular pH, the pore size of the nanotubes was found to be increasing with anodization potential (Sample 10, 11, 12). However, the pore size was independent of the pH at a particular potential.

Figure 1 shows an illustrative cross-sectional image of a 25°C sample anodized at 10 V. **Figure 2** shows the glancing angle x-ray diffraction (GAXRD) patterns of Sample 17, the longest nanotube sample, annealed at different temperatures up to 580°C. The crystallization occurs in

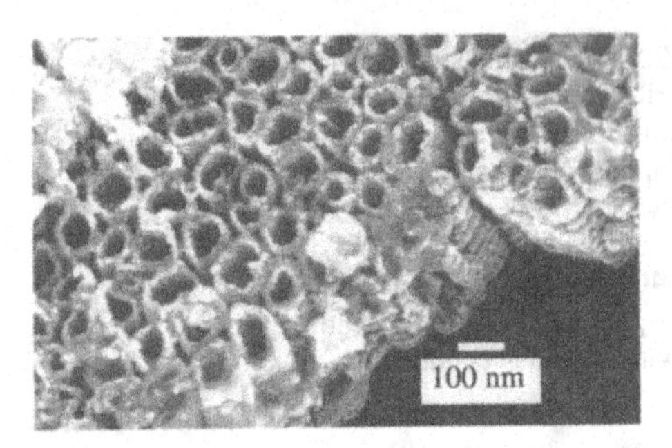

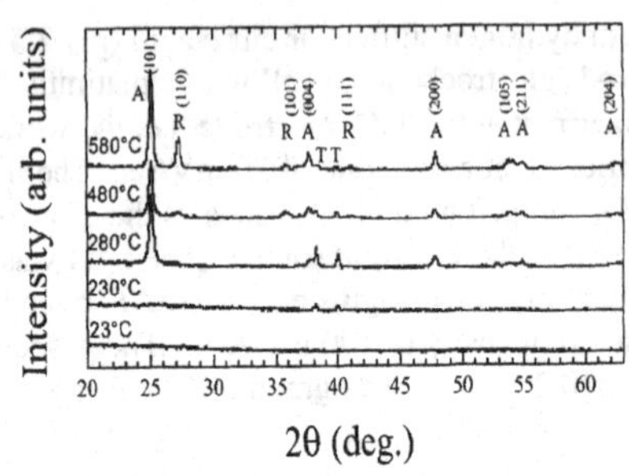

Figure 1. Cross-sectional FESEM image of 10 V titania nanotube arrays anodized at 25°C
Figure 2. GAXRD patterns of samples 17 (pH=4.5, 25V) annealed at different temperatures.

anatase phase at a temperature near 280°C. Electrolyte concentration or pH has no apparent influence on the crystallization temperature of the nanotubes. As the temperature increases the crystallinity increases as more amorphous regions became crystalline and co-existence of rutile phase with the anatase phase can be seen at 480°C. On increasing the annealing temperature further to 580°C, rutile concentration increases as evident from the higher ratio of anatase 101 peak to rutile 110 peak. No sign of nanotube disintegration was observed in the FESEM images of the 580°C annealed samples.

TEM images and selected area diffraction (SAD) patterns of Sample 17 after annealing at 480°C were taken. The studies indicate that the nanotube walls contain only anatase crystals and they have not undergone transition to rutile phase. The TEM data in conjunction with the GAXRD patterns indicate that anatase crystals are formed at the nanotube-Ti substrate interface region as a result of the oxidation of the metal at elevated temperatures and in the nanotubes. At a temperature between 480°C and 580°C, the bigger anatase crystals at the nanotube-Ti substrate interface region are converted into rutile. The constraints imposed by the nanotube walls, however, make critical radii needed for rutile nucleation very large [10]. This prevents the anatase phase at the nanotube walls from undergoing transition to rutile phase. Therefore, nanotubes, annealed at temperatures between 480°C and 580°C, can be considered as anatase crystallites stacked in cylindrical shape on a rutile foundation.

Using an electrolyte comprised of 0.5% HF + acetic acid mixed in a 7:1 ratio, titania nanotube arrays were anodized at 10 V at four different electrolyte bath temperatures, 5°C, 25°C, 35°C and 50°C. The resulting wall thicknesses (WT), and lengths (L), are as follow (in nm). 5°C: WT=34, L=224. 25°C: WT=24, L=176. 35°C: WT=13.5, L=156. 50°C: WT=9, L=120. The pore diameter is essentially the same (22 nm) for the 10 V anodized titania nanotube arrays fabricated at these different temperatures. As the wall thickness increases with decreasing anodization temperature the voids in the interpore areas fill; with the tubes becoming more interconnected the discrete tube-like structure approaches a nanoporous structure in appearance.

Figure 3 shows the photocurrent density versus potential of the titania nanotube-array electrodes, measured in 1 M KOH electrolyte as a function of anodization bath temperature. At 1.5 V the photocurrent density of the 5°C anodized sample is more than three times the value for the sample anodized at 50°C. The lower anodization temperature also increases the slope of the photocurrent-potential characteristic. For crystalline semiconductors, under certain limitations the squared photocurrent is a linear function of the applied bias. In **Figure 4** squared photocurrent is plotted against electrode potential, the traces closely approximate a straight line at low values of

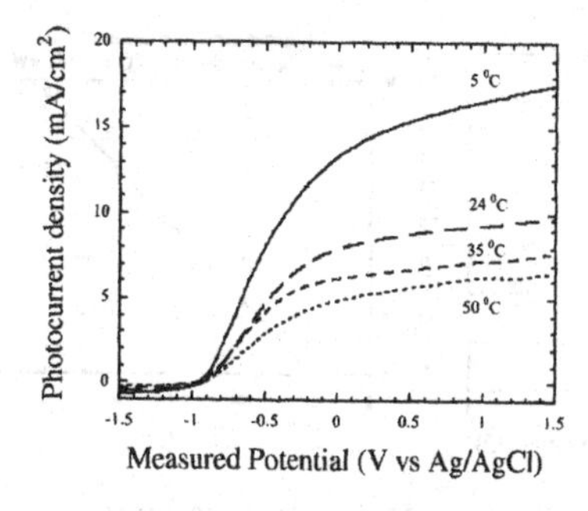

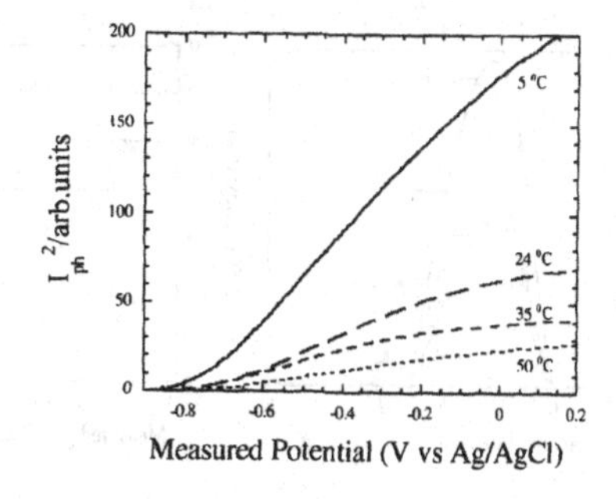

Fig. 3 Variation of photocurrent density versus measured potential [vs. Ag/AgCl] for 10 V samples anodized at four anodization bath temperatures, i.e. 5°C, 25°C, 35°C and 50°C.

Fig. 4 Squared-photocurrent vs. potential for 10 V samples as a function of bath temperature.

applied bias indicative that photogenerated charges being separated by the electric field of the depletion layer [11]. At higher anodic polarization the squared photocurrent-potential plot deviates from linearity for all the samples due to band-bending [11]. As can be seen in **Figure 4**, the sample anodized at 5°C adheres to linearity over the largest bias range; the quantum efficiency of this sample, at 337 nm, is approximately 95%.

The titania nanotube array architecture results in a large effective surface area in close proximity with the electrolyte thus enabling diffusive transport of photogenerated holes to oxidizable species in the electrolyte. Separation of photogenerated charges is assisted by action of the depletion region electric field. Minority carriers generated within a 'retrieval' length from the material surface, that is a distance from the surface equal to the sum of the depletion layer width and the diffusion length, escape recombination and reach the electrolyte. The relevant structure sizes of the titania nanotube arrays, i.e. half the wall thickness, are all smaller than 20 nm which is less than the retrieval length of crystalline titania, hence bulk recombination is greatly reduced and the quantum yield enhanced.

Due to light scattering within a porous structure incident photons are more effectively absorbed than on a flat electrode [12]. However while bulk recombination is reduced by the nanotube architecture, photogenerated minority carriers can be trapped by surface states. Hence recombination between separated electrons and holes is sensitive to the band-bending potential that the electrons have to overcome to reach the surface where (unwanted) recombination takes place. In general, within the nanometer regime, as the size of the particle or grain decreases so does its ability to sustain a significant amount of band-bending. As a consequence, we can anticipate that the titania nanotube-array with 34 nm wall-thickness has greater band-bending than a similar array with 9 nm walls. The enhanced band-bending of the thicker walls decreases surface recombination rates thus increasing the photocurrent.

Photoconversion efficiency [4], at 320 nm to 400 nm, as a function of potential for the photoanodes made as a variation of temperature is shown in **Figure 5**; A maximum conversion efficiency of 6.8% is obtained for nanotubes anodized at 5°C; a normalized hydrogen generation rate of 36 mL/W•hr was achieved for this sample. **Figure 6** shows the photo-conversion efficiency of Sample 17, 22 nm diameter 6 μm long nanotube array, of approximately 12.5%. **Figure 7** shows a quantitative description of the hydrogen evolution as a function of time for a photoelectrochemical cell with Pt mesh cathode and photoanode consisting of Sample 17; a

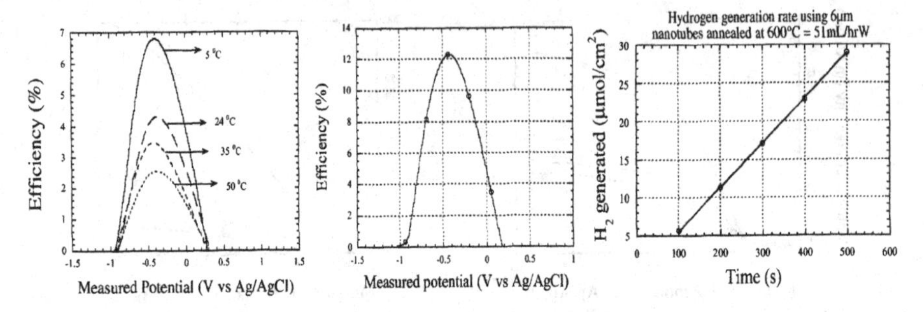

Fig. 5 Photoconversion efficiency as a function of measured potential [vs. Ag/AgCl] for 10V samples anodized at four temperatures [i.e. 5°C, 25°C, 35°C and 50°C].

Fig. 6 Photoconversion efficiency for Sample 17, Table I.

Fig. 7 Hydrogen photoproduction using Sample 17 as a photoanode with a Pt mesh cathode.

normalized hydrogen generation rate of 51 mL/W•hr is achieved. A nanotube sample of Sample 17 recipe but 4.4 μm length showed a photoconversion efficiency of ≈ 11.8%, with a hydrogen gas evolution rate of 44 mL/hr•W. In all cases gas chromatographic analysis was used to verify that the volume ratio of the evolved hydrogen and oxygen was 2:1, which confirmed water-splitting.

CONCLUSIONS

The ability to achieve longer nanotubes increases with pH so long as the electrolyte remains acidic. Increasing pH reduced both the nanotube growth rate, and the rate of dissolution, with more time required for a growth-dissolution equilibrium to be achieved. This made the length of the nanotubes grown in weak acidic electrolytes more time dependent. For a 20 hr anodization, we have fabricated a nanotube array 6.0 μm in length in an electrolyte of 0.1mol/L KF, 1.0mol/L of H_2SO_4 and 0.2 mol/L of citric acid at pH 4.5, a factor of 10 longer than anodic oxidation fabricated nanotube arrays previously reported [1,9]. For Sample 17, a 6 μm nanotube array, a normalized hydrogen generation rate of 51 mL/W•hr is achieved, to the best of our knowledge the best hydrogen generation rate of any titania based photoelectrochemical cell.

REFERENCES

[1] O.K. Varghese, M. Paulose, C.A. Grimes, E.C. Dickey, *J. Mater. Res.* 18, 156 (2003).
[2] O.K. Varghese, G.K. Mor, C.A. Grimes, M. Paulose *J. Nanosci. Nanotech.*, 4, 733 (2004).
[3] G.K. Mor, O.K. Varghese, M.V. Pishko, C.A. Grimes, *J. Mater. Res.* 19, 628 (2004).
[4] Mor, G. K.; Shankar, K.; Varghese, O. K.; Grimes, C. A. *J. Mater. Res.* 19, 2989 (2004).
[5] B.C. Yang, M. Uchida, H.M. Kim, X.D. Zhang, T. Kokubo, *Biomaterials*, 25, 1003 (2004).
[6] Y.T. Sul, C.B. Johansson, Y. Jeong, T. Albrektsson, *Med. Eng. Phys.* 23, 329 (2001).
[7] D. Gong, C.A. Grimes, O.K. Varghese, …, E.C. Dickey, *J. Mater. Res.* 16, 3331 (2001).
[8] R. Beranek, H. Hildebrand, P. Schmuki, *Electrochem. & Solid State Lett.* 6, B12 (2003).
[9] G.K. Mor, O.K. Varghese, M. Paulose, C.A. Grimes, *J. Mater. Res.* 18, 2588 (2003).
[10] Y. Ohya, H. Saiki, T. Tanaka and Y. Takahashi, *J. Am. Ceram. Soc.* 79, 825 (1996).
[11] Pleskov, Y. V.; Krotova, M. D. *Electrochimica Acta* 38, 107 (1993).
[12] Marin, F.I.; Hamstra, M.A.; Vanmaekelbergh, D., *J. Electrochemical Soc. 143*, 1137 (1996).

Mater. Res. Soc. Symp. Proc. Vol. 836 © 2005 Materials Research Society L1.11

Performance of a solid-state Photoelectrochromic window

Ursa Opara-Krasovec [1], Anneke Georg[2], Andreas Georg[3] and Marko Topic[1]

[1] Laboratory of semiconductor devices, University of Ljubljana, Faculty of electrical engineering, Trzaska 25, SI-1000 Ljubljana, Slovenia
[2] Freiburg Materials Research Centre, Freiburg, Germany
[3] Fraunhofer Institute for Solar Energy Systems, Freiburg, Germany

ABSTRACT

The photoelectrochromic device reported in this paper is a switchable window, which combines a dye-sensitized solar cell with an additional electrochromic layer. The coloring process takes place under open circuit conditions and illumination, bleaching is achieved under short circuit conditions. Solid PE devices in a size up to 10x10 cm^2 employing a solid Li^+ ion conductor are built and characterized. The photopic transmittance of the solid PEC device is varied between 62 % (bleached state) and 1.6 % (coloured state). The corresponding coloring time under 1 sun illumination is about 10 minutes and about 10 minutes are needed for bleaching.

INTRODUCTION

Switchable windows are used to control daylighting and thermal conditions in modern buildings [1]. They represent a new class of glazing that provide a reversible modulation of the optical transmittance, reflectance, absorptance, and emittance. There is no scattering of visible light in any state of the device and the outdoor scene remains clearly visible. The photoelectrochromic device reported in this paper is a new type of switchable window, which combines a dye-sensitized solar cell [2] with an additional electrochromic layer. Electrochromic layers change their transmittance reversibly when electrons and cations are injected. In photoelectrochromic systems the dye solar cell provides the energy for the coloration of the electrochromic layer. Thus, the transmittance of the photoelectrochromic device can be decreased under illumination and can be increased again when illuminated or in the dark. An external voltage supply is not required. Applications of these devices include, for example, switchable sunroofs in cars or smart windows in buildings.

Layer configuration of a photoelectrochromic device is presented on Fig. 1. A sol-gel route followed by dip-coating deposition is used for the preparation of photoactive (dye-sensitized TiO_2) and the electrochromic (WO_3) layers. Pt layer is sputtered on the opposite TCO electrode. Between both electrodes is an electrolyte containing Li^+ ions and a redox couple (I^- and I_3^-) [3,4].

During illumination (upper part of Fig.1), a dye molecule absorbs a photon of the incident light. Then an electron is rapidly injected from the excited state of the dye into the conduction band of the TiO_2 and diffuses to the WO_3. Ionized dye molecules are reduced by I^- in the electrolyte according to the reaction: $3I^- \rightarrow I_3^- + 2e^-$. Lithium ions intercalate into the WO_3 and keep the charges balanced. Upon reduction WO_3 colors blue.

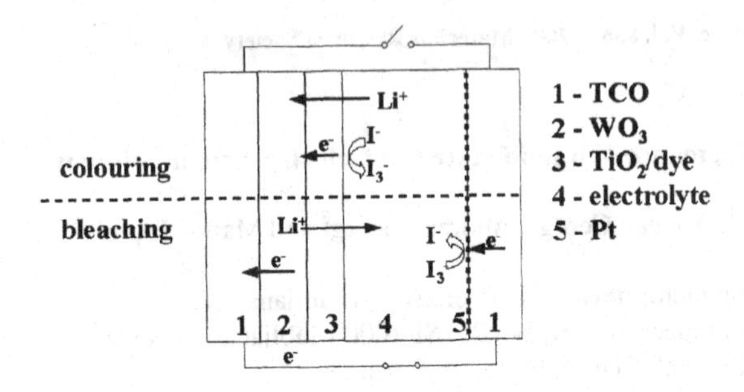

Fig. 1 Layer configuration in photoelectrochromic device (TCO: transparent conductive oxide). The thickness of the layers is not to scale. The upper part of the figure shows coloring processes taking place in the PE device (open circuit conditions and illumination). Bleaching is achieved under short circuit conditions and is presented in the lower part of the figure.

If electrons are allowed to flow via an external circuit from the WO_3 via a TCO layer to the Pt electrode (lower part of Fig.1, external switch closed), then the platinum catalyses the reverse reaction, i.e. the reduction of I_3^- to I^-. Lithium ions leave the WO_3, and the WO_3 is bleached fast. This process occurs also during illumination. If the external switch is open, electrons can leave the WO_3 only by loss reactions. This process is very slow. More detailed information about the system can be found in Refs. 3 and 4.

An alternative photoelectrochromic configuration was first published by Bechinger et al. [5]. The coloring and the bleaching are competing processes, because the bleaching is possible only via loss reactions. Therefore, either fast coloring and bleaching with a small transmittance change [5], or a large transmittance change with slow bleaching are achievable. In our new PE device, the materials can be optimized for coloring and bleaching independently, so it simultaneously allows fast coloring and bleaching and high contrast [3,4].

Solid PE devices in a size 10x10 cm^2 employing a solid Li^+ ion conductor, based on organically modified silane network having incorporated I^- and I_3^-, were built and characterized [6]. The visible transmittance of such solid PE device is 62 % in the bleached state and 1.6 % in the colored state. Time for coloring and bleaching is about 10 minutes. The performance of the PE devices aged at room temperature and exposed to different accelerated conditions (cycling, UV and visible illumination, elevated temperature: 65 and 85°C is also discussed. Thicker layers of TiO_2 will lead to a window, which can be used as a solar cell as well as switched in its transmittance. Using the same layer configuration and a thick WO_3 layer it is possible to build a photovoltaically self-charging battery [7].

EXPERIMENTAL

Layer preparation:

Tungsten oxide and TiO_2 layers were successively dip-coated onto the TCO-coated glass substrate. The detailed information about the preparation of electrodes is found in Refs. [8,9]. The thickness of WO_3 and TiO_2 layers was 800 and 100 nm, respectively. The layers were than left ethanol solution of dye [cis-bis (isothiocyanato) bis (2,2'-bipyridyl-4,4'-dicarboxylato) ruthenium(II)dye] (Solaronix). The Pt layers were sputtered with a thickness of 2 nm. We used

TCO (F:SnO_2)-coated glass substrates from Flabeg with 8 Ω/ . All other chemicals were from Fluka, purum or puriss.

Electrolyte preparation and assembling of PE devices:

As an electrolyte an ormosilane network was synthesized. All chemical reagents are commercially available (Fluka, Aldrich). According to the procedure reported in Ref. [10], equimolar amounts of 3-isocyanatopropyltriethoxysilane and 0,0'Bis (2-Aminopropyl-polyethyleneglycol) (M_w 4000 g/mol) were reacted to form a precursor composed of poly(propylene) oxide bis-end capped with triethoxysilane groups. Hydrolization/condensation reactions were catalyzed with acetic acid. The concentration of LiI in the electrolyte was 0.5M and an appropriate amount of I_2 was added to obtain 0.005M concentration. In addition, a cationic surfactant was added to the ormosil in a mass ratio (surfactant/ormosil) of 1:3. More detailed information can be found in Ref. 6. A polymer foil (Surlyn from Dupont) on each electrode with a thickness of 25 μm served as a spacer and a glue between both electrodes. Electrolyte was doctor-blade printed onto the electrodes. The electrodes were than pressed together and heated up to 90°C for few minutes. The size of the PEC samples was 5x5 cm^2 or 10x10 cm^2.

Characterization of the photoelectrochromic devices

For characterizing the kinetics of the coloring and bleaching processes during illumination and bleaching in the dark, we developed the set-up which measures the transmittance in the same way in both cases [4]. A halogen lamp illuminates the photoelectrochromic sample, and the intensity of the transmitted light is detected by a silicon photodiode. The light intensity of the halogen lamp on the surface of the photoelectrochromic cell corresponds to 1 sun (1000W/m^2), taking into account the mismatch factor of dye solar cells. The two electrodes of the photoelectrochromic device are connected via a variable shunt resistance and a switch. For all configurations, the TiO_2 layer is always directed towards the lamp, so that the coloring of the WO_3 does not alter the light intensity on the TiO_2. The shunt resistance was chosen to be 10Ω, which is similar to the resistance of the TCO layer. With this construction, the voltage in open-circuit state (Uoc) and the current in short-circuit (Isc) state were measured.

The measured value of the transmittance is a folding over the spectra of the halogen lamp, the photodiode and the filter. In order to calculate the visible transmittance out of the measured value of the transmittance, the set-up was calibrated. More detailed information about the set-up and calibration measurements can be found in Ref. 6.
Transmittance spectra were measured with Perkin-Elmer 330 Spectrometer.

DISCUSSION

The structural analysis of the sol-gel made WO_3-TiO_2 layers confirmed their complex nano-structure [4,8,9]. The size of the particles in the WO3 layers is around 30 nm, while the particles in the TiO_2 layer are smaller (10 nm). The WO_3 particles consist of a crystalline monoclinic core (m-WO_3), which is surrounded by 5-10 nm thick amorphous phase [9]. Because of the preparation process, TiO_2 and SiO_2 are left inside the WO_3 layer, mainly situated in the amorphous phase. The high porosity of the WO_3 and TiO_2 layers allows a penetration of the electrolyte into the pores even for polymeric ionic conductors [6].

Fig. 2 shows photos of a photoelectrochromic device in bleached (left, short circuit) and in colored state (right, after illumination in a sun-simulator). The corresponding spectra of the transmittance are presented on Fig.3. The visible (solar) transmittances changes from 62% (41%) to 2% (1%) in roughly 10 min under illumination of 1 sun. The transmittance in the colored state depends strongly on the thickness of the WO_3 layer. It should be noted that the thickness of the Pt-layer and the amount of the dye are small enough to allow a transmittance of 62% in bleached state for the visual spectrum, and 41% for the solar spectrum. The main losses of transmittance are due to the TCO layers (especially in the infra-red region) and the redox-electrolyte, which can be made thinner.

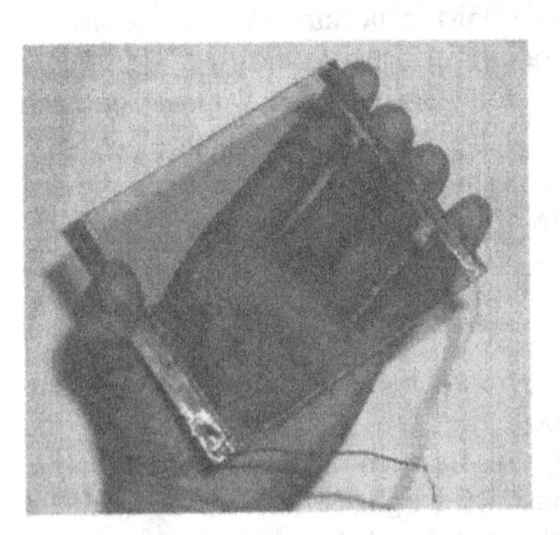

Fig. 2 Sample of a photoelectrochromic device in bleached state (left, short circuit) and colored by illumination equivalent to one sun (right, open circuit).

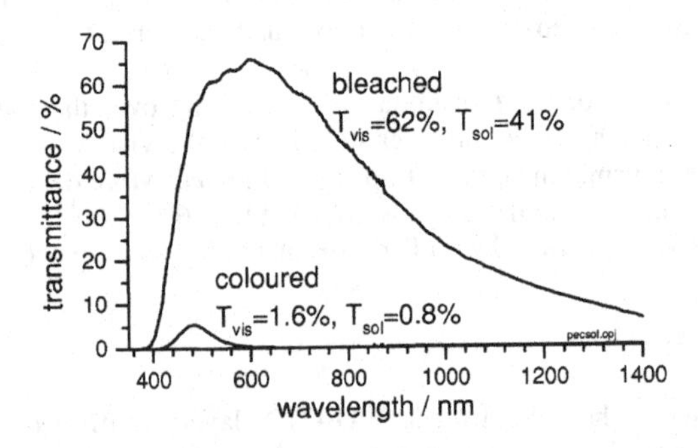

Fig. 3 Transmittance spectra of the PE device in the colored and bleached states. The visible (solar) transmittances changes from 62% (41%) to 2% (1%) in roughly 10 min.

The photoelectrochromic device, as described in this paper, colors on illumination with open circuit and it bleaches in the dark with short circuit within about 10 minutes. It is possible to adjust the electrolyte in that way, that the device bleaches with short circuit under illumination or that it keeps its colors. A slow bleaching occurs with open circuit conditions in the dark (up to 100 hours). Thus, the only impossible process seems to be a coloring in the dark that is usually not required. However, if needed, the device still acts as an electrochromic device, i.e. it can be colored and bleached by applying an external voltage independent on the conditions of illumination.

The coloring and bleaching is shown in Fig. 4. Under illumination and open circuit conditions, the voltage of the device reaches about 0.5V. For short circuit, the current density can be measured with respect to the area of the device, and integrated to get the charge. The charge is roughly proportional to the optical density, the coefficient of proportionality being the coloration efficiency.

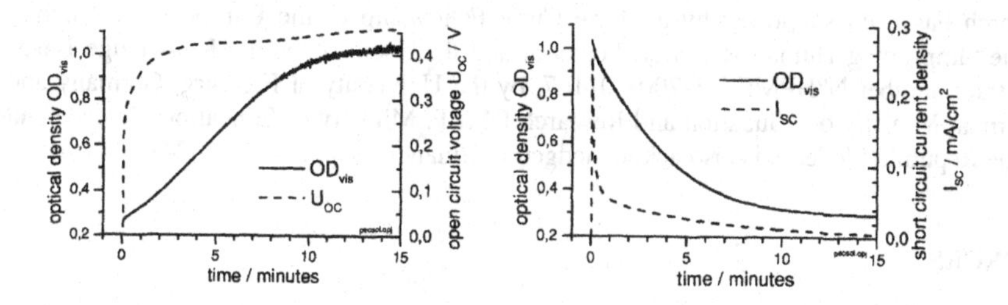

Fig. 4 Left: Coloring in open circuit illuminated by an intensity of 1 sun (1000W/m^2),
Right: Bleaching in short circuit in the dark.

Ageing of the photoelectrochromic devices

The change of the optical properties of the photoelectrochromic systems is negligible when the devices are kept under room temperature conditions (checked up to two months). The samples were exposed to different ageing conditions (visible illumination, UV illumination and to higher temperatures (65°C, 85°)). Upon cycling and UV illumination the most stressful was found to be UV illumination. The samples colored, but did not bleach under short circuit conditions. From the spectral measurements of the colored devices we observed that the concentration of I_3^- decreases and the only way to bleach the samples is to apply the voltage. Ageing the samples at higher temperature (85°C) influenced mainly the bleaching kinetic, bleaching time was increased and devices did not bleach to their full extent. Devices were also cycled up to 1000 cycles. During the first few cycles the coloration kinetic improves and stays constant. The bleaching kinetic becomes slower with cycling, after 200 cycles the time needed to achieve complete bleaching is doubled, while further cycling leads to the uncompleted bleaching of the PE systems. The results showed that the main problems left are: the increase of the charge transfer resistance between the electrolyte and Pt electrode, slow diffusion of the I_3^- species in the electrolyte layer and the decrease of the I_3^- concentration in the electrolyte. To overcome these problems better sealing is needed and an UV filter should be used (either as a layer or included in the electrolyte) to maintain the performance of the devices.

CONCLUSION

A photoelectrochromic device was introduced, which combines an electrochromic layer of WO_3 with a dye solar cell. The layers show a complex nano-structure. The high porosity allows a penetration of the electrolyte into the layers of WO_3 and TiO_2 even for polymeric ionic conductors. The device can be switched under illumination as well as in the dark. For a cell with solid electrolyte the visual (solar) transmittances changes from 62% (41%) to 2% (1%) in roughly 10 min. The results of the ageing test are encouraging, but the improvement of the sealing and an introduction of an UV filter seem to be unavoidable.

ACKNOWLEDGMENT

This research has been supported by a Marie Curie Fellowship of the European Community programme "Improving Human Research Potential and the Socio-economic Knowledge Base" under contract number N°HPMF-CT-2001-01167, by the University of Freiburg, Germany and by the German Ministry of Education and Research BMBF. Ministry of Education, Science and Sport of the Republic Slovenia is also acknowledged for financial support (Z2-3524).

REFERENCES

1. C. G. Granqvist, *Handbook of Inorganic Electrochromic Materials*, (Elsevier, Amsterdam, 1995).
2. B. O'Regan and M. Grätzel, Nature (London), **353,** 737 (1991).
3. A. Hauch, A. Georg, S. Baumgaertner, U. Opara Krašovec, B. Orel , *Electrochim. acta* **46,** 2131 (2001).
4. A. Hauch, A. Georg, U. Opara Krašovec, B. Orel, *J. Electrochem. Soc.*, **149,** H159 (2002).
5. Clemens Bechinger, Suzanne Ferrere, Arie Zaban, Julian Sprague, Brian A. Gregg, *Nature*, **383**, 608 (1996).
6. U. Opara Krašovec, Anneke Georg, Andreas Georg, V. Wittwer, J. Luther[3,4] and M. Topič, *Solar Energy Mat. Sol. Cells*, **84**, 369 (2004).
7. A. Hauch, A. Georg, U. Opara Krašovec, B. Orel, *J. Electrochem. Soc.*, **149,** A1208 (2002).
8. U. Opara Krašovec, R. Ješe, B.Orel, J. Grdadolnik and G. Dražić, *Monatsh. Chem.*, **133,** 1115 (2002).
9. U. Opara Krašovec, Anneke Georg, Andreas Georg, G. Drazic and M. Topič, *J Sol-Gel Sci. and Techn.* (submitted, 2004).
10. E. Stathatos, P. Lianos, U. Lavrenčič Štangar and B. Orel, *Adv. Mater.*, **14,** 354 (2002).

Nanoparticle-Hybrid Solar Cells

Mater. Res. Soc. Symp. Proc. Vol. 836 © 2005 Materials Research Society L2.4

Time Evolution of Photoconductivity in TiO_2 Electrodes Fabricated by a Sol Gel Method

Zhibin Xie [a 1], Victor M Burlakov[a], Bernard M. Henry[a], Kiril R. Kirov[a], Christopher R. M. Grovenor[a], Hazel E. Assender[a], G. Andrew D. Briggs[a], Mitsuru Kano[b] and Yusuke Tsukahara[b]
(a): Department of Materials, University of Oxford, Parks Road, Oxford OX1 3PH, UK
(b):Technical Research Institute, Toppan Printing Co., Ltd., Takanodai-Minami 4-2-3, Sugito-machi, Kitakatsushika-gun, Saitama 345, Japan
(1): Corresponding author Email: zhibin.xie@materials.ox.ac.uk Tel: +44 1865 273699
Fax: +44 1865 273789

ABSTRACT

We report here on the time evolution of photoconductivity under continuous illumination in nanocrystalline TiO_2 samples prepared by a sol gel method, and also on the conductivity decay once the illumination is switched off. We observe strong dependence of the photoconductivity on the illumination intensity for both processes. It is found that the conductivity decay after high-intensity illumination is slower than after low-intensity illumination, and we have attempted to explain these experimental results using a model involving hole trapping-detrapping processes.

INTRODUCTION

The wide gap semiconductor TiO_2 is often used as an electron acceptor in composite solar cells (CSC's) consisting of a conjugated polymer acting as a photon absorber and hole conductor. Photogenerated excitons in the conjugated polymer dissociate at the interface of the conjugated polymer and titania, injecting electrons into the latter [1]. The overall performance of CSC's is strongly determined not only by how efficiently the electrons are transferred to the electron accepting TiO_2, but also on how they pass through the porous nanocrystalline TiO_2 thin film and are collected at an anode. Electron transport is heavily affected by both intrinsic imperfections in the TiO_2 crystal structure [2, 3] and by the characteristic thin film morphology [4]. High quantum efficiencies in CSC's require large organic-inorganic interface areas, which in turn suggest the need for highly porous titania. At the same time, large interface areas mean a high concentration of surface states, which may trap electrons and holes. To understand the interplay between interfacial trapping and the transport properties through the nanocrystalline TiO_2 thin films, we have studied the photoconductivity (PC) of this material in the time domain. In this paper we present results on the microstructure of nanocrystalline TiO_2 thin films prepared by a sol gel method, and how the PC properties depend on various illumination conditions under vacuum.

EXPERIMENTAL DETAILS

PC samples with $ITO/TiO_2/Au$ configuration were formed on pre-patterned ITO/glass substrates. Before depositing titania, the substrates were thoroughly ultrasonically cleaned in iso-propanol and acetone for 10 minutes, and then dried with nitrogen gas. TiO_2 layers were produced by spin coating a sol similar to the one used in Grätzel cells[5]. Titanium (IV) isopropoxide was added to a 0.1M solution of nitric acid, and was heated until peptidisation of the TiO_2 occurred. A surfactant, Triton-X-100, was subsequently added to the solution. The filtered solution was spin coated onto the cleaned ITO substrate at 2000rmp for 1 minute. The as-deposited sample was dried at 100°C for 5 minutes, and then annealed at 450°C for 1 hour on a hotplate in air. Transparent thin films of thickness 140nm were obtained by three sequential depositions. Finally Au electrodes (45 nm thick) were deposited by thermal evaporation at a rate of 20nm/min. in a vacuum of order 10^{-6} mbar. The active area of each device was 0.063 cm^2.

The crystal structures present in the TiO_2 films were identified by X-ray diffraction (XRD) using CuKα radiation. The surface morphology of the films was studied by a JEOL6500F SEM and Park Scientific Instruments AFM in the tapping mode. Electrical measurements were carried out both in the dark and under illumination using a Keithley 2400 source meter with an ITC Oxford Instruments temperature controller. A KH Steuernegel 1200 Solar Simulator was employed as a light source. The samples were mounted in a cryostat with pressure maintained between 10mtorr and 15mtorr by a rotary pump. For all PC measurement samples, a forward bias of 0.5 V was applied to the gold electrode. A range of meshes and neutral density filters were used to control the illumination intensity, which was measured with a calibrated silicon photodiode at the sample position before device testing.

RESULTS AND DISCUSSION

Microstructural characterisation

It was established by X ray diffraction that the as-deposited films are predominately amorphous. It was also found that annealing in air has a strong influence on the crystal structure of the films. Anatase was observed to form at 400°C, and rutile at 550°C. For all PC samples, partially crystalline anatase layers were formed by a heat treatment at 450°C.

As shown in Figure 1a, AFM images showed that the surface morphology of the films consists of particles with diameter ranging from 20nm to 30nm, with average surface roughness of 4.5nm. The SEM image shown as figure 1(b) indicates that some particles formed agglomerates of size 100nm-400nm.

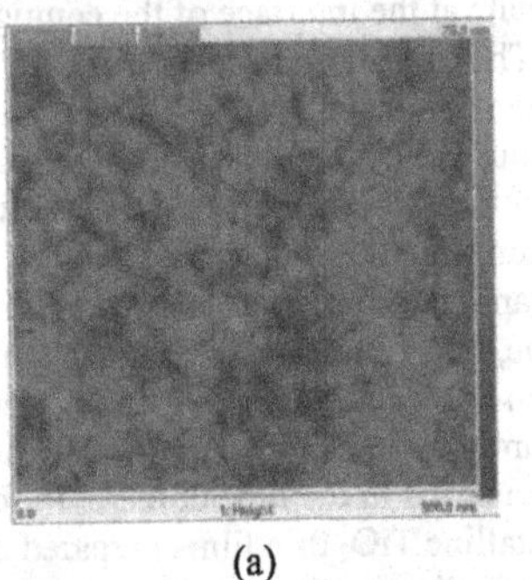

(a)

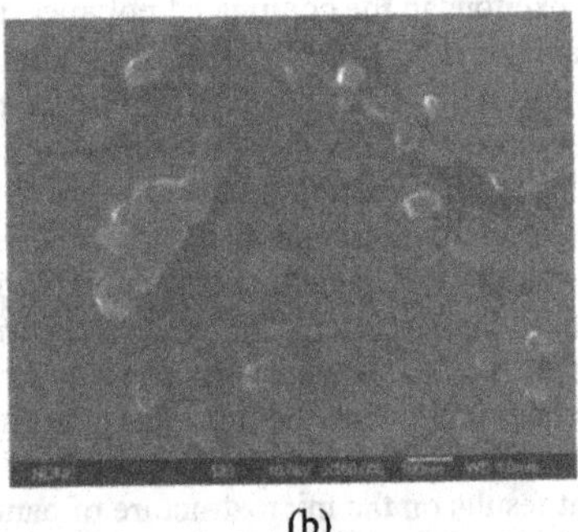

(b)

Figure 1: (a) AFM images (0.5 μm x 0.5 μm) and (b) SEM images of a sol gel TiO_2 layer

Photoconductivity

All current-voltage (I-V) curves measured in the dark under vacuum show diode-like behaviour due to rectifying properties of the TiO_2/Au interface. The shapes of the I-V curves at direct bias are characteristic of the presence of space charge effects, modified by a high concentration of traps. In the zero bias limit, the curve is linear and conduction is dominated by the intrinsic free carrier density. We obtained a dark conductivity value at room temperature of $1.8E\text{-}11\Omega^{-1}m^{-1}$ in good agreement with Studenikin *et.al.*[6]

It is well established that the PC of anatase is very sensitive to the presence of oxygen and water [7]. To avoid such influence as much as possible all our measurements were carried out in vacuum. The samples were heated at 380K in vacuum for 2 hours to remove traces of absorbed water before PC measurements. The time dependent PC behaviour of nanocrystalline TiO_2 obtained under an illumination intensity of 1 mW/cm^2 (1% Sun) is shown in Figure 2. The PC behaviour is characterised by a very slow increase during illumination and subsequently a slow decay after switching off the light. It can also be seen from Figure 2 that the PC in the second illuminated period rises more rapidly than in the first.

The PC behaviour strongly depends on the illumination intensity. As shown in Figure 3a, when the light intensity decreases it takes a much longer time for PC to approach saturation. In practice it is nearly impossible under these weak illumination conditions to reach saturation, and the normalized PC shown in Figure 3b illustrates the very strong dependence of conductivity on the light intensity. Switching off the illumination results in slow PC decay (see Fig. 3c), the mechanism of which seems to include a memory effect, i.e. the decay rate also depends on the previous light intensity. However, in contrast to the increase in PC under illumination, the initial relative decay is faster if the sample was illuminated with a lower light intensity, as shown in Figure 3d. It is indeed the light intensity rather than the exposure time, which is responsible for the memory effect in the PC decay. In Figure 4, two different illumination times under the same light intensity result in PC decay rates within the range of sample-to-sample variation.

Golego *et al* have observed similar long decay times for PC in both Nb-doped TiO_2[8] and ZnO[9] thin films, from which they have extracted relaxation times ranging from 10^2 to more than 10^4 seconds. They attribute this decay behaviour to carrier trapping and recombination in grain boundary states in their nanocrystalline films, but the precise mechanisms that lead to such long characteristic time constants are not identified. Using a similar exponential fitting process on our PC decay data, we have also extracted relaxation times in the range 10^2 - $2x10^3$ seconds. The strong dependence of PC behaviour with time has been previously modelled using a broad spectrum of charge carrier traps in nanocrystalline TiO_2 [10]. The higher the light intensity, the quicker the traps are filled and consequently the quicker PC saturation is achieved. More difficult to explain is the intensity dependence of PC decays. Consider a trap density *g(E)*, and assume that only holes are trapped[7]. Then the PC can be expressed as

$$\sigma_p = K \int_{E_{min}}^{E_{max}} n_h(E,t) \cdot g(E) dE \tag{1}$$

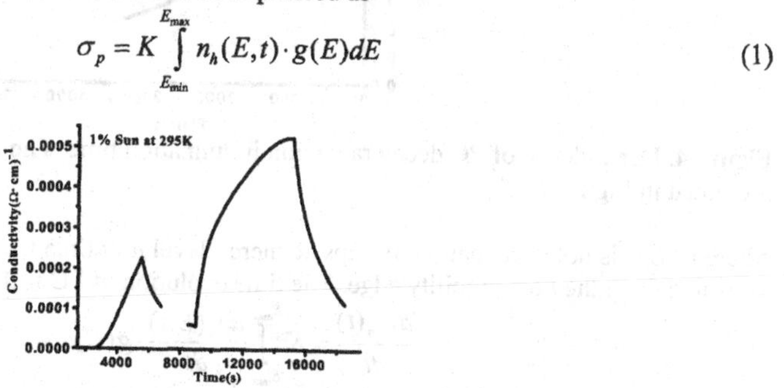

Figure 2. Typical PC curve for nanocrystalline TiO_2 at 295K in vacuum under 1% Sun illumination intensity

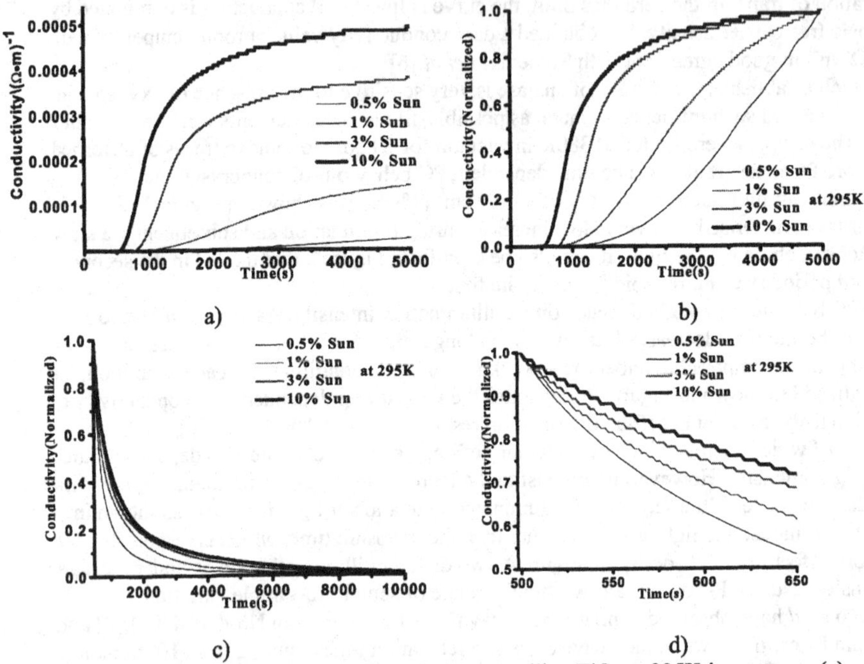

Figure 3. Light intensity effects on PC of nanocrystalline TiO_2 at 295K in vacuum; (a) conductivity rise (b) normalized conductivity rise (c) normalized conductivity decay (d) normalized initial decay

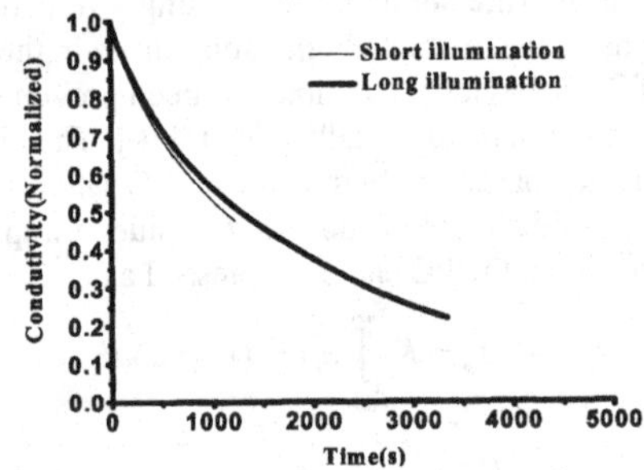

Figure 4. Dependence of PC decay rate upon illumination time. The results correspond to those presented in Figure 3.

where $n_h(E,t)$ is hole occupation of traps at energy level E at time t, and E is the trap energy measured from the hole mobility edge. The time evolution of PC is then given by

$$\frac{d\sigma_p(t)}{dt} = K \int_{E_{\min}}^{E_{\max}} \frac{dn_h(E,t)}{dt} \cdot g(E)dE \tag{2}$$

Taking into account that the evolution of the occupation is controlled by the generation rate G_E and the thermal ionisation rate of traps at temperature T with attempt frequency f

$$\frac{dn_h(E,t)}{dt} = G_E - f \cdot n_h(E,t) \cdot \exp\left(-\frac{E}{kT}\right) \quad (3)$$

Eq. (2) is reduced to

$$\frac{d\sigma_p(t)}{dt} = K \int_{E_{min}}^{E_{max}} \left[G_E - f \cdot n_h(E,t) \cdot \exp\left(-\frac{E}{kT}\right)\right] \cdot g(E)dE \quad (4)$$

Assuming that each trap has a single energy level, in which case *g(E)* represents the concentration of traps with energy *E*, and that filling or emptying of a trap does not depend on the occupation of other traps we can express G_E as

$$G_E = G/Z, Z = \int_{E_{min}}^{E_{max}} g(E)dE \quad (5)$$

where G is the overall generation rate which is proportional to incident light intensity I_l. Substituting Eq. (5) into Eq. (4) we have

$$\frac{d\sigma_p(t)}{dt} = K \int_{E_{min}}^{E_{max}} \left[\frac{G}{Z} - f \cdot n_h(E,t) \cdot \exp\left(-\frac{E}{kT}\right)\right] \cdot g(E)dE = KG - Kf \int_{E_{min}}^{E_{max}} \left[n_h(E,t) \cdot \exp\left(-\frac{E}{kT}\right)\right] \cdot g(E)dE \quad (6)$$

which shows some dependence of PC on light intensity.

To model PC decay, we assume that the illumination is switched off at time t_0. The hole occupation numbers then can be obtained by solving Eq. (6)

$$n_h(E,t_0) = \exp\left(\frac{E}{kT}\right) \cdot \left(KG - \frac{d\sigma_p(t_0)}{dt}\right) / KfZ, \quad (7)$$

and relative rate of PC decay at the initial stage of the decay process can be described by substituting Eq. (7) into Eq. (4) with $G_E=0$

$$\frac{1}{\sigma_p}\frac{d\sigma_p(t)}{dt} = -\frac{fZ}{\int_{E_{min}}^{E_{max}} \exp\left(\frac{E}{kT}\right) g(E)dE} \quad (8)$$

According to Eq. (8) the initial relative decay rate of PC is independent of the light intensity in contrast to our experimental results. Such inconsistency suggests that our assumption about lack of correlations between various trap states is not valid, and that to understand the possible mechanisms involved in their correlations requires further investigation.

The PC behaviour is related to the test temperature as well. As shown in Figure 5(a), the PC excitation rate increases with increasing test temperature. The higher the temperature, the higher the frequency of interaction between holes and traps, and as a result, the sooner it reaches steady state. As illustrated in Figure 5(b), at higher temperatures, the initial relative PC decay rate is faster. Similar temperature effects were also reported recently by some other researchers [11], but Nelson *et al* did not find this phenomenon [12]. Eq. (8) predicts a strong temperature dependence of the initial relative PC decay rate, in qualitative agreement with our experiments

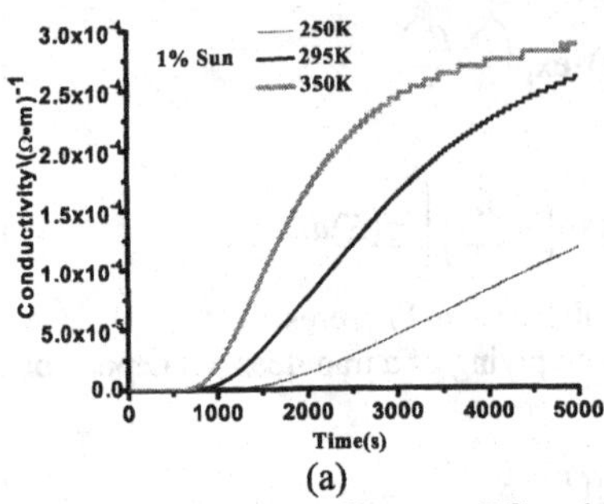

(a)

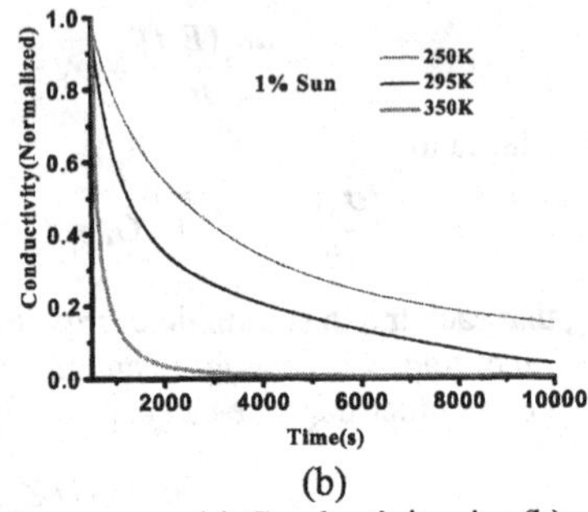

(b)

Figure 5: Temperature effects on PC under 1% under vacuum (a) Conductivity rise (b) Normalized conductivity decay

CONCLUSIONS

We have performed investigations of photoconductivity in nanocrystalline TiO_2 samples fabricated by a sol gel method. Our results confirm the long-term transient processes in PC on both switching on and switching off the illumination. We observe that both the rate of increase in PC under illumination and the rate of PC decay in the dark are strongly dependent on the illumination intensity and test temperature. The rate of PC decay in the dark is slower after high-intensity illumination than after low-intensity illumination. A model of the intensity-dependent PC decay requires further development, possibly taking into account correlations between charge carrier traps.

ACKNOWLEDGEMENTS

We would like to thank Professor G.D.W. Smith, FRS, for the provision of laboratory facilities. We are also grateful to Dr. M. J. Carey and Mr. R. Turner for their help in setting up the PC experiments. The financial support of the Toppan Printing Company Ltd, Japan is gratefully acknowledged.

REFERENCES

1. C.Grant, A M Schwartzberg, G P Smestad, J Kowalik, L M Tolbert and J Z Zhang *Synthetic Metals* **132** 197 (2003)
2. J Nelson, S A Haque, D R Klug, and J R Durrant *Physical Review B* **63** 205321(2001)
3. G. Franco, *et al Journal of Physics and Chemstry B* **103** 692(1999)
4. A C Arango, *et al Advanced Materials* **12** 1689 (2000)
5. M. Gratzel, Patent US5350644 (1994)
6. S A Studenikin, N Golego, M Cocivera *J. Appl. Phys.* **84** 5001 (1998)
7. J Nelson, A M Eppler, I M Ballard, *J. Photochem.& Photobio. A* **148** 25 (2002)
8. N Golego, S A Studenikin, M Cocivera *Journal of Materials Research* **14** 698 (1999)
9. S A Studenikin, N Golego, M Cocivera *Journal of Applied Physics* **83** 2104 (1998)
10. B. Dulieu, J Bullot, J Wery, M Richard, L Brohan., *Physical Review B* **53** 10641(1996)
11. A. Brajsa, *et al* Pomoni, *Opitcal Materials* **26** 151-153 (2004)
12. A M Epper, I M Ballard, J Nelson, *Physica E* **14** 197-202 (2002)

Mater. Res. Soc. Symp. Proc. Vol. 836 © 2005 Materials Research Society L2.5

Fabrication of Nanorod Arrays for Organic Solar Cell Applications

Susan Huang, Harry Efstathiadis, and Pradeep Haldar
College of Nanoscale Science and Engineering, The University at Albany - SUNY
Hee-Gyoun Lee
Korean Polytechnic University, Siheung, Korea
Brian Landi, and Ryne Raffaelle
NanoPower Research Laboratory, Rochester Institute of Technology

ABSTRACT

We report on a novel use of nanorod arrays for organic based solar cell devices. A metal foil with copper nanorods attached to the surface was developed by electrodepositing copper from a copper sulfate solution into an anodic alumina oxide (AAO) template that had been coated with a metal on one side. The AAO membrane was dissolved in NaOH leaving behind an aligned array of copper nanorods. This nanorod array was evaluated to explore the possibility of increasing the power conversion efficiency of organic solar cells. Nanorod array characteristics were investigated by focus ion beam, scanning electron microscopy, and x-ray diffraction spectroscopy. A solar cell device was made by applying a polymer layer of poly(2-methoxy-5-(3',7'-dimethyloctyloxy)-1,4-phenylene-vinylene) (MDMO-PPV) mixed with 6,6 phentl-C61-butyl acid-methylester (PCBM) onto the copper nanorod array and sandwiching it with a film of poly(3,4-ethylenedioxythiophene): poly(styrene-sulfonate) (PEDOT:PSS) applied onto a indium tin oxide coated glass substrate.

INTRODUCTION

Fabricating solar cells from organic materials is a promising method for inexpensive, flexible photovoltaics. Current efficiencies of organic solar cells are still less than 4% but there is a large effort on developing methods of increasing efficiencies. One of these methods is to embed nanomaterials into photoactive polymer matrix. Nanomaterials that have been incorporated into organic solar cells include chemically modified fullerenes [1,2], single wall carbon nanotubes [3], and nanorods [4] to increase exciton dissociation and carrier transport. Organic solar cells with colloidal nanorods embedded in the polymer matrix have been demonstrated to have relatively high efficiencies of 1.7% [5]. We plan to explore the effect on device efficiency of polymer solar cells with embedded aligned nanorod arrays. Metallic [6] and semiconducting [7] nanorods have been previously fabricated by electrodeposition into a template. Electrodeposition offers a simple alternative solution for high yield production of nanoparticles and nonorods compared to other methodologies developed for the synthesis of metal nanostructures. The use of nanorods is expected to increase the efficiency of the device because each nanorod could provide a direct path to the electrode. Metallic nanorod arrays have been prepared by electrodeposition into anodic alumina oxide (AAO) templates [8,9] for organic solar cell applications and characterized by scanning electron microscopy (SEM), x-ray diffraction spectroscopy (XRD) and focus ion beam

(FIB) microscopy. These nanorod arrays have been incorporated into the polymer poly(2-methoxy-5-(3',7'-dimethyloctyloxy)-1,4-phenylene-vinylene) (MDMO-PPV) for preliminary materials characterization.

EXPERIMENTAL DETAILS

Aligned copper and silver nanorod arrays were fabricated by electrodeposition of metal into AAO templates in a two-electrode cell configuration. Commercial AAO templates with a cited pore size of 200 nm and diameters ranging from 14 mm to 47 mm were employed as substrates. A copper plate was used as the anode and the AAO membrane was used as cathode. Both electrodes were immersed in an electrolyte while no surfactants or additives were utilized. One side of the AAO template was coated with a 400 nm layer of Ag using sputter deposition in order to obtain a conductive electrode for electrodeposition. The Ag coated AAO was attached to a piece of indium tin oxide (ITO) coated glass with the silver side in contact with the ITO. For Cu nanorods, a 0.014M solution of copper sulfide ($CuSO_4$) was used with a Cu counter electrode. A current density of 3-4 mA/cm^2 was used to deposit the nanorods. For Ag nanorods, a commercial Ag plating solution and a pure Ag counter electrode was used. The current density for Ag deposition was between 1-2 mA/cm^2. Cu and Ag electrodeposition was performed for 5 min, 10 min, 20 min and 60 min into AAO membranes. The resulting nanorods were freed from the AAO membrane by dissolving the AAO in 0.5M NaOH for 45 min. The resulting nanorod arrays were rinsed in dionized water. A schematic of the fabrication process for the aligned nanorods is shown in Figure 1.

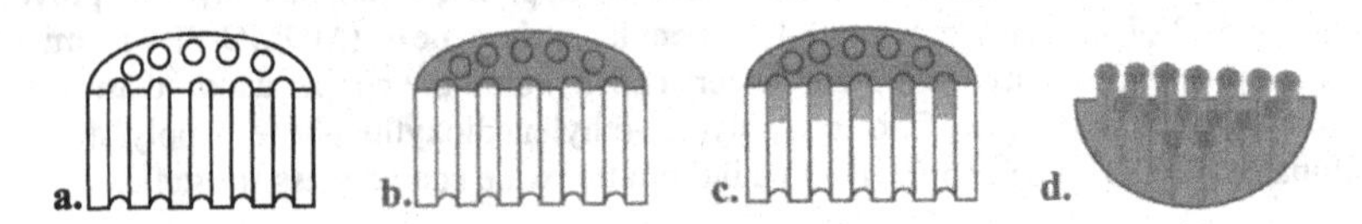

Figure 1. Metallic nanorods were fabricated by using a) an AAO template and b) sputtering Ag on one side. Then c) metal was electrodeposited into the pores of the AAO and finally d) the AAO was removed in NaOH leaving an aligned nanorod array.

RESULTS AND DISCUSSION

For all the films grown, several analytical techniques were employed to perform thorough material characterization. AAO membrane pore size and nanorod, nanotube dimensions was measured by SEM and FIB microscopy. Crystal structure was investigated by XRD analysis carried out on an x-ray diffractometer equipped with a Cu K_α x-ray source. Normal angle (Bragg-Bretano) incidence geometry was used and collected Cu and Ag patterns were compared, respectively, to the Cu and Ag reference patterns from the standard Joint Committee for Powder Diffraction Standards (JCPDS) powder diffraction file (PDF). Cross sectional SEM and FIB images shown in Figure 2, indicate that Cu electrodeposition resulted in aligned nanorods for the 60 min deposition time. The mean diameter of the nanorods was approximately 225 nm, which is slightly larger than the cited diameter. The length of the nanorods vs. time was measured for electrodeposition times of 5 min, 10 min, 20 min and 60 min. For the 5 min deposition run, there is no evidence of nanorods. The growth of the nanorod seems to be increasing linearly up to 20 min but plateaus for longer deposition times (Figure 3). This is most likely due to depletion of the plating solution. A typical XRD spectrum of these nanorod arrays is shown in Figure 5 and indicates that the nanorods were composed of polycrystalline Cu and the presence of CuO [10].

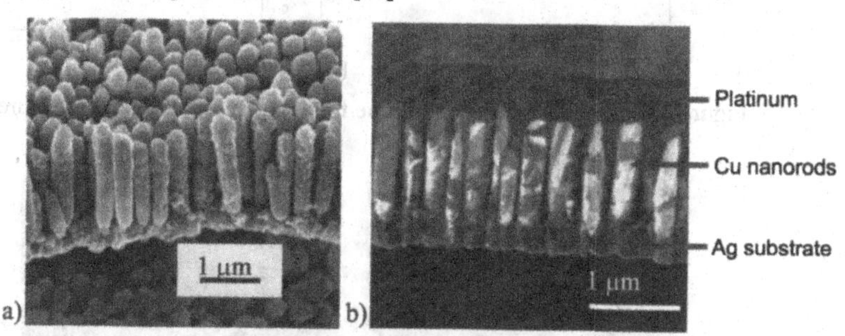

Figure 2. a) SEM image of Cu nanorods, 60 min deposition and b) FIB image of Cu nanorods

Ag electrodeposition consistently resulted in the formation of nanotubes rather than nanorods as can be seen from the cross-sectional SEM images in Figure 4. This is perhaps due to the concentration of the solution, the deposition time and current density. The nanotube length vs. time also shows that the length increases with time up to 20 min and then nanotube growth slows down dramatically (Figure 3) similar to the Cu nanorods. An XRD pattern (Figure 5) of the nanotubes shows that they were composed of polycrystalline Ag.

Additional work is under way to investigate how the electrodeposition parameters such as bath temperature, current density, voltage, and additives control the shape, length, and diameter of the nanostructures. The possible function of additives to kinetically control the growth rates of various Ag and Cu faces by interacting with these faces through adsorption and desorption is under investigation. The characteristics of the nanostructures on cell performance will also be investigated.

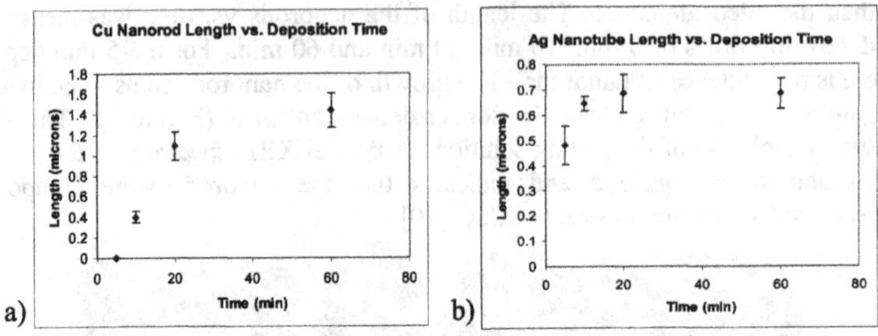

Figure 3. Length as a function of time for a) Cu and b) Ag nanostructures.

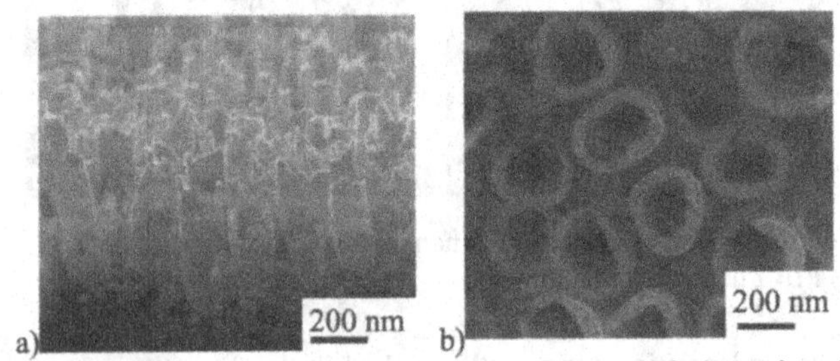

Figure 4. Aligned Ag nanotube array from side (a) and plane view (b).

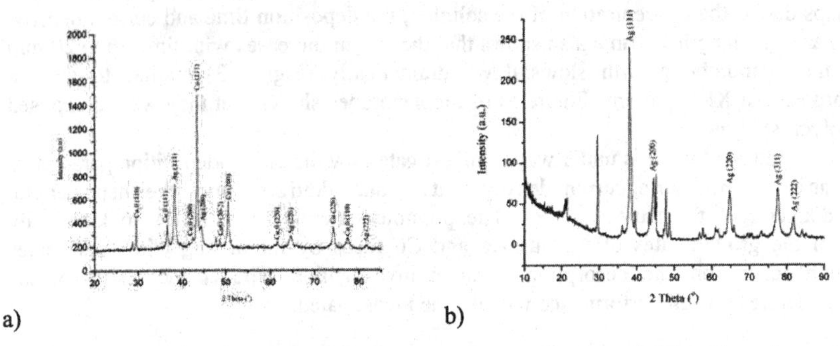

Figure 5. XRD of a) Cu nanorods indicates presence of polycrystalline Cu and Cu_2O and b) Ag nanotubes show presence of polycrystalline Ag.

CONCLUSIONS

Aligned metal nanostructures were fabricated using a simple electrodeposition method into AAO membranes. Both Cu nanorods and Ag nanotubes of ~ 200 nm diameter were synthesized using controlled current electrodeposition. Additional work will be carried out to investigate electrodeposition parameters that control the shape, length, and diameter of the nanostructures. Preliminary investigations of incorporating aligned nanorods into the photoactive polymer of an organic solar cell are also under way.

ACKNOWLEDGMENTS

This work was supported by the Energy and Environmental Technology Center (E2TAC) at the College of Nanoscale Science at Engineering (CNSE) at the University at Albany - SUNY . The authors would like to thank Dr. P. Choi and O. Paschos at CNSE for film characterization and their suggestions in this work.

References

[1] C. J. Brabec, N. S. Sariciftci, and J. C. Hummelen, *Adv. Funct. Mater.*, **2001**, *11*, 15.
[2] S. E. Shaheen, C. J. Brabec, N. S. Sariciftci, F. Padinger, T. Fromherz, and J. C. Hummelen, *Appl. Phys. Lett.*, **2001**, *78*, 841.
[3] E. Kymakis and G. A. J. Amaratunga, *Appl. Phys. Lett.*, **2002**, *80*, 112.
[4] W. U. Huynh, X. G. Peng, and A. P. Alivisatos, *Adv. Mater.*, **1999**, *11*, 923.
[5] W. U. Huynh, J. J. Dittmer, A. P. Alivisatos, *Science*, **2002**, *295*, 2425.
[6] A. J. Yin, J. Li, W. Jian, A. J. Bennett, and J. M. Xu, *Appl. Phys. Lett.*, **2001**, *79*, 1039.
[7] D. S. Xu, D. P. Chen, Y. J. Xu, X. S. Shi, G. L. Guo, L. L. Gui, and Y. Q. Tang, *Pure Appl. Chem.*, **2000**, *72*, 127.
[8] C. A. Foss Jr., M. J. Tierney, and C. R. Martin, *J. Phys. Chem.*, **1992**, *96*, 9001.
[9] A. J. Yin, J. Li, W. Jian, A. J. Bennett, and J. M. Xu, *Appl. Phys. Lett.*, **2001**, *79*, 1039.
[10] A. K. Srivastava,, P. Tiwari, A. Kumar, R. V. Nandedkar, **2004**, 86, 22.

Mater. Res. Soc. Symp. Proc. Vol. 836 © 2005 Materials Research Society L2.8

Quantum Dot-Single Wall Carbon Nanotube Complexes for Polymeric Photovoltaics

Brian J. Landi[1], Stephanie L. Castro[2], Chris M. Evans[1], Herbert J. Ruf[1], Sheila G. Bailey[3], and Ryne P. Raffaelle[1]

[1]NanoPower Research Laboratories, Rochester Institute of Technology, Rochester, NY 14623, USA
[2]Ohio Aerospace Institute, Brookpark, OH 44142, USA
[3]NASA Glenn Research Center, Cleveland, OH 44135, USA

ABSTRACT

The ability to dissociate the photo-generated excitons and transport the resulting charge carriers are the major impediments in improving the efficiency of polymeric solar cells. In order to simultaneously address both of these issues, we have investigated the use of quantum dot-single wall carbon nanotube (QD-SWNT) complexes as a suitable nanomaterial dopant in these devices. The formation of CdSe-SWNT complexes occurred through covalent attachment of carboxylic acid-functionalized SWNTs with CdSe-aminoethanethiol (AET) quantum dots. An additional synthetic approach was evaluated using both electrostatic and covalent attachment schemes for $CuInS_2$-mercaptoacetic acid (MA) quantum dots and amine terminated SWNTs. The efficacy of each approach is discussed, including the necessary transmission electron microscopy (TEM) and optical absorption spectroscopy data to probe the interactions between nanomaterials. The potential effects of charge transfer between components may have important implications in the efficiency of these materials for polymeric photovoltaic devices.

INTRODUCTION

Polymeric solar cells are currently being investigated as an alternative photovoltaic technology, due to the potential reduction in processing cost, improved scalability, and opportunity for lightweight, flexible devices [1]. The use of nanomaterials as an additive in these polymer systems is currently an area of very active research, with most efforts directed at semiconductor quantum dots (QDs), fullerenes, and single wall carbon nanotubes (SWNTs) [1-3]. Selection of the appropriate nanomaterial should maximize exciton dissociation and promote efficient carrier transport in the device. Additionally, the potential exists for a multiple-junction device that is tailored to the air mass zero (AM0) spectrum since conducting polymers, QDs, and SWNTs each absorb in a different spectral region. To better facilitate this process of photon absorption by the components and dissociation of the excitons by the highest electron affinity material nearest the exciton, we are evaluating the use of QD-SWNT complexes as an ideal nanomaterial.

Quantum dot-single wall carbon nanotube (QD-SWNT) complexes represent a class of materials that can encompass both high electron affinity and high electrical conductivity. The electron affinity of the QDs can be tailored based upon the selection of semiconductor material used as well being influenced by the nanocrystal size [4, 5]. For example, CdSe QDs have a range of electron affinities reported from 3.5-4.5 eV, while bulk $CuInS_2$ ranges from 4.1-4.9 eV [5, 6]. In addition, the low percolation threshold of SWNTs coupled with their extraordinary electrical conductivity (10^4 S/cm for metallic (10,10) SWNTs) allow for significant enhancement of electron transport at even very low doping levels (<1% w/w) [7]. Although a mixture of QDs and SWNTs in the polymer may also prove worthwhile, coupling of the QDs to SWNTs [8-10] would presumably provide the most efficient combination of exciton dissociation, charge transfer, and carrier transport. In this report, the synthetic details associated with the formation of CdSe-SWNT and $CuInS_2$-SWNT complexes have been outlined, including the appropriate chemical functionalization and surface ligand exchange steps required for covalent attachment. Optical absorption spectroscopy is performed to probe any differences in potential electronic interactions between nanomaterials for electrostatic and covalent attachment schemes.

EXPERIMENTAL DETAILS

Synthesis of the single wall carbon nanotubes (SWNTs) was performed using an Alexandrite laser vaporization process, previously described in detail [11]. The raw soot was purified using conventional nitric acid and thermal oxidation steps, to achieve SWNT mass fractions of >95% w/w in the overall sample [11]. CdSe quantum dots were synthesized from CdO as the precursor, following the established protocol [12]. $CuInS_2$ quantum dots were synthesized by thermal decomposition of the molecular single-source precursor $(PPh_3)_2CuIn(SEt)_4$ in the presence of hexanethiol (ht) at 200 ºC in dioctylphthalate [13]. The reaction solution is mixed for several hours at the elevated temperature and then cooled to room temperature. The nanocrystals precipitate with addition of a (1:1) volumetric mixture of methanol:acetonitrile and subsequent wash steps with methanol complete the extraction of high quality $CuInS_2$ nanocrystals [13].
A summary of the synthesis details for each of the QD-SWNT attachment schemes is depicted in Figure 1. The initial steps for either QD route is appropriate surface ligand exchange on the QDs and preparation of carboxylic acid-functionalized SWNTs performed by ultrasonication in a 4:1 mixture of concentrated H_2SO_4:H_2O_2 for 2.5 hours [12]. The formation of CdSe-SWNT complexes occurred through covalent attachment of carboxylic acid-functionalized SWNTs with CdSe-aminoethanethiol (AET) quantum dots. Ligand exchange of the TOPO with AET proceeded by stirring the CdSe-TOPO quantum dots in neat AET at 100 ºC for 24 hours [10]. The product was rinsed with 0.5M NaOH and resuspended in N,N-dimethylformamide (DMF). Activation of the carboxylic acid-functionalized SWNTs was performed using a ten-fold excess of 1-ethyl-3-(3-dimethylaminopropyl) carbodiimide hydrochloride (EDC) and sulfo-N-hydroxy succinimide (sNHS) for 30 minutes prior to coupling with the CdSe-AET quantum dots in a DMF solution which stirred for 2 hours and was quenched with ethanolamine. The reaction mixture was centrifuged and the precipitate rinsed with DMF. The final product was dried at 70 ºC in vacuo for one hour and is represented by (1) in Figure 1. In the case of the $CuInS_2$ quantum dots, ligand exchange of the hexanethiol (ht) occurred with mercaptoacetic acid (MA) by stirring with pyridine overnight [14]. In order to couple both carboxy-terminated species, (i.e. $CuInS_2$-MA and SWNTS-COOH) a coupling reaction (activated with dicyclohexylcarbodiimide (DCC)) was employed using ethylenediamine (en) to react with the carboxylic acid-terminated SWNTs initially to produce amine-terminated SWNTs. The approach to electrostatic attachment involved stirring the $CuInS_2$-MA quantum dots with the amine terminated SWNTs in DMF for six hours [14]. This reaction product is depicted in Figure 1 by (2). As described for the CdSe covalent coupling, the carboxylic acid-terminated $CuInS_2$ quantum dots can be activated with

EDC and sNHS under equivalent conditions. The resulting product was reacted with the amine-terminated SWNTs to produce a covalent amide linkage as shown by (3) in Figure 1.

Characterization of the SWNTs, quantum dots, and QD-SWNT complexes during synthesis and chemical modification was performed by both conventional spectroscopy and microscopy. Optical absorption measurements were made with a Perkin Elmer Lambda 900 spectrometer. Transmission electron microscopy (TEM) was done with a Phillips CM200 at an operating voltage of 200 kV using lacey carbon grids.

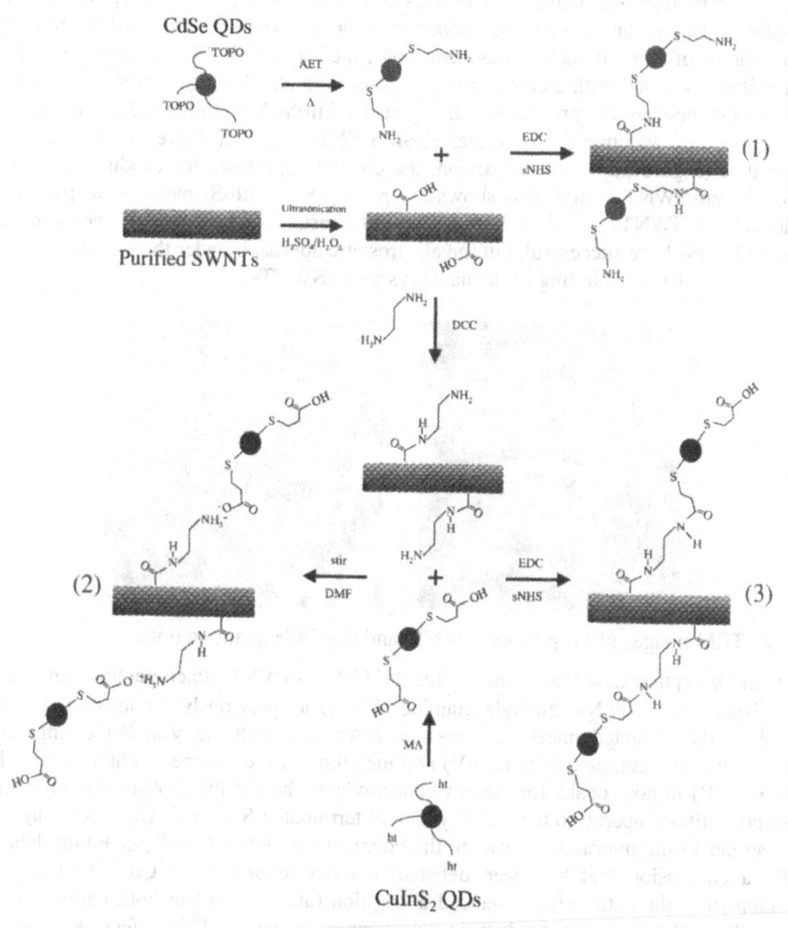

Figure 1. Reaction schemes for the attachment of QDs to SWNTs resulting in (1) covalent attachment of CdSe –AET-SWNT complexes, (2) electrostatic $CuInS_2$-MA-en-SWNT complexes, and (3) covalent $CuInS_2$-MA-en-SWNT complexes.

DISCUSSION

Each of the reaction schemes in Figure 1 was performed utilizing purified single wall carbon nanotubes (SWNTs) represented by the TEM image in Figure 2a. The predominance of highly ordered bundles with minimal sidewall impurities and lack of metal catalysts is typical for these purified materials. Similarly, Figure 2b shows a TEM image of high quality CdSe nanocrystals: typical of a ~3.5 nm diameter distribution prepared by the colloidal synthesis. The attachment results from TEM are demonstrated in Figure 3 for each of the attempted schemes. As reported previously for product (1) in Figure 1 (CdSe-AET-SWNT complexes), [12] the TEM analysis clearly shows the attachment of CdSe nanocrystals along various sidewalls of the SWNTs. The covalent nature of this interaction has been confirmed using FT-IR spectroscopy and various Raman shifts associated with electronic interactions have also been observed [12]. The current attempt for the electrostatic product (2) in Figure 1 ($CuInS_2$-MA-en-SWNT complexes) has not shown nanocrystal attachment, but rather clean SWNT bundles (Figure 3c) and aggregates of quantum dots (Figure 3d). In comparison, the covalent approach for product (3) in Figure 1 ($CuInS_2$-MA-en SWNT complexes) shows the presence of $CuInS_2$ nanocrystal particles along the sidewalls of SWNTs (Figures 3e-f). The TEM results suggest that the covalent schemes outlined in Figure 1 are successful, but the electrostatic approach under the prescribed conditions fails to promote efficient binding of the nanocrystals to SWNTs.

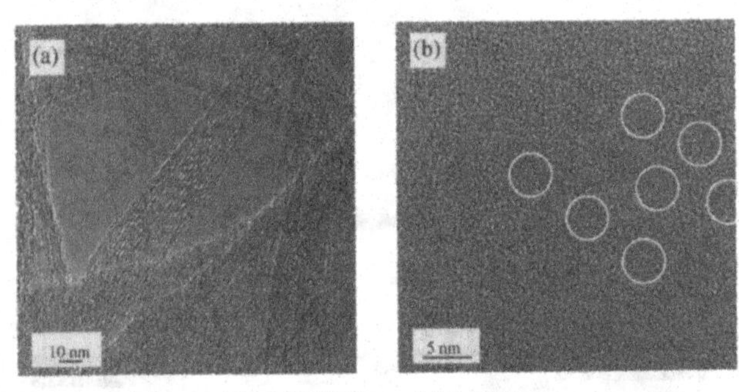

Figure 2. TEM images of (a) purified SWNTs and (b) CdSe quantum dots.

Optical absorption data was acquired for the $CuInS_2$-SWNT attachment schemes using 2.5 μg/mL dispersions in N,N-dimethylacetamide (DMA) as previously demonstrated [11]. The results show the expected interband transition associated with the Van Hove singularities in SWNTs for the semiconducting (~1.2 eV) and metallic (~1.8 eV) types. There is a slight blue-shift (~10 meV) in both peaks for the covalent product whereas the electrostatic product shows no energetic shift compared to the carboxylic acid-terminated SWNTs. This result suggests that there is an electronic interaction between the covalently attached $CuInS_2$ quantum dots and the SWNTs: a conclusion that has been debated in other reports using CdSe [8-10]. Another observation from the data is the enhanced absorption (above 2 eV) in both $CuInS_2$ attachment samples, albeit the covalent product is significantly greater. This effect is presumably a manifestation of the quantum dots' absorption in this range,[13] however, the effects of quantum dot electronic interaction on the π-plasmon of the SWNTs has yet to be investigated.

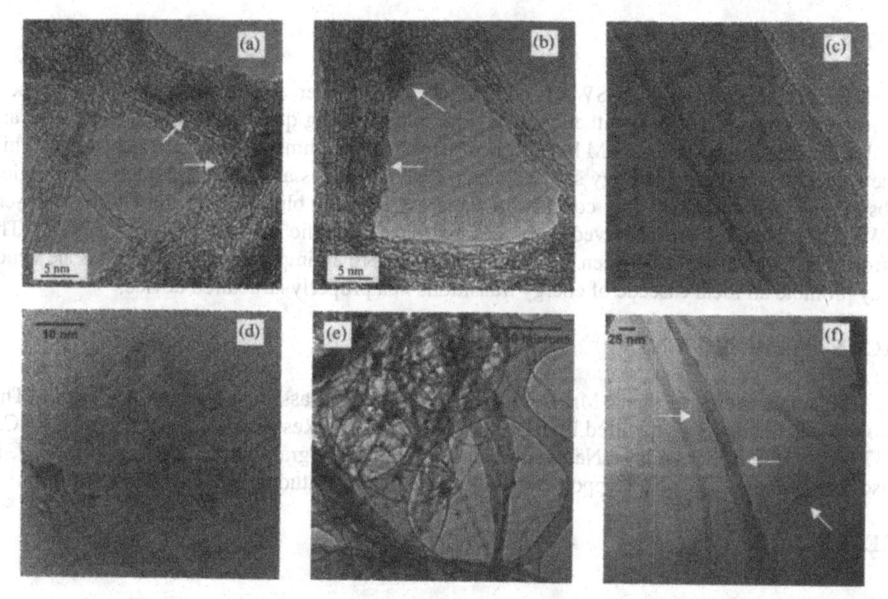

Figure 3. TEM images for (a and b) reaction scheme involving covalent attachment of CdSe-AET-SWNT complexes from product (1) in Figure 1; (c and d) reaction scheme involving electrostatic attachment of $CuInS_2$-MA-en-SWNT complexes from product (2) in Figure 1; (e and f) reaction scheme involving covalent attachment of $CuInS_2$-MA-en-SWNT complexes from product (3) in Figure 1.

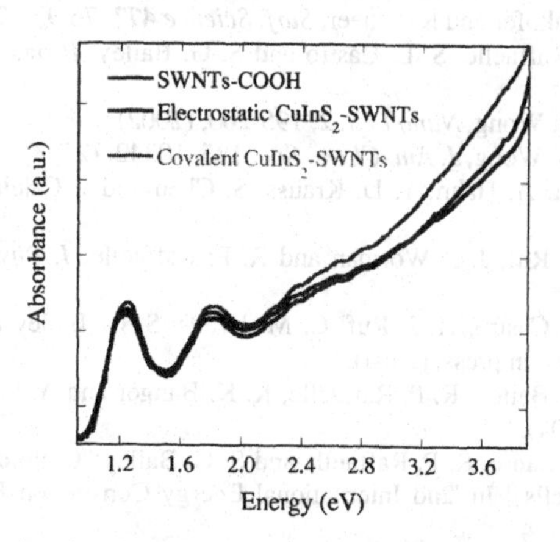

Figure 4. Optical absorption data from 2.5 μg/mL DMA dispersions for the carboxylic acid-terminated SWNTS (SWNTS-COOH), electrostatic $CuInS_2$-SWNT attachment product (2) from Figure 1, and covalent $CuInS_2$-SWNT attachment product (3) from Figure 1.

CONCLUSIONS

The development of QD-SWNT complexes for polymer photovoltaic devices has been investigated using multiple synthetic routes with two differing quantum dot moieties (CdSe and CuInS2). The results using TEM have shown successful attachment from a covalent route, while the electrostatic binding strategy showed clean SWNT bundles and QD aggregates. The optical absorption data supported this conclusion since a ~10 meV blue-shift in the CuInS2-MA-en-SWNT complexes were observed over the starting carboxylic acid-terminated SWNTs. The effects of charge transfer between select nanomaterials are an important fundamental issue which may promote an ideal cascade of energy transitions in a properly structured device.

ACKNOWLDEGEMENTS

The authors wish to thank Mr. David Hull for his TEM assistance during this work. This project was financially supported by BP Solar, NASA Glenn Research Center (grant nos. NCC3-937 and NAG3-2595) and the National Science Foundation (grant no. ECS-0233776). B. J. L. also acknowledges financial support from a NASA Graduate Student Research Fellowship.

REFERENCES

1. V. Dyakonov, *Thin Solid Films* **451-452**, 493-497, (2004).
2. C. J. Brabec, S. E. Shaheen, C. Winder, N. S. Sariciftci and P. Denk, *Appl. Phys. Lett.* **80**, 1288-1290, (2002).
3. E. Kymakis, I. Alexandrou and G. A. J. Amaratunga, *J. Appl. Phys.* **93**, 1764, (2003).
4. L. E. Brus, *J. Chem. Phys.* **79**, 5566-5571, (1983).
5. E. Kucur, J. Riegler, G. A. Urban and T. Nann, *J. Chem. Phys.* **119**, 2333-2337, (2003).
6. R. Hunger, C. Pettenkofer and R. Scheer, *Surf. Science* **477**, 76-93, (2001).
7. B. J. Landi, R. P. Raffaelle, S. L. Castro and S. G. Bailey, *Prog. Photovot: Res. Appl.*, in press, (2004).
8. S. Banerjee and S. S. Wong, *Nano Lett.* **2**, 195-200, (2002).
9. S. Banerjee and S. S. Wong, *J. Am. Chem. Soc.* **125**, 10342, (2003).
10. J. M. Haremza, M. A. Hahn, T. D. Krauss, S. Chen and J. Calcines, *Nano Lett.* **2**, 1253-1258, (2002).
11. B. J. Landi, H. J. Ruf, J. J. Worman and R. P. Raffaelle, *J. Phys. Chem. B* **108**, 17089-17095, (2004).
12. B. J. Landi, S. L. Castro, H. J. Ruf, C. M. Evans, S. G. Bailey and R. P. Raffaelle, *Sol. Energ. Mat. & Sol. Cells*, in press, (2004).
13. S. L. Castro, S. G. Bailey, R. P. Raffaelle, K. K. Banger and A. F. Hepp, *J. Phys. Chem. B* **108**, 12429-12435, (2004).
14. S. L. Castro, B. J. Landi, R. P. Raffaelle and S. G. Bailey "Colloidal CuInS2 Nanoparticles for Polymeric Solar Cells." In '2nd International Energy Conversion Engineering Conference'. (Providence, RI, 2004).

Mater. Res. Soc. Symp. Proc. Vol. 836 © 2005 Materials Research Society L2.10

Nanoporous Si – Organic Composite Photovoltaics

I. A. Levitsky, W. B. Euler[1], N. Tokranova[2], B. Xu[2], and J. Castracane[2]

Emitech, Inc., Fall River, Massachusetts 02720, U.S.A.
[1]Department of Chemistry, University of Rhode Island, Kingston, Rhode Island 02881, U.S.A.
[2]College of Nanoscale Science and Engineering, University at Albany (SUNY), New York 12203, U.S.A.

ABSTRACT

We report novel composite photovoltaics fabricated from nanoporous Si filled with Copper Phthalocyanine (CuPC) and its derivatives, including discotic liquid crystal CuPC. Porous n-type Si (PSi) was prepared electrochemically (pore size 10-20 nm) and filled with CuPC solution by impregnation under pressure or spin casting (liquid crystal CuPC) onto a porous surface followed by heating to the temperature of the mesophase transition. For porous layer with the thickness more than 1500-2000 nm, the conversion efficiency was up to 3% under a tungsten lamp with solar filters (30 mW/cm^2). The porous structure and pore filling were examined by SEM and XPS techniques. Photocurrent spectra of the CuPC-PSi cell demonstrate two bands that correspond to the absorbance of the protonated form of CuPC. A comparative analysis between CuPC and liquid crystal CuPC has been made and conclusions about possible mechanisms of charge transfer and transport are drawn. The solar cell optimization is discussed as well.

INTRODUCTION

The study and development of organic-inorganic composite solid solar cells is a rapidly growing area in the material science for photovoltaics (PV). For the past several years, organic-inorganic composite solar cells demonstrated enhanced performance due to nanoscale fabrication of inorganic components (nanodots [1], nanorods[2]), the use of Si and III-V type semiconductors [3,4] and novel types of dyes and hole transport materials for solid dye synthesized solid cells (DSSCs) [5]. DSSCs based on mesoporous TiO_2 have been an attractive alternative to the purely organic devices because of their huge interfacial area and high electron mobility in mesoporous titania [5,6]. Meanwhile, employment of other nanoporous inorganic matrices filled with organic materials could lead to fabrication of novel hybrid photovoltaics systems with efficient photoinduced charge transfer (PCT) and fast charge transport.

In this paper we demonstrate a solar cell based on n-type nanoporous silicon (PSi) filled with copper phthalocyanine (CuPC) and its derivatives (including a discotic liquid crystal form). The composite materials were characterized by SEM, XPS and Vis-NIR spectroscopy. The distinctive feature of such a device is the contribution of the semiconductor component to light absorption and energy conversion, which is impossible for

TiO_2 based DSSCs. Also, the nanoporous PSi structure provides a large interfacial area between the organics and the semiconductor, in contrast to a flat heterojunction [3, 4], and reduces reflection as compared with bulk Si.

EXPERIMENTAL DETAILS

The highly doped n-type Si (100 orientation, resistivity = 0.025 Ωcm) has been used for PV cell fabrication. It was observed in study [7] that only n-type Si provides PCT from CuPC to semiconductor. The back side of a Si wafer was treated with HF/ethanol solution to remove SiO_2 followed by evaporation by Al (20 nm) or Cr/Au (20 nm/150 nm) to form an ohmic contact. Nanoporous Si layers (500 – 15000 nm) were prepared by Si anodization (50 mA/cm^2) in HF/ethanol solution (15%,) followed by plasma etching to ensure that all pores were opened (Fig. 1). Some samples were processed without plasma etching to provide references. CuPC (Aldrich) and its derivatives CuPC1 (Aldrich) and CuPC2 (discotic liquid crystal, Polymer Source, Inc.) were chosen as organic fillers (Fig. 2, a). The choice of CuPC2 was motivated by the possibility of increasing the hole mobility along the quasi -1D columns similar to the LC coronen derivatives [8]. A DSC measurement of CuPC2 (Fig. 2, b) showed the LC mesophase in the range of 67-250 °C, consistent with previously reported data [9]. It is important for fabrication of the solid solar cell that LC phthalocyanines can preserve their columnar structure in the solid phase after cooling [10].

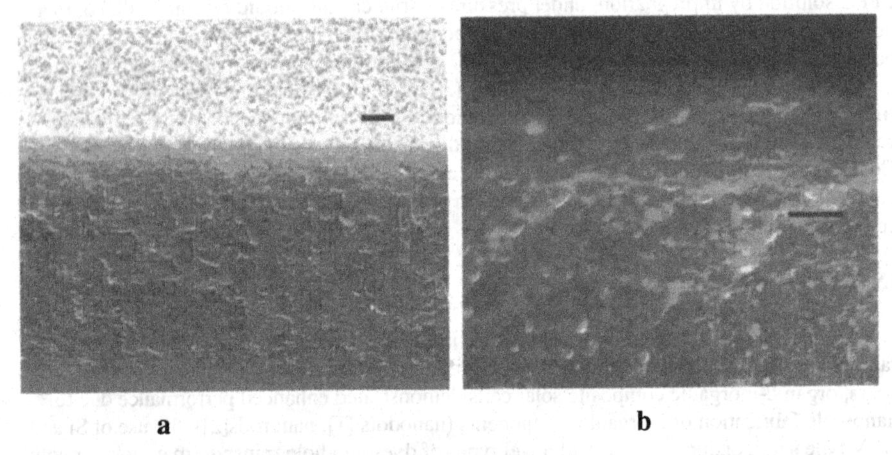

Figure 1. **a:** SEM image of Si surface and cross-section after electrochemical etching: 15% HF/ethanol solution, 20 s, 50 mA/cm^2, the porous layer depth is 1800nm, 300nm scale (black bar); **b:** the same sample after of being filled with CuPC /H_2SO_4 (cross-section only).

PSi was filled by dropping a solution (~10^{-2} M) of CuPC (concentrated H_2SO_4), CuPC1 ($CHCl_3$), or CuPC2 ($CHCl_3$) onto the surface followed by the mechanical pressing (7.5 kg/cm^2) for 30 s and rinsing with water (CuPC) or chloroform (CuPC1, CuPC2). LC CuPC2 was also spin-cast onto a PSi substrate followed by heating to 80 °C (above the solid-

LC transition) and left to cool slowly to ambient temperature. The top electrode (anode) was deposited onto the organic/PSi surface by sputtering gold (~20 nm) or ITO (80-100 nm) through a shadow mask. The active area of the cell was about 0.05 cm^2.

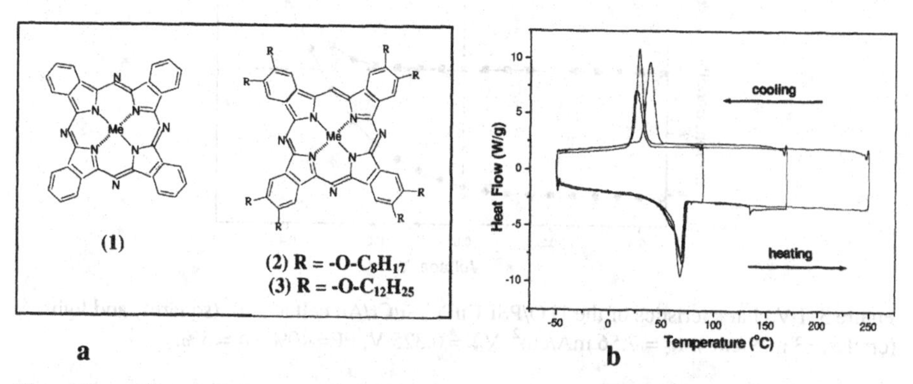

Figure 2. **a**: Chemical structure of CuPC (1), CuPC1 (2) and Liquid Crystal CuPC2 (3); **b**: DSC thermogram (first, second and third cycles, heating at 10 °C/min) for LC-CuPC. The temperatures of the endothermic peaks (68 °C) associated with the solid–discotic LC transition. The LC mesophase is stable up to 250 °C (in accordance with Ref. [9]).

The devices were tested under dark and illuminated conditions using a 250 W halogen-tungsten lamp (QTH 6334, Oriel) equipped with a KG4 heat absorbing filter. Proximity to the standard solar simulator (class B) was estimated by the value of conversion efficiency of the calibrated silicon photodiode, which was less than 25% of NREL data. The light intensity was measured by an Oriel 70260 broadband power meter and was varied by circular neutral filters. *I-V* characteristics were detected by a Keithley 236 source-measure unit connected to a PC. A Perkin-Elmer Lambda 900 spectrophotometer was used to record Vis-NIR absorption spectra. X-ray Photoelectron Spectroscopy (XPS) analysis has been performed using a Thermo VG scientific Theta Probe.

DISCUSSION

We did not find any sizable PV effect for samples filled with CuPC1 and LC CuPC2 compounds. PV response of the PSi-CuPC1 cell was low, but higher than that for reference cells without organics. The best conversion efficiency was ~ 0.02% (30 mW/cm^2) and could be slightly improved after 15 hours annealing at 100 °C. Contrary to our expectation to observe a strong PV response in the PSi filled with the LC CuPC2, the device conversion efficiency (< 0.01%) was comparable with that of reference cells.

Relatively high conversion efficiency (up to 3%) was observed for PSi filled with CuPC. Figure 3 shows I-*V* characteristics for an ITO/PSi-CuPC/Cr/Au cell at 33mW/cm^2 light power. It was found that *I-V* curves of PSi samples that were not treated by plasma etching to open all pores exhibited a "kink" behavior leading to a decrease of the V_{oc} and I_{sc} values. A similar "kink" was observed in other reports [11-13] and was explained recently

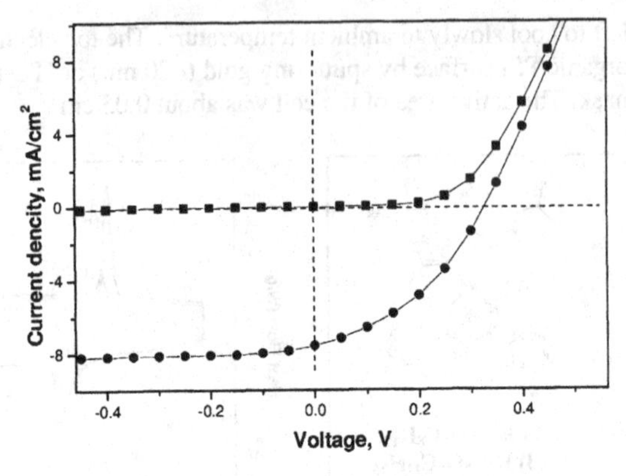

Figure 3. I-V characteristics of the ITO/PSi-CuPC/Si/Cr/Au cell at dark (squares) and light (circles, 33 mW/cm^2); I_{sc} = 7.56 mA/cm^2, V_{oc} = 0.325 V, FF= 40% , η = 3%.

in a theoretical study [14]. We can suggest that samples with a partially opened pores or with poorly developed pore structure cannot provide a sufficiently dense filling of CuPC molecules to form a "conductive wire" for hole transport to the anode. This is consistent with the theoretical model [14], which indicates that the *I-V* curve degradation ("kink" feature) occurs as a result of a decrease of the hopping coefficient: poor pore filling leads to isolation of the CuPC aggregates making them insulated from each other.

In order to confirm that Si pores are really filled with organic molecules and to estimate the depth of pores filling XPS analysis has been applied to the PSi-CuPC samples. It was found that the atomic concentration for Cu-2p3, N-1s and C-1s (all organic components) decreases exponentially as $\sim exp(-x/d)$ from the surface to the bulk with an approximately the same d value of 80-100 nm for each atom. The trace of these atoms was clearly observed even at the 200 nm depth. This result is consistent with the SEM image of the filled PSi (Fig.1, b), where most of the organic material is seen in the near-surface region. Thus, we can conclude that PSi indeed is filled with organic molecules, however the effective length of the CuPC penetration is less than the pores depth. Such a circumstance allows us to significantly improve the PV device performance in case of the complete pore filling. The fact that the η value strongly depends on the filling pressure supports the above assumption. For example, without any pressure PV efficiency η = 0.0007%; at a pressure of 2.4 kg/cm^2 $\eta \sim$ 0.02%; and at a pressure of 7.5 kg/cm^2, η = 2-3 %.

The significant difference between low (CuPC1, CuPC2) and high (CuPC) conversion efficiency of hybrid devices can be understood in terms of the interfacial area between organic and inorganic components. Likely, the electron transfer from CuPC1 to Si is slow or nonexistent due to relatively long alkyl chains (Fig 2, a), which can prevent close contact of the CuPC1 molecular core with PSi. This is supported by the fact that the LC CuPC2 based cell has the lowest η value, comparable with that of the references. Thus, the critical issue becomes an average distance between a CuPC core to the PSi surface and LC organization cannot compensate the low PCT caused by the long alkyl chains.

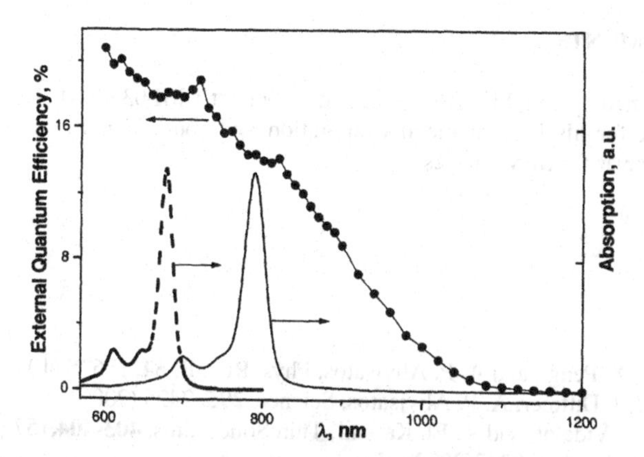

Figure 4. External quantum efficiency of the Au/Psi-CuPC/Si/Al cell. Absorption spectra of CuPC in concentrated H_2SO_4 solution (solid line) and CuPC1 in chloroform solution (dashed line). Concentration is 10^{-5} M. Shift of CuPC spectrum with respect to CuPC1 spectrum is the result of CuPC protonation in H_2SO_4 solution.

Contribution of the CuPC to the PV effect can be seen in the spectral dependence of external conversion efficiency, which exhibits two peaks (720 and 820 nm), matching the absorption bands of the protonated CuPC form in H_2SO_4 solution (Fig. 4). A small shift to longer wavelengths (~20 nm) can be associated with aggregate formation of the protonated CuPC inside the nanopores. The long wavelength edge of the spectrum corresponds to the band gap of Si (1.1 eV) but not pSi (~ 2 eV). Hence, the absorbance of the porous layer is low or the porosity in the n-type Si is not sufficient to change the band gap. Replacement of the ITO anode by Au does not change the V_{oc} value (~0.3 V, 30mW/cm^2), but decreases the backward dark current. Independence of V_{oc} on electrode workfunction (ITO, 4.5-4.7 eV and Au, 5.1 eV) can be associated with the fact that organic and hybrid PV cells do not require a built-in electrostatic potential for charge separation, unlike inorganic devices [14,15].

CONCLUSIONS

We demonstrated that nanoporous Si filled with copper phthalocyanine can provide an efficient PV effect with conversion efficiency up to 3% under white illumination (30 mW/cm^2). X-ray Photoelectron Spectroscopy data for CuPC based cells shows that the trace of organic molecules exists at a 200 nm depth. A critical issue for the pore filling is the pressure applied to the organic solution. Both CuPC and PSi contribute to the photocarrier generation. Further improvements in the device performance are expected with pore diameter/depth/structure optimization, increase of the filling pressure and fabrication of thin, free-standing PSi membranes.

ACKNOWLEDGMENTS

This work is supported by an NSF SBIR grant under contract DMI-0338664. The authors are grateful M. Platek for his help in metal evaporation and sputtering, R. Moore for XPS analysis and A. Gracias for SEM images.

REFERENCES

1. N.C.Greenham, X. Peng,, and A. P. Alivisatos, Phys. Rev. B. **54,** 17628 (1996).
2. W. U. Huynh, J. J. Dittmer, A. P. Alivisatos, Science, **295,** 2425 (2002).
3. J. Ackermann, C. Videlot, and A. El. Kassmi, Thin Solid Films, **403-404,**157 (2002).
4. F. Garnier, J. Optics. A **4,** S247(2002).
5. M. Grätzel, J. Photochem. Photobiolog. C :Photochem Rev. **4**(2003)145.
6. Q.-B. Meng, K. Takahashi, X.-T. Zhang, I. Sutanto, T. N. Rao, O. Sato, A. Fujishima, H. Watanabe, T. Nakamori, and M. Uragami, Langmuir **19**,3572(2003).
7. A. S. Komolov, and P.J. Moller, Synth.Met. **128**,205(2002).
8. L. Schmidt-Mende, A. Fechtenkotter, K. Mullen, E. Moons, R.H. Friend, and J.D. MacKenzie, Science **293** 1119 (2001).
9. C. Piechocki, J. Simon, A. Skoulios, D. Guillon, and P. Weber, J. Am. Chem. Soc. **104,** 5245(1982).
10. D. Markovitsi, I. Lecuyer, and J. Simon, J. Phys. Chem. **95,**3620 (1991).
11. P. Peumans, and S. R. Forrest, Appl. Phys. Lett. **79**, 126(2001).
12. M. Y. Song, J. K. Kim, K. J. Kim,and D. Y. Kim, Synth. Met.**137**, 1387(2003).
13. P. Ravirajan, S. A. Haque, J. R. Durrant, D. Poplavskyy, D.D.C. Bradley and J. Nelson, J. Appl. Phys. **95**,1473(2004).
14. J. Nelson, J. Kirkpatrick and P. Ravirajan, Phys.Rev B **69**,035337(2004).
15. B. A. Gregg, J. Phys. Chem. B **107**, 4688(2003).

Polymer-Based Devices

Mater. Res. Soc. Symp. Proc. Vol. 836 © 2005 Materials Research Society L3.2

High Efficiency P3HT/PCBM Solar Cell

Kanzan Inoue[ab], Ross Ulbricht[a], Pallavi C. Madakasira[ab], Miaoxin Zhou[c], Sergey B. Lee[a], John Ferraris[ac], Anvar A. Zakhidov[ab]

a) NanoTech Institute, The University of Texas at Dallas, 2601 N. Floyd Rd. Richardson, TX 75083, USA.
b) Department of Physics, The University of Texas at Dallas, 2601 N. Floyd Rd. Richardson, TX 75083, USA.
c) Department of Chemistry, The University of Texas at Dallas, 2601 N. Floyd Rd. Richardson, TX 75083, USA.

ABSTRACT

We report a nearly twofold increase of short circuit current: from $I_{sc} \sim 10$ mA/cm^2 to $I_{sc} = 16$-20 mA/cm^2 in P3HT/PCBM solar cells (SC) employing freshly prepared regio-regular poly(3-hexylthiophene) (RR-P3HT) without special purification . The power conversion efficiency is enhanced to $\eta \geq 4\%$ as compared to our best $\eta=3.8\%$ in SC with commercial polymer despite the decreased filling factor (FF= 0.42, as compared to best FF = 0.59). We used our earlier found [1] procedures with optimal post heat treatment temperatures and time for our polymer SC. We also discovered a strong correlation between the device preparation procedures and performance. The optimal phase separation of PCBM and RR-P3HT into a bi-continuous network structure occurs after quite long solution stirring times (enhanced homogenization) and surprisingly very short annealing time at optimal temperature. We also found that the optimal concentration of PCBM in a RR-P3HT matrix is rather low, only c~35 wt%, contrary to high c~80 wt% in PPV based SC.

INTRODUCTION

Efficiencies of organic solar cells have improved dramatically since the introduction of the concepts of a bulk-heterojunction and bi-continuous interpenetrating network of acceptor and donor materials. The optimal nanoscale phase separation of the donor and acceptor into a bi-continuous network is critical for the performance of solar cells because of the short exciton diffusion length (~10 nm) in organic materials. The effect of the postproduction treatment on polymer/[6,6]-phenyl-C61 butyric acid methyl ester (PCBM) organic solar cells has been studied by several groups [2-5]. However, the mechanism behind this effect is yet to be understood clearly. In our earlier papers [1] we demonstrated the effects of the postproduction heat treatment and solution homogenization time on the performance of bulk heterojunction solar cell based on commercial RR-P3HT and PCBM over a range of different temperatures and durations, as well as, explored the physical phenomenon behind the improvement during the treatment.

In the present paper, we demonstrate, that "freshly prepared" polymer even without special purification shows much better photocurrent in optimally processed solar cells, as compared to the RR-P3HT from commercial sources (e.g. American Dye Source Co.). Here "fresh" means stored in argon glove box to avoid oxidation and used within a few days after synthesis. The effect of the solution homogenization time on the performance of the device is also reported in this paper. Long stirring of freshly synthesized, but intentionally unpurified P3HT, creates I_{sc} current increase from a typical range of 10-11 mA/cm^2 to 16-20 mA/cm^2, but the filling factor decreases from 0.59 to 0.42. We discuss here possible explanations of this enhancement of photocurrent phenomena as due to moderate p-doping of SC by catalyst residuals after synthesis, which creates better barrier with hole collecting ITO.

EXPERIMENTAL DETAILS

Device fabrication

ITO coated glass substrates (<15 Ω/sq with ~85% light transmission) and ITO on PET were obtained from Delta Technologies Ltd. EL-grade PEDOT-PSS was purchased from Bayer AG. RR-P3HT and PCBM were purchased from American Dye Source. All materials were used as received without further purification. We fabricated four devices on each substrate, each having an area of ~9 mm^2. The ITO coated glass substrate is etched and cleaned before being plasma-treated for five minutes (90 seconds for flexible substrates) under O_2 gas. A layer of PEDOT:PSS is then spin coated onto the substrate at 6100 rpm creating a 30-35 nm layer.

The sample is then dried by being heated at ~120°C for 100 minutes (60 minutes at 110°C for flexible substrates) in a glove box. The photoactive material solution is dispersed by a magnetic stirrer for days until is determined to be the best condition. The solution is then spin-coated onto the sample at 700 rpm creating a 50-60 nm layer using a toluene solution consisting of roughly 1:2 ratios of PCBM and RR-P3HT, respectively. The final layer is made up of 65% RR-P3HT and 35% PCBM. An aluminum cathode is then deposited under high vacuum (<10^{-6} torr) at an initial deposition rate of 0.4 Å/s and gradually increasing to 1.0 Å/s with a 450 sec ramp time to create a final thickness of 1000 Å. A thin 6 Å LiF layer is deposited prior to Aluminum deposition for flexible devices. A surface profiler (AMBIOS XP-1) was used to measure film thickness. The finalized device is then annealed on a hot plate in a glove box at the desired temperature for the desired amount of time.

The absorption spectra were measured on a Perkin-Elmer Lambda 900 UV-VIS-NIR Spectrophotometer. The current-voltage characteristics were recorded with a Keithley 236 source-measure unit. A solar simulator (150W Xenon lamp with AM0 and AM1.5 filters from Spectra-Physics and focusing lens) with light intensity calibrated at 100 mW/cm^2, was used as the light source for solar cell efficiency measurements. The reported efficiency measurement was not corrected for spectral mismatch.

Synthesis of regio- regular poly-3-hexylthiophene (RR-P3HT)

The standard synthetic procedure of McCullough [7,8] was followed, changing only the amount of reagents.
Synthesis is made in three steps:
Step 1 (Synthesis of 3-hexylthiophene (compound 1), according to literature [6]. To a suspension of Mg (1.22g, 50.2mmol) in 10ml of anhydrous ether, hexylbromide (9.40g, 50.7mmol) was added dropwise in 35ml of ether. After complete disappearance of Mg, 3-bromothiophene (8.89g, 52.9mmol) and 30mg Ni (dppp) Cl_2 were added. After stirring and heating (40 to 50 °C) for 48 hours, the reaction mixture was poured into a mixture of crushed ice and diluted HCl (2N) and extracted from ether. The combined ether layers were dried over $MgSO_4$. After removal of the solvent, the residue was distilled in vacuum to yield compound 1 as a colorless oil (4.32g, 51.2%).
Step 2 (Synthesis of 2,5-dibromo-3-hexylthiophene (compound 2, according to literature[7]. (a) 3-hexylthiophene (4.23g, 25.1mmol) was dissolved in 75ml dry THF. N-Bromosuccinimide (9.04g, 25.1mmol) was added to the solution over a period of 5 min. The solution was stirred at

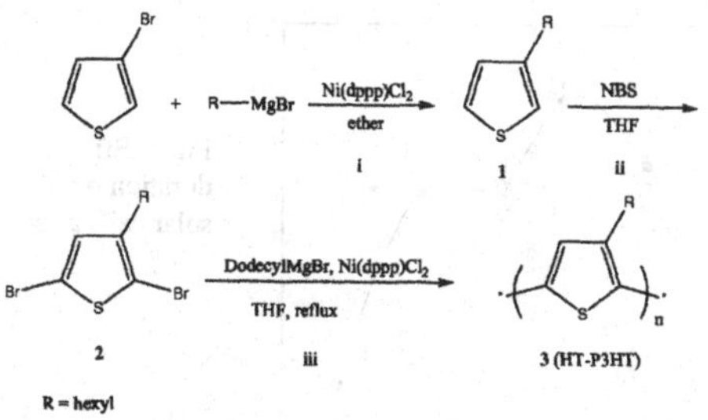

Fig. 1 Synthetic route for regio-regular poly(3-hexylthiophene).

room temperature for 24 hr. The solvent was removed in vacuo and hexane was added to precipitate succinimide. The mixture was filtered through a silica plug and the solvent was removed in vacuo. The residue was distilled in vacuum to yield compound 2 - yellowish oil (6.58g, 80.4%).

Step 3 (Synthesis of HT-Poly-3-hexylthiophene (compound 3), according to literature [8]. Compound 2 (1.45g, 4.45mmol) was dissolved in 40ml of dry THF. To this solution was added dodecylmagnesium bromide (4.50ml, 1.0M solution in diethyl ether). The mixture was heated to reflux for 1h and then Ni(dppp)Cl_2 (12mg) was added. After 3hr at reflux, the reaction was poured over 225 ml MeOH and then filtered through a Soxhlet thimble, which was then subjected to Soxhlet extraction with MeOH, hexanes and chloroform. The polymer was recovered from the chloroform fraction by reprecipitation by MeOH and filtration. The solid was further dried in vacuum over for 3 hr to yield HT-P3HT with green metallic luster (0.33g, 45%), 90 % HT-coupled as estimated by methylene protons directly adjacent to aromatic ring).

RESULTS AND DISCUSSION

The ability of forming nano-scale phase separation and bi-continuous networks largely depends on the dispersion of the photoactive solution. A more evenly dispersed solution yields better device performance. A well dispersed solution shows roughly 10-20 % higher filling factor coefficients in comparison to the not-so-well dispersed solution when both solutions are used to fabricate devices under the same conditions. However, heavy usage of an ultra/bath sonicator can damage the polymer or PCBM resulting in a decrease of the charge carrier mobility and photogeneration efficiency, and consequently, the efficiency of the device. We used a magnetic stirrer to continuously stir the solution in glove box at room temperature. A mild stirring speed was used (300-500prm). This way the solution was never heated up while being dispersed. Fig.2 indicates that the estimated optimal stirring duration is 9 to 10 days. The improvement of the spin coated film quality was observed with the naked eye as the stirring duration approached its maximum. After 10 days of stirring, the device performance decayed quickly, although the film quality continued to be good. We haven't found a feasible explanation of such a rapid decrease of performance once the optimum stirring duration is surpassed.

Fig. 3 shows the effect of the duration of the postproduction heat treatment at different temperatures on the efficiency of the devices fabricated on glass substrate, measured under simulated sun light AM1.5 (100 mW/cm^2).

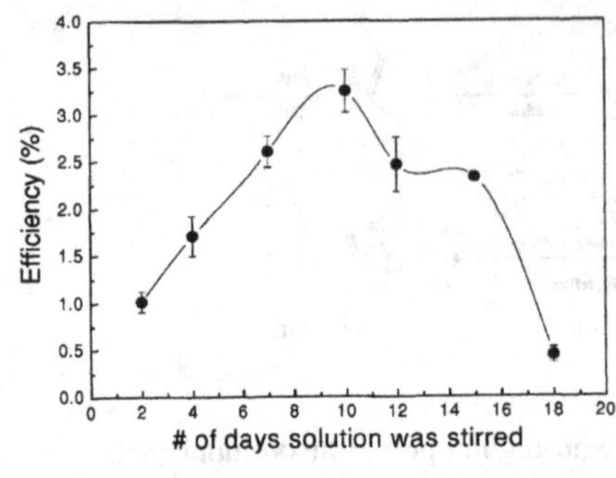

Fig.2. Effects of the solution stirring duration on glass-base P3HT/PCBM solar cells after annealing at 96 °C

The devices show significant improvement in efficiency after being annealed for a few minutes. However, for longer durations of heat treatment, they behave very differently at different temperatures. At 75 °C and 90 °C the efficiency was stabilized after a 3 min heat treatment and no further changes in efficiency were observed. Whereas, the efficiencies started decreasing as annealing time increased at 120 °C and 150 °C. Especially at 150 °C the efficiency was observed to increase significantly after only a 20 sec heat treatment and started decreasing rapidly for longer durations of heat treatment. A similar test was done for the flexible solar cells and 96 °C was determined to be the optimal temperature for flexible RR-P3HT/PCBM devices. Fig. 4 shows the effect of annealing duration on flexible solar cells at 96 °C. The decrease in device efficiency with respect to the duration of annealing is much more rapid as compared to devices on glass.

The results of thermal treatment on efficiency of ITO/PEDOT:PSS/RR-P3HT_PCBM/Al devices indicate that at high temperatures the PCBM molecules can diffuse more easily within polymer matrix and form large nanocrystals. We have fabricated bulk-heterojunction solar cells on

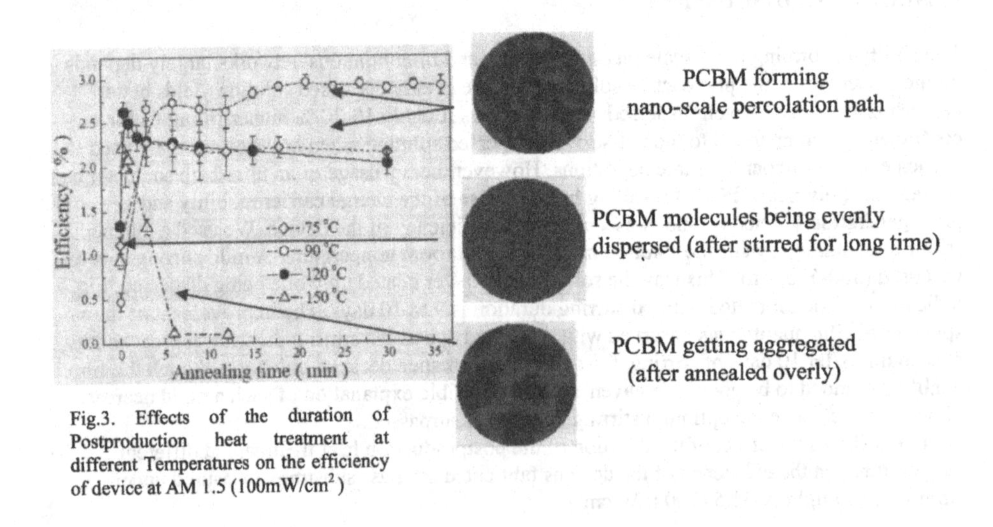

Fig.3. Effects of the duration of Postproduction heat treatment at different Temperatures on the efficiency of device at AM 1.5 ($100 mW/cm^2$)

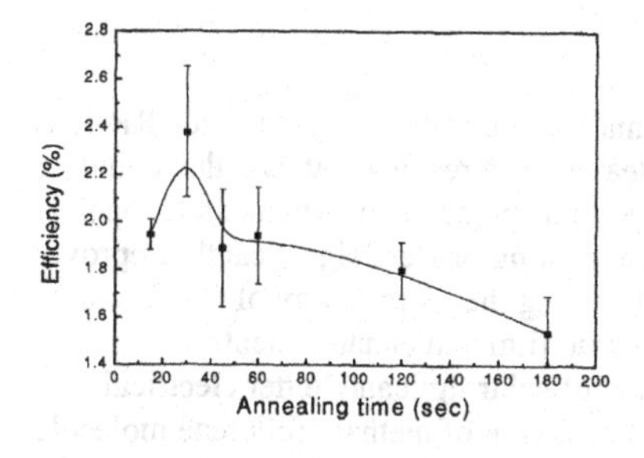

Fig.4 P3HT/PCBM solar cells on plastic substrates annealed at 96 °C for 5 minutes.

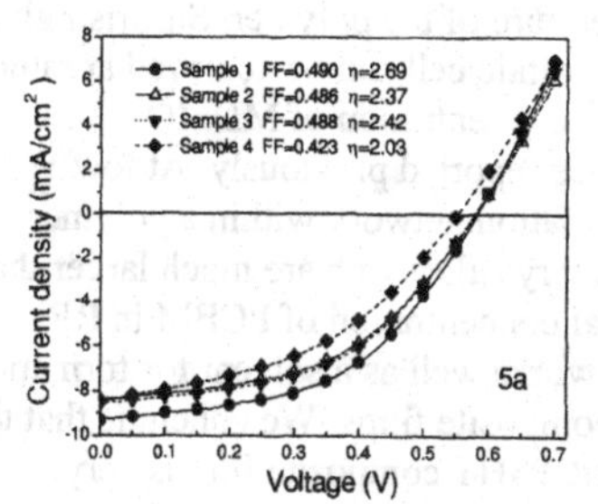

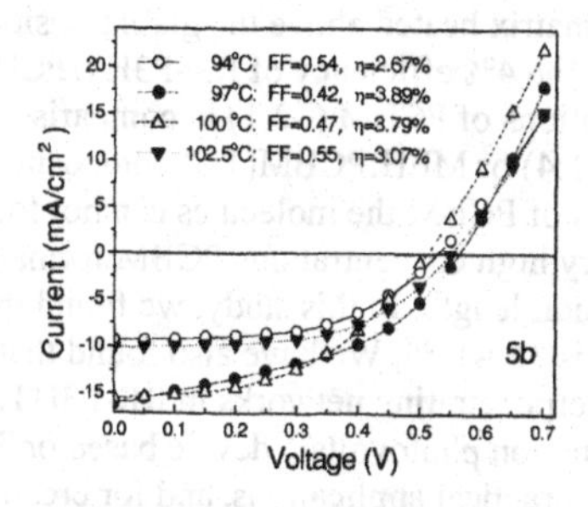

Fig. 5 a) ITO/PEDOT: PSS/P3HT_PCBM/LiF/Al solar cells on plastic substrates annealed at 96°C for 30 sec b)ITO/PEDOT:PSS/P3HT-Hm_PCBM/Al solar cells on glass substrates annealed for 5 min at various temperatures. All measurements were taken under simulated solar light AM 1.5 100mW/cm².

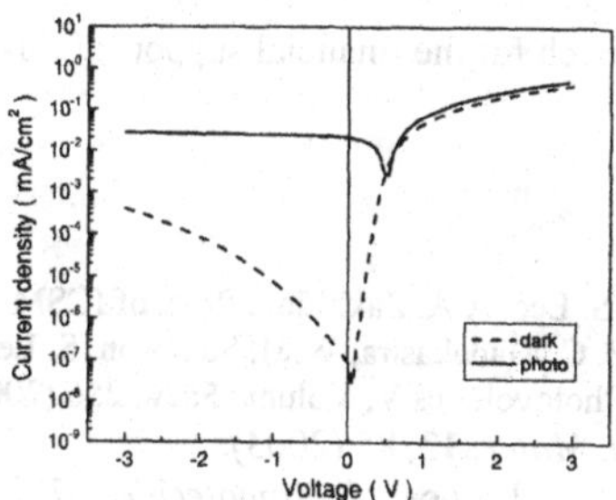

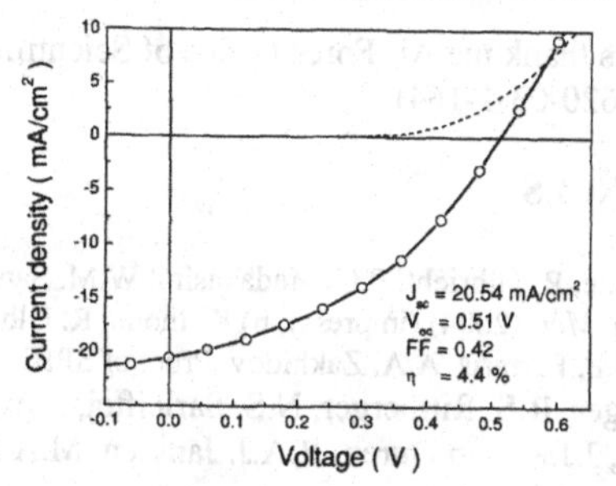

Fig.6. Current-voltage characteristics of the best ITO/PEDOT:PSS/PCBM+RR-P3HT/LiF/Al device in dark and under simulated solar light AM 1.5 100 mW/cm²

flexible substrates using optimized heat treatment parameters (Fig.5a). These solar cells are extremely light weight and have a thickness of less than 0.5mm. These plastic-based photovoltaic devices have a much higher degradation rate as compared to the glass base devices. Therefore, protection layers and an encapsulation are essential for the applications. Fig 5b compares the devices performances with different annealing temperatures. The devices were made with a solution after optimized stirring. Clearly the temperature range of 97 °C to 100 °C yields a surprisingly higher I_{sc} and thus higher efficiencies than at rest of the annealing temperatures. Fig. 6a and 6b show the current-voltage characteristics of our best devices fabricated on glass substrates under dark and simulated solar light AM 1.5 100 mW/cm².

CONCLUSION

In summary, we have presented a study of solar cells made from in-house synthesized RR-P3HT polymer, which was intentionally unpurified so as to leave some residual catalyst that could act as a moderate p-dopant. Magnetic stirring process helps to disperse the polymers and suspend PCBM molecules in the polymer matrix. The thermal annealing process significantly improves the efficiency of RR-P3HT/PCBM solar cells by ameliorating the morphology of the organic active layer in several ways; 1) densification of the organic film and enhancement of intermolecular interaction leading to a higher absorption of solar light and better electrical transport; 2) optimized phase segregation by enhanced diffusion of methanofullerene molecules in a polymer matrix heated above the glass transition temperature of the polymer. Surprisingly, we found that the 4% efficiency of RR-P3HT/PCBM photovoltaic cell can be achieved at rather low concentrations of PCBM (~2:1) in comparison to a high concentration of MDMO-PPV:PCBM (1:4) or MEH:PCBM(1:5) composite film device reported previously. At low concentrations of PCBM the molecules can not form a percolation network within a polymer matrix. At very high concentrations PCBM forms large nanocrystals which are much larger than exciton diffusion length. In this study, we found that optimal concentration of PCBM in RR-P3HT matrix is 35 wt. %. We have also found that toluene works well as a solvent for formation of optimal interpenetrating networks in RR-P3HT:PCBM composite films. We conclude that the bulk heterojunction photovoltaic device based on PCBM/RR-P3HT composite film is very promising for practical applications, and for creating flexible solar cells on plastic substrates.

ACKNOWLEDGMENTS

The authors thank the Air Force Office of Scientific Research for the financial support of this work (F49620-03-1-164).

REFERENCES

1. a) K. Inoue, R. Ulbricht, P.C. Madakasira, W.M. Sampson, S. Lee, A.A. Zakhidov, Proc. of ICSM 2003, *Synth. Met.* (2004) (in press); b) K. Inoue, R. Ulbricht, P. C. Madakasira, W.M. Sampson, S. Lee, J. Gutierrez, J.P. Ferraris, A.A. Zakhidov, Proc. of SPIE – Org. Photovoltaics V, Volume **5520**, 256 (2004).
2 F. Padinger, R.S. Rittberger, N.S. Sariciftci, *Adv. Funct. Mater.* **13,** 85 (2003).
3 X. Yang, J.J.K. van Duren, R.A.J. Janssen, M.A.J. Michels, J. Loos, *Macromolecules* **37,** 2151 (2004).
4. J.J.K. van Duren J. Loos, F. Morrissey, C.M. Leewis, K.P.H. Kivits, L.J. van IJzendoorn, M.T. Rispens, J.C. Hummelen, R.A.J.Janssen, *Adv. Funct. Mater.* **12,** 665 (2002).
5. N. Camaioni, G. Ridolfi, G. Casalbore-Miceli, G. Possamai, M. Maggini *Adv. Mater.* **14,** 1735-38 (2002).
6. Pham,C.V.; Mark, H.B., Jr.; Zimmer, H. *Synth. Commun.* **16,** 689-696 (1986).
7. Loewe, R.S.; Khersonsky, S.K.; McCullough, R.D. *Adv. Mater.* **3**, 250 (1999).
8. Loewe, R.S.; Ewbank, P.C.; Liu, J.; Zhai, L.; McCullough, R.D. *Macromolecules* **34,** 4324-4333 (2001).

Mater. Res. Soc. Symp. Proc. Vol. 836 © 2005 Materials Research Society L3.7

Plasticized Conjugated Polymers: A Possible Route to Higher Voltage Solar Cells

Paul Wentzel, Aurelien Du Pasquier
Energy Storage research group
Department of Ceramics and Materials Engineering
Rutgers, The State University of New Jersey
10, Knightsbridge Road, Piscataway NJ 08854

ABSTRACT

The introduction of propylene carbonate (PC) into the poly (3-alkyl)thiophenes (P3AT) poly(octyl)thiophene (P3OT) and poly(hexyl)thiophene) (P3HT) is studied. In both cases, we observe a diminution of their melting temperature, which demonstrates a plasticizing effect. We also observe an increase of photovoltage and photocurrent in sandwich devices M//P3AT+PC//ITO-PET where M=Al, Mg and ITO-PET is Indium thin oxide coated on polyester. We attribute this effect to the introduction of oxygen from the PC. When mixed with an electron acceptor PCBM, the photocurrents only slightly increase and the photovoltages are similar. In this case, the presence of oxygen is detrimental to the system.

INTRODUCTION

Blends of conjugated polymers and fullerene derivatives have been used successfully as photoactive layers to build organic solar cells. This is because they create efficient dissociation centers for excitons. This effect relies on the ultra fast electron transfer from conjugated polymers to fullerenes [1,2,3], with long-lived charge separation [4] and transport of the separated holes and electrons to metallic electrodes of different work functions. Such donor-acceptor bulk heterojunction results in charge separation quantum efficiency near unity. In order to improve the film formation and the homogeneity of the blends, the use of a plasticizer such as polystyrene or poly (vinyl) carbazole has been investigated. It was found that no decrease in performance was observed up 10 Wt.% added plasticizer [5]. Further, other conjugated plasticizer such as poly-3-butyl-co-3,4-dibutylthiophenes have been found to improve homogeneity and dispersion of the fullerene C_{60} in P3OT polymer, as evidenced by SEM [6]. There are other potential benefits to plasticizers, which include facilitating the fabrication of solar cells by lamination [7] and increasing the diffusion coefficient of small molecules in the blend, which can act as redox shuttles. In his context, we have investigated the addition of propylene carbonate on the thermal and photochemical properties of poly (alkyl) thiophenes blended or not with a fullerene derivative.

EXPERIMENTAL DETAILS

Regioregular poly (3-octylthiophene) was purchased from Aldrich. Regioregular poly (3-hexyl) thiophene was kindly provided by Plextronics. [6,6]- phenyl-C61-butyric acid methyl ester) (PCBM) was purchased from American Dye Source. The polymers and PCBM were dissolved at room temperature in chloroform to make 5 g/L solutions.

Various amount of 5 g/L propylene carbonate (PC hereafter) solution in chloroform were added to their solution, resulting after casting and solvent evaporation in plasticized films with PC contents ranging from 0 to 100 Wt.%. Differential scanning calorimetry (DSC hereafter) measurements of those films were performed on a DuPont 910 DSC system. Plastic solar cells were built by using the following procedure: ITO-PET transparent conducting substrates (Sheldahl, 40 ohm/□) were cut to 1.5*3 cm^2 size and cleaned in methanol. A thin layer of poly(ethylene) dioxythiophene doped with poly(styrene) sulfonate (PEDOT: PSS hereafter) (Baytron P, Bayer) was spin coated on the central portion.

For making the contacts between the plastic solar cell and the potentiostat, a sample-holder was home-built. It consisted of a cylinder with a quartz window at one end, and a 1 cm^2 metal piston pushed by a vise at the other end. A spring between the piston and the vice ensured that the same pressure was applied for all measurement. For testing of different metals and different photoactive films, the piston was either aluminum or magnesium.

The polymer solutions were drop-cast on the base of the piston, and pressed against the conductive face of the PEDOT: PSS coated ITO-PET substrate. For I-V measurement, an EG&G 273 potentiostat-galvanostat was used, under control of Corrware software (Scribner Associates). A white halogen bulb was used as light source. The light intensity could be varied between 1 and 165 mW/cm^2, and was measured with a solar cell calibrated to the solar spectrum (Daystar, Inc).

RESULTS

DSC measurements on P3OT and P3HT plasticized with PC

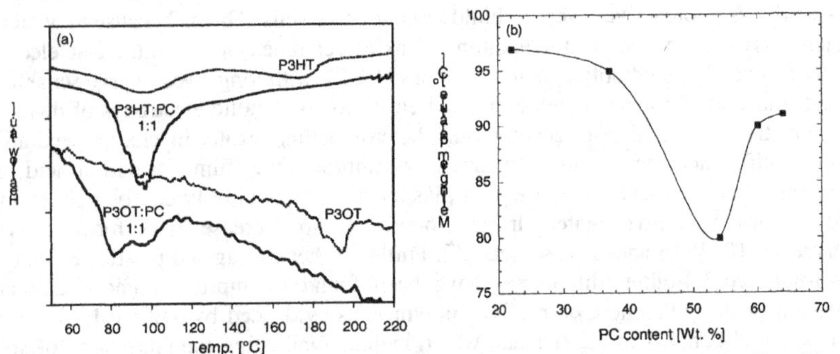

Figure 1. (a) DSC traces of P3HT and P3OT plasticized or not with PC at 1:1 weight ratio, (b) melting temperature of P3OT as a function of PC contents.

DSC measurements were performed on P3HT and P3OT films drop cast from 5g/L solutions in chloroform containing various amounts of PC. At 50 wt.% PC contents, the melting temperature is reduced from 180°C to 100°C for P3HT, and from 200°C to 80°C for P3OT (Fig. 1a). We also observed, in the case of P3OT, a minimum of the melting temperature at 54 wt.% PC contents (Fig. 1b). We attribute this as being the maximum plasticizer amount that the system can accept. Larger amounts result in phase separation, which causes the melting temperature to rise again.

I-V measurements on ITO// P3OT+PC// M = Al, Mg films

When P3OT films are used in a Schottky diode configuration PET-ITO//P3OT+x%PC//M, the rectification ratio in the dark increases with an increase in PC content. Under constant white illumination of 165 mW/cm^2, the open-circuit voltage increases from 610 to 800 mV with an increase in PC from 22 to 54 Wt.%. At the same time, short-circuit currents increase from 1.3 to 2.4 μA/cm^2 (Fig. 2). The results suggest that there is an optimum PC content, because PC presence also results in higher dark currents and lower fill factor, which is not desirable. Also, higher PC contents are not desirable because the mechanical stability of the plasticized P3OT films would be compromised. In the range studied, the films remained freestanding.

When the aluminum contact is replaced by a magnesium contact, the open circuit voltage increases up to 1.8 Volts and the short-circuit current up to 5 μA/cm^2 for a 60 Wt.% PC content in P3OT (Table I).

PC content, wt%	V_{oc} (V)	I_{sc} (μA/cm^2)	Metal contact	Fill Factor
22	0.61	1.29	Al	0.32
37	0.78	1.49	Al	0.26
54	0.80	2.41	Al	0.23
60	0.63	2.02	Al	0.24
60	1.885	5.02	Mg	0.30

Table I: Effect of PC content and metal contact on the characteristics of Al/P3OT+PC/PEDOT: PSS/ITO-PET plastic solar cells

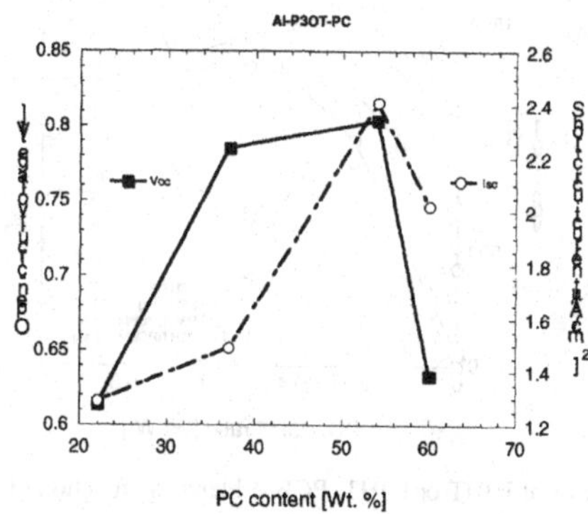

Figure 2. Effect of PC content on I_{sc} and V_{oc} of Al//P3OT+PC/PEDOT: PSS//ITO-PET plastic solar cells

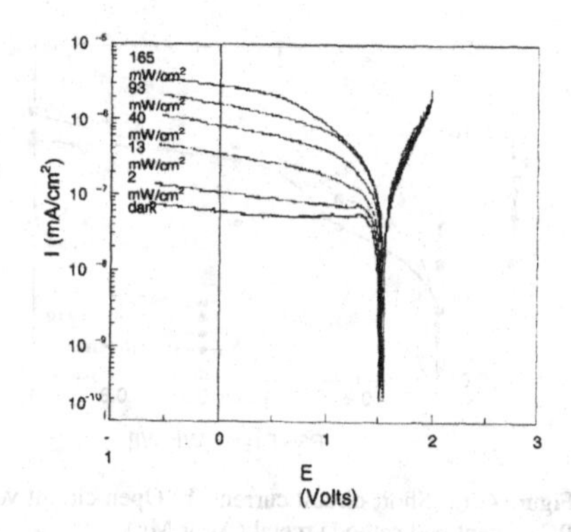

Figure 3. I-V curves at various light intensities of Mg//P3OT+PC/PEDOT: PSS//ITO-PET plastic solar cells

Addition of PCBM to the P3HT-PC system

We used a P3HT: PCBM weight ratio of 1:3 as suggested by the literature [8]. We measured the I-V curves of P3HT: PCBM: PC films at various PC contents as indicated on Table II. The I-V curves were measured against either Al or Mg cathodes.

P3HT [Wt]	PCBM [Wt]	PC [Wt]
1	3	0
1	3	0.25
1	3	0.5
1	3	1
1	0	0.5

Table II: Weight compositions of P3HT: PCBM: PC films tested

At low PC contents, we observe higher photocurrents with PCBM, but when the PC contents increases, the photocurrents are similar or slightly higher to those observed without PCBM (Fig. 4). This suggests that PCBM only contributes marginally to the photocurrent in this system. The observation of open-circuit voltages suggests the same conclusion: at low PC content, the V_{oc} values are higher with PCBM than without. But at high PC content, the V_{oc} values depend solely on the cathode metal, and are no longer fixed by the HOMO-LUMO difference between P3HT and PCBM. The effect of light intensity on V_{oc} in the case of the Mg cathode (not shown here) suggests a similar conclusion: without PC, V_{oc} values decrease to 750 mV at increasing light intensity. This indicates that the voltage is fixed by the HOMO-LUMO difference between P3HT and PCBM, as reported in the literature [9]. But in the presence of PC, the V_{oc} values increase up to 1300 mV. Further, in the absence of PCBM, even higher V_{oc} values of 1450 mV are measured for the 1:0:0.5 composition.

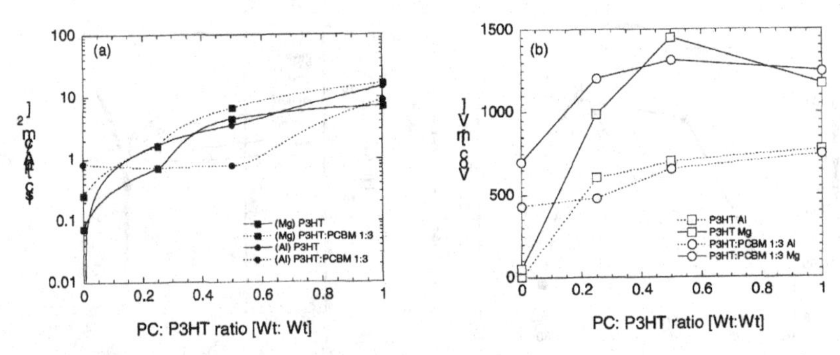

Figure 4. (a) Short-circuit current (b) Open-circuit voltage of P3HT or P3HT: PCBM blends as function of PC content and cathode metal (Al or Mg)

DISCUSSION

In the Mg//P3OT+PC/PEDOT: PSS//ITO-PET system, the study of short-circuit current I_{sc} dependency to light intensity I indicates a linear relationship $I_{sc} \sim I^{\alpha}$ with the exponent α passing through a maximum when the melting temperature of P3OT is lowest. An example of I-V curves obtained at various light intensities is depicted on Figure 3. Figure 5a shows the dependence of the short circuit current on the incident light intensity. Exponents α close to 1 are expected for devices, where both electron and hole transport are comparably efficient and bimolecular recombination not significant [10].

The maximum α of 0.73 is lower than the values observed in bulk heterojunction devices. This can indicate that electron transport is lower than hole transport, because of the absence of electron acceptor to efficiently separate the photogenerated excitons in the polymer. When plotting the scaling exponent α as function of the PC contents, the values are maximum at about 50 Wt. % PC, which suggests that this is the optimum PC contents (Fig. 5b). This confirms the results initially observed with an Al cathode.

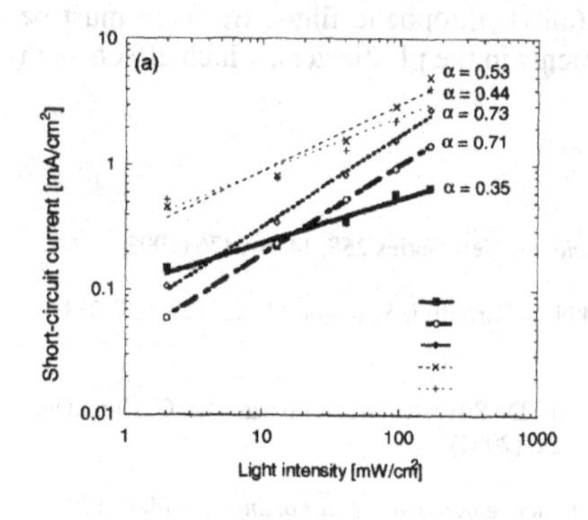

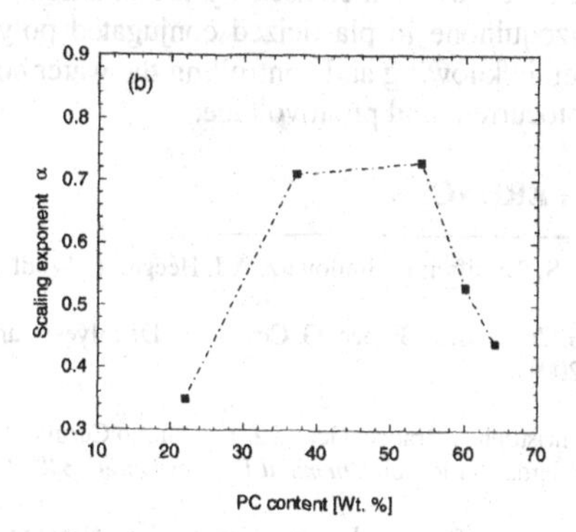

Figure 5. (a) Logarithmic plot of short-circuit current as function of light intensity for Mg//P3OT+PC/PEDOT: PSS// ITO-PET plastic solar cells at various PC contents (b) Scaling exponent α of the $I_{sc} \sim I^{\alpha}$ dependency of sort-circuit current to incident light for various PC contents in Mg//P3OT+PC/PEDOT: PSS//ITO-PET plastic solar cells.

For both polymers, the presence of PC increases short-circuit current by at least one order of magnitude, and open-circuit voltage increases by changing the cathode metal from Al to Mg. To explain this effect, we ruled out the possibility of a structural reason by extracting the PC in the films with ether, which resulted in a decrease of open-circuit voltage and short-circuit current. We assume that the presence of PC in the film increases its oxygen content and/or mobility, either from water electrolysis or direct dissolution from the atmosphere. Oxygen can be easily transported in the plasticized polymer, and can act as an efficient electron transfer shuttle, as it has already been observed with liquid electrolytes [11]. This effect enhances the photocurrent of the polymers, but also hinders electron transfer to PCBM when it is present.

The presence of oxygen can also explain the increase in photovoltage, by forming oxides at the surface of Al and Mg. Although care was taken to sand the surface of the electrodes before each measurement, such oxides could reform spontaneously in the presence of water traces contained in the plasticizer.

The overall values of photocurrent that we are measuring are low, because of the thickness of the films we are working with (~ 10 μm). This was necessary for using mechanical contacts, since evaporated contacts are not possible with a plasticizer such as PC. With a boiling point of 240°C, PC would be evaporated at the vacuum required to deposit Al contacts.

CONCLUSION

PC is a good plasticizer for P3OT and P3HT. This can be advantageous for building solar cells by lamination. PC also causes higher open-circuit voltages than PCBM. We attribute this to the formation of oxides at the metal cathode. An increase in photocurrents in the P3OT and P3HT alone is also observed, which we attribute to electron transfer from polymer to the oxygen. We conclude that short-circuit current could be further increased by the addition of other electron transfer molecules such as benzoquinone in plasticized conjugated poly (alkyl) thiophene films. But care must be taken in knowing and controlling the water contents in the plasticizers, which affects both photocurrent and photovoltage.

REFERENCES

1. N.S. Sariciftci, L. Smilowitz, A.J. Heeger, F. Wudl, *Science*, New Series **258,** 1474-1476 (1992)

2. G. Zerza, C. J. Brabec, G. Cerullo, S. De Silvestri and N. S. Sariciftci, *Synthetic Metals* 1**19**, 637-638 (2001)

3. Christoph J. Brabec, Gerald Zerza, Giulio Cerullo, Sandro De Silvestri, Silvia Luzzati, Jan C. Hummelen and Serdar Sariciftci, *Chemical Physics Letters* **340**, 232-236 (2001)

4. N.S. Sariciftci, A.J. Heeger, *Handbook of Organic conductive molecules and Polymers,* Willey (1997)

5. F. Padinger, C. J. Brabec, J. C. Hummelen, R. A. J. Janssen and N. S. Sariciftci, *Synthetic Metals* **102**, 1285-1286 (1999)

6. N. Camaioni, M. Catellani, S. Luzzati and A. Migliori, *Thin Solid Films* **403-404**, 489-494 (2002)

7. M. Granstrom, K. Petritsch, A. C. Arias, A. Lux, M. R. Andersson & R. H. Friend, *Nature* **395**, 257-260 (1998)

8. Maher Al-Ibrahim, H.-K.H. -Klaus Roth, Uladzimir Zhokhavets, Gerhard Gobsch and Steffi Sensfuss *Solar Energy Materials and Solar Cells* **85**, 13-20 (2005)

9. Christoph J. Brabec, Antonio Cravino, Dieter Meissner, N. Serdar Sariciftci, Thomas Fromherz, Minze T. Rispens, Luis Sanchez, and Jan C. Hummelen, *Adv. Funct. Mater.***11**, 374-380 (2001)

10. D. Gebeyehu, C.J. Brabec, F. Padinger, T. Fromherz, J.C. Hummelen, D. Badt, H. Schindler, N.S. Sariciftci, *Synthetic Metals* **118,** 1-9 (2001)

11. L. Micaroni, C. N. Polo da Fonseca, F. Decker and M. -A. De Paoli, *Solar Energy Materials and Solar Cells* **60**, 27-41 (2000)

Mater. Res. Soc. Symp. Proc. Vol. 836 © 2005 Materials Research Society L3.9

Polymer solar cells: screen-printing as a novel deposition technique

Tom Aernouts, Peter Vanlaeke, Jef Poortmans, Paul Heremans
Polymer and Molecular Electronics, MCP Division
Interuniversity Micro-Electronics Centre (IMEC)
Kapeldreef 75, B-3001 Leuven, Belgium

ABSTRACT

Screen-printing is studied as deposition technique for conjugated material based layers. Photovoltaics based on the principle of bulk donor-acceptor heterojunction are tested using a blend of poly(2-methoxy-5-(2'-ethyl-hexyloxy)-1,4-phenylene vinylene) (MEH-PPV) mixed with the C_{60}-derivative (6,6)-phenyl C_{61}-butyric acid methyl ester (PCBM). First, different solution concentrations of the donor MEH-PPV material and of the blend are subjected to rheology measurements. Addition of the acceptor (PCBM) to a donor material based solution induces a decrease of the solution viscosity. However, the overall flow behaviour of the blend remains similar to that of the MEH-PPV based solution. Secondly, it is shown that specific printer settings have to be used to obtain active layers that are suitable for opto-electronic applications. Finally, devices with an overall energy conversion efficiency of 1.25% under standardized simulated solar illumination (AM1.5G; 100mW/cm^2) have been obtained showing that screen-printing can be a suitable technique for the deposition of the active layer of polymer solar cells.

INTRODUCTION

Solar cells with conjugated materials as active layer have the potential to compete with standard Si-based photovoltaics when production costs are taken into account. The possible use of low-cost substrates as well as the necessity of only a very thin organic active layer leads to a substantial materials cost reduction. Moreover, chemical modifications of the materials such that they become soluble in common solvents make it possible to exploit solution based processes to deposit the active layer, reducing the production costs even further. A common technique to process such conjugated materials is spin coating. In this way, a thin, homogeneous film is easily applied onto large substrates. However, no direct patterning of the deposited layer can be obtained by this technique. Using printing techniques to process organic materials offers the opportunity to obtain patterned films directly onto the substrate. Inkjet printing has in this way already shown its suitability to prepare multi-color displays [1]. The high resolution and the small features that come in reach makes inkjet printing even suitable for fabrication of full plastic transistors. Nevertheless, processing organic solar cells probably does not require this high precision and may need a process with higher throughput. Therefore, screen-printing as a linear casting technique can be interesting as a novel deposition technique to process organic photovoltaics. Moreover, the patterning of the film in one step will even facilitate the production of solar cell modules and the integration of the photovoltaic device into several application tools like smart cards, mobile phones ...

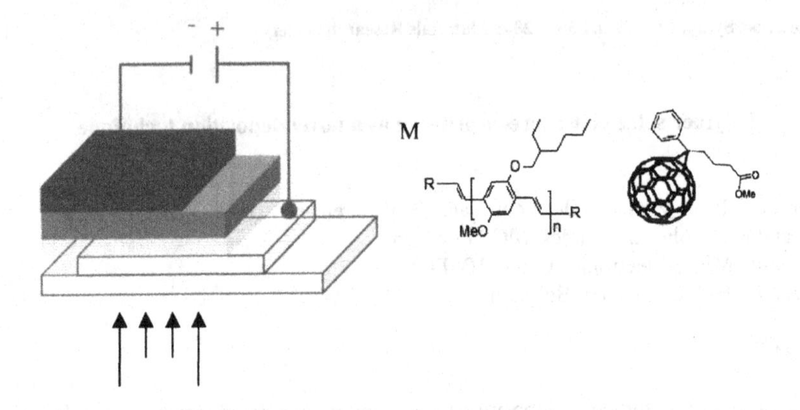

Figure 1: Schematic representation of bulk donor/acceptor heterojunction solar cell with the chemical structure of the materials in the active layer

One of the most promising concepts is that of the bulk donor-acceptor (D/A) heterojunction solar cell, based on the principle of ultrafast (subpicosecond) photo-induced charge transfer from a conjugated polymer to a fullerene (C_{60}) molecule [2]. Blending the donor and acceptor material an interpenetrating bi-continuous network is formed such that generated charges can be transported towards their respective contacts. IndiumTinOxyde (ITO) covered with a thin layer of poly(ethylenedioxythiopene)/poly(styrene-sulfonate) (PEDOT/PSS) is commonly used as transparent electrode whereas evaporated metals like Ca or LiF/Al form the backside contact (see Fig. 1). In this way, energy conversion efficiency well over 3% has been achieved under simulated solar illumination [3]. Processing is almost fully independent of the substrate material and can be done on glass as well as on e.g. poly(ethylene therephtalate) (PET) foils. The latter offers the opportunity to produce fully flexible solar cells due to the mechanical flexibility of the organic active layer.

To study the feasibility of screen-printing as a deposition technique for the active layer, poly(2-methoxy-5-(2'-ethyl-hexyloxy)-1,4-phenylene vinylene) (MEH-PPV) is used as donor material, whereas the acceptor is the C_{60}-derivative (6,6)-phenyl C_{61}-butyric acid methyl ester (PCBM). Chemical structures of these compounds are given in Fig. 1. Both materials can be dissolved in common organic solvents such that processing of the blend from solution is possible. Rheology measurements on polymer based solutions as well as on blends are performed to see what the effect is of the addition of PCBM to the donor based solution. However, optimising and studying the influence of printing parameters is also important. This is first done with a single component solution containing only the donor material. In a later stage, screen-printing of D/A blends is done, to achieve photovoltaic devices. It is shown that screen-printing can be a suitable deposition technique for the active layer of polymer solar cells. Moreover, the direct patterning of the deposited layer will facilitate the production of photovoltaic modules.

RESULTS

Rheology measurements

Rheological measurements were carried out on a Carri-Med Rheometer (CSL^2 500) with a cone/plate measuring system such that the shear rate γ is constant over the full surface of the geometer. This equipment allows studying the viscosity η as function of the shear rate γ and the temperature T. We have examined solutions with different concentrations of the MEH-PPV polymer as well as a blend of the donor and acceptor material in a ¼ weight ratio. The solvent was chosen to be in all cases chlorobenzene (CB) as it was proven to be beneficial for the performance of the actual solar cell [4].

Fig. 2 depicts the result of a rheological measurement for different solutions at a constant shear rate of 100/s but whereby the temperature was gradually varied from 10°C to 55°C. For all these solutions it can be observed that the viscosity increases when the temperature is lowered. However, no strong increase is seen, showing that no gelation is appearing in this temperature range for any of these solutions. Taking the three uppermost curves into account, the variation of the viscosity for different polymer concentrations is shown. The 1% (w/v) concentrated MEH-PPV solution has a viscosity ranging from $7x10^{-2}$ Pa.s at 10°C to $4x10^{-2}$ Pa.s at 55°C. These values triple when the concentration is doubled and even increase almost 12 times for 3% concentrated solutions. It was not possible to study the behaviour of higher concentrated MEH-PPV solutions due to limited solubility of the polymer in chlorobenzene.

The lower curve in Fig. 2 shows the behaviour of a 1% polymer solution to which PCBM is added in a ¼ ratio by weight. Also in this case, the viscosity increases when the temperature is raised without any appearance of gelation in the applied temperature range. Remarkably, the values vary from $3x10^{-2}$ Pa.s at 10°C to $1.6x10^{-2}$ Pa.s at 55°C being lower than what was obtained for the pure 1% MEH-PPV solution. So, the viscosity decreases despite the addition of solid material.

Fig. 3a shows the result of a measurement on the 1% polymer solution whereby the shear rate was varied. This was done for different temperatures ranging from 10°C to 55°C. It can

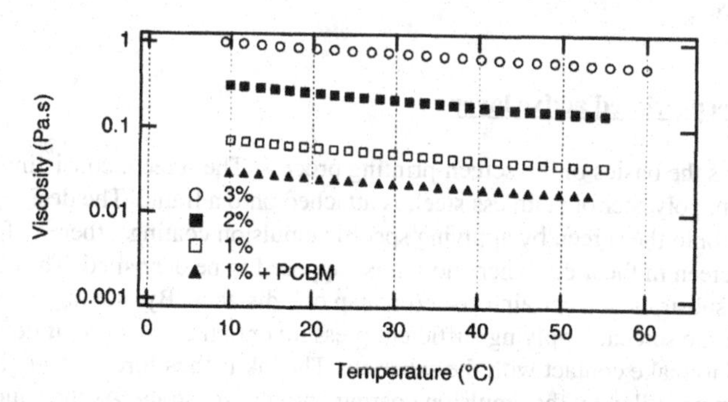

Figure 2: Viscosity measured as function of temperature for different polymer solution concentrations and for a donor/acceptor blend solution

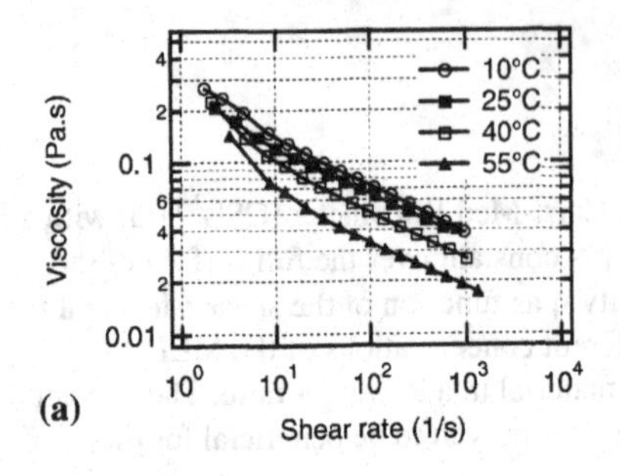

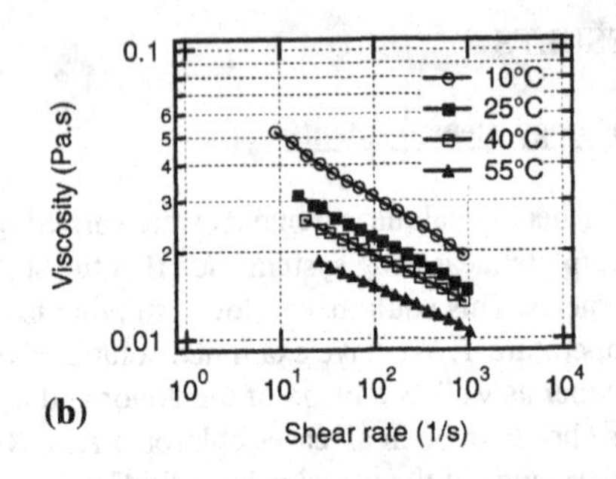

Figure 3: Viscosity as function of applied shear rate for different temperatures for **(a)** 1% MEH-PPV solution and **(b)** 1% MEH-PPV solution with additional PCBM in a ¼ weight ratio

again be observed that the viscosity increases when the temperature is lowered. Furthermore, there is a clear dependence of the viscosity upon the shear rate in such a way that η decreases for higher γ values. This type of flow behaviour is often called shear-thinning or pseudo-plastic and can be caused by a strong interaction between the polymer chains in the solution.

A similar measurement is performed on the 1% concentrated solution after the addition of PCBM in a ¼ weight ratio. The result is given in Fig. 3b, showing also in this case the same temperature dependence as measured before. If we compare the values of the blend with those obtained for the pure polymer solution, we see that the viscosity is lower for the blend, under similar measurement conditions. This confirms the observation of the previous measurement. Due to the fact that PCBM is a rather small molecule with respect to the long polymer chains of MEH-PPV it appears to have a weakening effect on the pure solution. In this way the interaction between the polymer chains is reduced leading to a decrease of viscosity. However, it can be observed from Fig. 3 that for the blend the viscosity still depends on the applied shear rate. So, the addition of PCBM decreases the viscosity of the solution but the overall pseudoplastic flow behaviour remains.

Devices with screen-printed active layer

Fig. 4a depicts the basics of the screen-printing process. The screen, consisting of woven wires of e.g. nylon, polyester or stainless steel, is attached onto a frame. The desired printing pattern is defined onto the screen by applying specific emulsion coatings, thereby filling the openings in the screen in the areas where no ink is supposed to be deposited. The screen is then placed above the substrate at a certain offset (or snap off) distance. By moving the squeegee, the ink is spread over the screen. Applying sufficient pressure onto the squeegee, it deflects the screen downward to make contact with the substrate. The ink is then forced through the open areas of the screen not filled by the emulsion coating onto the substrate. As the squeegee passes a given point, screen fabric tension snaps the screen back, leaving the ink behind.
The experiments described further on were all carried out using a manually operated SP002-R screen printer (ESSEMTEC) with a polyurethane squeegee (shore hardness 85), set at an angle

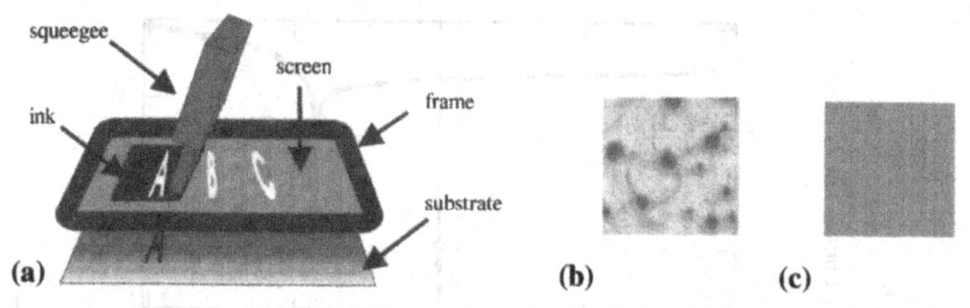

Figure 4: (a) Basic scheme of the screen-printing process, and digital scans (0.8cm by 0.8cm) of screen-printed layers of MEH-PPV with **(b)** low speed and **(c)** high speed of the squeegee movement

of 45°. Polyester screens where commercially bought (KOENEN Gmbh) with an open printing pattern of 4.5cm by 4.5cm. The printing itself is done in ambient atmosphere. Due to the photovoltaic application we are studying, the substrates were always glass sheets (5cm by 5cm) covered with ITO on which a 30nm thin layer of PEDOT/PSS is spin coated.
Concerning the machine set-up, we focused on variables like the printing speed, the snap off distance and the squeegee pressure. Screens with different mesh sizes, i.e. the number of wires per cm, were also studied. The influence of these different printer settings on the film formation and thickness was examined for solutions containing only the donor material MEH-PPV.

It turned out that the speed of movement of the squeegee had the most serious influence on the film morphology. Photographs of such MEH-PPV layers printed at low and at high speed are given in Fig. 4. The film after printing with low speed of the squeegee movement is non-homogeneous with very uneven spreading of the material over the substrate. Areas with almost no material present are contrasting with places of thick droplet-like zones. On the other hand, when a sufficiently high printing speed is applied a much more homogeneous film is formed (see Fig. 4b). Even coverage of the substrate and good spreading of the material is observed in this case. Therefore, all subsequent results are obtained by printing at high speed of the squeegee movement. The effect of the snap off distance, the squeegee pressure and the mesh size on the final print result was already reported elsewhere.

In accordance with these experiments, we have fabricated photovoltaic devices with a screen-printed active layer. We used a donor/acceptor blend based on a 1% MEH-PPV solution with additional PCBM in a ¼ weight ratio. The ITO film on the glass substrates was patterned beforehand by UV-photolithography. Prior to the printing of the polymer solution, a thin PEDOT/PSS layer was spin coated on these patterned substrates. To finalize the device structure, a metallic backside contact of LiF/Al was evaporated in high vacuum (~10^{-8} Torr) on top of the active layer through a shadow mask. In this way, solar cells were fabricated with active areas ranging from 0.09cm^2 to 0.47cm^2, all with very similar performance.

Typical current-voltage (I-V) characteristics of such device are presented in Fig. 5, as well in dark as under standardized simulated solar illumination (AM1.5G, 100mW/cm^2). For the dark IV-curve clear diode behaviour is observed with a rectification ratio of over 10^3 at 2V. Under illumination a short circuit current density J_{sc} of 3.4 mA/cm^2, an open circuit voltage V_{oc} equal to 845 mV and a fill factor FF of 44% is obtained. This results in an overall energy conversion efficiency of 1.25%. This is almost comparable to standard spin coated PPV/C_{60} based solar

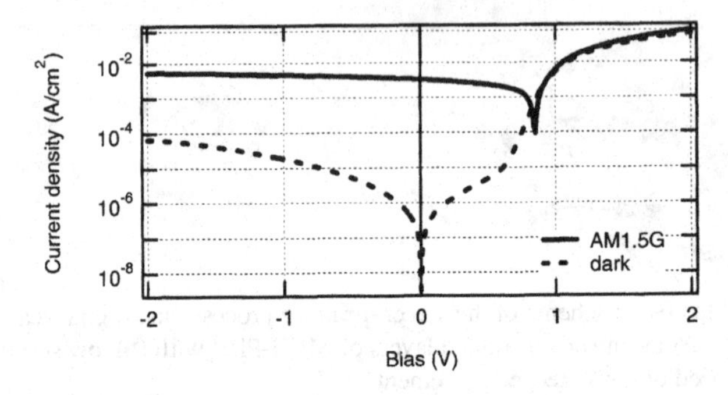

Figure 5: Current-Voltage characteristic of a bulk donor/acceptor heterojunction solar cell with screen-printed active layer in dark and under standardized simulated solar illumination (AM1.5G; 100mW/cm^2)

cells. It is believed that further improvements can be achieved when printing conditions for the D/A blend are optimised.

CONCLUSIONS

We have shown that screen-printing can be a suitable deposition technique for the active layer of polymer based photovoltaic devices. Rheology measurements pointed out that the behaviour of MEH-PPV based solutions is influenced by the polymer concentration. The addition of PCBM decreased the viscosity of the solution but the overall pseudoplastic flow behaviour remained. Also the influence of several printing parameters on the film formation and thickness was studied for MEH-PPV based solutions. It was shown that the printing speed had a serious influence on the film morphology. Furthermore, it was possible to fabricate solar cells with a screen-printed organic active layer. Energy conversion efficiency of 1.25% was achieved under standardised simulated solar illumination. Better adaptation of the printing conditions to the rheology of the formulated blends is however still necessary. It is believed that in this way the performance of the polymer solar cell with screen-printed active layer can be further improved. An important advantage of this process over e.g. spin coating is the direct patterning of the deposited layer. This will facilitate the production of organic photovoltaic modules and the integration of these devices into specific application tools.

REFERENCES

1. E. Haskal *et al*, SID Digest Tech. Pap. 33 (2002) 776
2. N.S. Sariciftci, L. Smilowitz, A.J. Heeger, F. Wudl, Science 258 (1992) 1474
3. F. Padinger, R.S. Rittberger, N.S. Sariciftci, Adv. Funct. Mater. 13 (2003) 85
4. S.E. Shaheen, C.J. Brabec, N.S. Sariciftci, Appl. Phys. 78 (2001) 841

Small Molecule-Based Devices

Mater. Res. Soc. Symp. Proc. Vol. 836 © 2005 Materials Research Society L4.6

Star-shaped heptamers of discotic dyes as new materials for photovoltaic devices

S. Holger Eichhorn, Nicholas Fox, and Bryan Bornais
University of Windsor, Department of Chemistry and Biochemistry, Windsor, Ontario, Canada

Abstract

Potentially n-type and p-type semi-conducting discotic liquid crystal dyes are linked together to star-shaped heptamers, which might self-organize into super-columns of separated p-type and n-type columnar stacks. Their synthesis, mesomorphism, and electronic properties will be discussed along with their potential use in photovoltaic devices.

Introduction

Organic photovoltaic devices (OPVDs) have attracted increasing attention from academia and industry because of their potential as a low cost alternative to silicon based devices. OPVDs are not expected to reach the efficiencies of their inorganic counterparts (up to about 20 %) but present efficiencies of only 3 % and low operating stability still are crucial limitations.

While high efficiencies are found for the charge carrier generation, the proportion of separated charge carriers remains low because of high recombination rates. It has been generally accepted that effective charge carrier separation takes place only at a donor-acceptor interface. Thus, high efficiencies are expected for materials with high interfacial areas between donor acceptor materials as well as good charge-carrier transport properties (large mean free paths of charge carriers). Optimization of both factors has been proven difficult despite tremendous recent advances in the morphological control of materials down to the molecular level.[1]

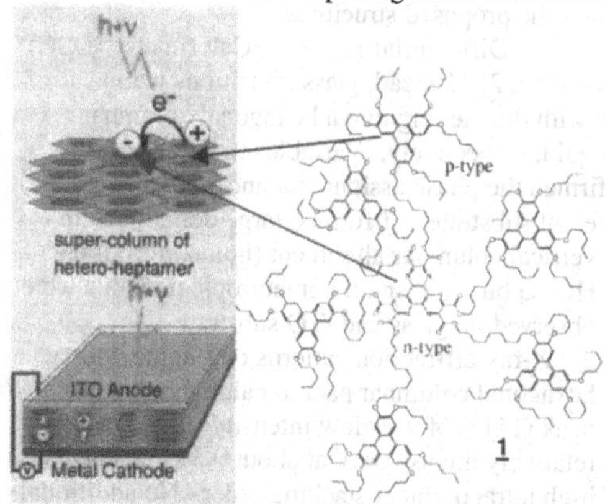

Figure 1 Simplified cartoon of a photovoltaic device containing a star-shaped hetero-heptamer self-organized and self-aligned into nano-separated columns of the same discotic entity

Presented here is a new attempt of nano-structuring an organic semiconductor based on self-organizing discotic liquid crystals (DLCs). DLCs based on flat polyaromatic cores might co-facially stack into columnar piles with stacking distances as small as 0.32 nm. This short distance allows for an intimate overlap of adjacent π-electron systems within a column that generates anisotropic charge carrier channels with mobilities as high as 0.1 $cm^2\ v^{-1}\ s^{-1}$.[2]

Discotic molecules with high electron affinity and with low ionization potential have been investigated as potential n-type and p-type semiconductors. High hole mobilities have been experimentally confirmed while electron mobilities seem to be at least a factor of 10^3 lower. Recent experimental and theoretical studies, however, have improved our understanding of the factors that govern charge carrier mobility in these systems and certainly provide new design criteria for synthetic chemists.[3]

Results and Discussion

The presented star-shaped hetero-heptamers 1 and 2 of p-type and n-type discotic units combine the advantages of oligomeric discotic liquid crystal with the alignment properties of small molecule discotics (Fig. 1 and 4). Oligomeric DLCs fully suppress crystallization and, instead, form anisotropic glasses of columnar stacks on cooling. They also stabilize the columnar mesophase and increase the intracolumnar stacking order. Linear oligomers, however, do not form large domains of uniform alignment on substrates and the columnar stacks tend to align parallel to a substrate (electrode).[4] A vertical alignment is required for high charge carrier mobility between the two electrodes of a photovoltaic device. The star-shaped design mimics the symmetry of a hexagonal columnar mesophase and an early paper by the Ringsdorf group[5] suggested a vertical alignment and monodomain formation of a similar homo-heptamer.

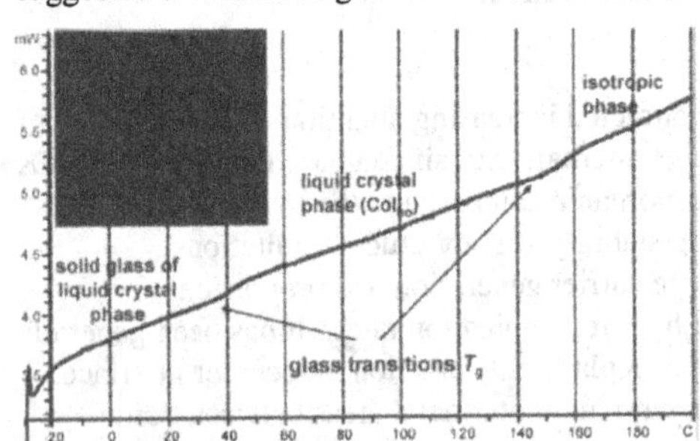

Figure 2 Polarized light microscope image of a partially aligned thin film of 1 on ITO (top left) and a DSC curve of the hetero-heptamer 1 (heating at 10 °C/min)

Several star-shaped hetero-heptamers have been prepared in our group based on hexaazatriphenylene hexamethylester and trisimidazole hexaamide as n-type core structures and hexaalkoxy triphenylenes as p-type ligands (Fig. 1 and 4). The monomer syntheses are based on established routes and the final fusions to the hetero-heptamers were achieved either under transesterification or transamidation conditions. All products were readily soluble in organic solvents and could be characterized by solution NMR. MALDI MS and elemental analysis also agreed with the proposed structures.

Differential Scanning Calorimetry (DSC) of 1 confirmed the absence of any crystallization (Fig. 2). Instead, glass transitions were observed at 40 °C and 155 °C, which coincided with the melting into a hexagonal columnar mesophase and the clearing into the isotropic liquid, respectively. Variable temperature polarized light microscopy measurements confirmed the phase assignment and were also employed for monitoring the alignment on different substrates. 1 formed large domains with vertical columnar alignment (homeotropic) on HOPG but only small homeotropic domains were observed on glass and ITO substrates.

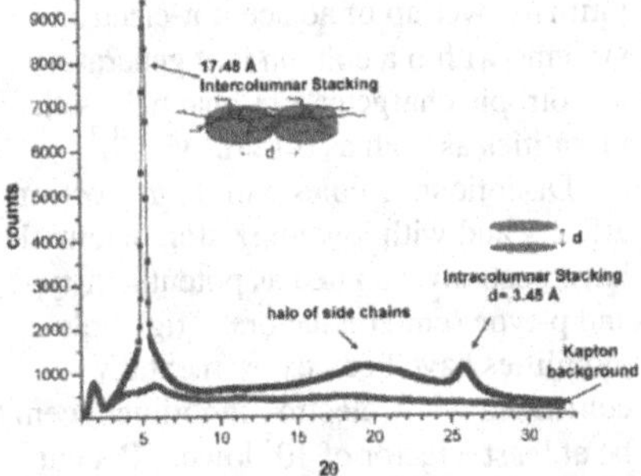

Figure 3 XRD pattern of the hetero-heptamer 1 at 25 °C

X-ray diffraction patterns of 1 agree with a hexagonal columnar packing although the cross peak (11) is of very low intensity (Fig. 3). A relatively intense peak at about 0.34 nm suggests high intracolumnar stacking order. No additional small angle peak for the formation of a super-column of the heptamer has been observed, probably because both discotic units have very similar sizes. Hexaamides based on a smaller trisimidazole core 2 (Fig. 4) are presently being investigated and the incommensurate sizes of the two discotics might lead to an observable peak for the super-column.

At this point, we have no conclusive experimental evidence for the formation of the super-columns. The super-column formation implies that all heptamers stack exactly on top of each other. Alternatively, the heptamers could also be shifted with regard to each other, which would not significantly alter the diffraction pattern of the individual hexagonally packed columnar stacks. High resolution X-ray diffraction measurements on aligned fibres are presently under way and would provide unambiguous evidence for the super-column formation if an additional small angle peak is observed.

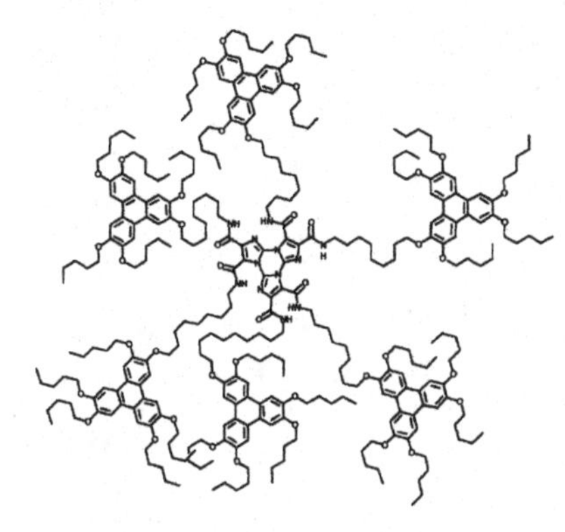

Figure 4 Hetero-heptamer 2 with trisimidazole hexaamide as central n-type core

Indirect evidence for the formation of the super-column might be provided by the comparison the properties of monomeric 6:1 mixtures of hexaalkoxytriphenylene and hexaalkyl substituted n-type cores with our heptamers. Comparative UV-VIS and fluorescence measurements on solutions and thin films, however, have been inconclusive in this regard. The red-ox properties are presently studied and so is the charge carrier mobility in thin films of these materials.

Incorporation of these oligomers into OPVDs is also planned as a collaborative effort. The formation of super-columns as depicted in Figure 1 would generate n-type and p-type channels right next to each other and create vast interfacial area. In addition, a vertical alignment of the semi-conducting channels should ensure an efficient charge separation. So, we expect the heptamer to give a superior performance in comparison to the 6:1 mixtures of monomers.

The macrocycles used in this study are not strong absorbers in the visible and IR spectral ranges and, therefore, not promising candidates for "real" OPVDs. Their sole purpose is the demonstration of the feasibility of this concept. Efforts have been started to incorporate strongly absorbing dyes such as phthalocyanines and perylenes.

References

[1] See for example recent review by H. Hoppe and N.S. Sariciftci J. Mater. Res. **19**, 1924 (2004);

[2] A. van de Craats P.G. Schouten, J.M. Warman J. Jap. Liq. Cryst. Soc. (EKISHO) **1**, 12 (1998); S.H. Eichhorn J. Porph. Phthalocyanines **4**, 88 (2000);

[3] V. Lemaur, D.A. da Silva Filho, V. Coropceanu, M. Lehmann, Y. Geerts, J. Piris, M.G. Debije, A.M. van de Craats, K. Senthilkumar, L.D.A. Siebbeles, J.M. Warman, J.-L. Bre´das, and J. Cornil, J. Am. Chem. Soc. **126**, 3271 (2004); X. Crispin, J. Cornil, R. Friedlein, K.K. Okudaira, V. Lemaur, A. Crispin, G. Kestemont, M. Lehmann, M. Fahlman, R. Lazzaroni, Y. Geerts, G. Wendin, N. Ueno, J.-L. Bre´das, and W.R. Salaneck J. Am. Chem. Soc. **126**, 11889 (2004);

[4] N. Boden, R.J. Bushby, A.N. Cammidge, A. El-Mansoury, P.S. Martin, Z.B. Lu, J. Mater. Chem., **9**, 1391 (1999)

[5] T. Plesnivy, H. Ringsdorf, P. Schumacher, U. Nuetz, S. Diele Liq. Cryst., **18**, 185 (1995);

Mater. Res. Soc. Symp. Proc. Vol. 836 © 2005 Materials Research Society L4.7

Controlling Organization in Photovoltaic Diodes from Discotic Liquid Crystals via Anode Surface Energy Alteration

Johanna P. Schmidtke[1]
Klaus Müllen[2]
Richard H. Friend[1]
[1]Cavendish Laboratory, University of Cambridge
Cambridge CB3 0HE, United Kingdom
[2] Max Planck Institut für Polymerforschung, Ackermannweg 10
55128 Mainz, Germany

ABSTRACT

We report the control of molecular ordering of discotic liquid crystals in thin, blended films used in photovoltaic diodes. The external quantum efficiency (EQE%) of photovoltaic diodes incorporating a crystalline hexabenzocoronene (HBC) derivative is improved by lowering the surface energy of the transparent anode surface with a short alkyl chain. Upon increasing the length of the functionalizing group on the anode, we find that the relative efficiency of the HBC component in the blend improves. Evidence of changed film morphology is also presented.

INTRODUCTION

Since the demonstration of an organic bilayer photovoltaic diode by Tang, the role of heterojunctions in photovoltaic diodes has been widely studied [1,2]. The role of morphology has been well described for polymer blend and small molecule photovoltaic diodes [3,4]. However, the role of surface properties in molecular organization of efficient photovoltaic systems is less well understood. Surface conditions are frequently used to manipulate the bulk organization of liquid crystals, including discotic liquid crystalline materials [5, 6].

Discotic liquid crystalline materials offer a combination of solubility, spectral absorption, self-organization, and high charge transport which is promising for optoelectronic applications. Discotic liquid crystals, including pthalocyanines and hexa-*peri*-hexabenzocoronenes (HBC), demonstrate strong self-organizing characteristics in solution, in thin films and at interfaces [7-9]. Due to strong π-π interactions, HBC derivatives form aggregates at concentrations as low as 10^{-9} M and demonstrate one-dimensional charge-carrier mobilities as high as 0.5 $cm^2 V^{-1} s^{-1}$ [10,11]. HBCs have also been effectively incorporated into field effect transistors and as hole-transporters in photovoltaic diodes [12, 13].

Here we report photovoltaic diodes fabricated from a blend of hexadodecyl-*peri*-HBC and a perylene derivative. We demonstrate control over of the molecular-scale organization of the film by altering the surface energy of the anode via functionalization with alkyl-trichlorosilanes and describe the effects on device performance.

EXPERIMENTAL DETAILS

The structures of the materials used in this study are shown in Figure 1. The hexadocdecyl-*peri*-HBC (HBC-C_{12}) was synthesized as reported in [14]. The perylene derivative (EPPTC) was purchased from Sensient Imaging Technologies GmbH and used without further purification. Trichlorosilanes were purchased from Aldrich. The photovoltaic diodes were prepared on glass/ITO substrates coated with 60 nm of poly(3,4-ethylenedioxythiophene) poly(styrenesulfonate) (PEDOT:PSS). The ITO/PEDOT:PSS anode was functionalized with alkyl-trichlorosilanes by following published method [15] and using a flat PDMS stamp fabricated on silicon. Photovoltaic active layers were spin cast in an N_2 environment onto the ITO/PEDOT:PSS anodes under nitrogen from a blend of 40:60 HBC-C_{12}:EPPTC in chloroform solution to give a 100 nm active layer. Atomic force microscopy images were obtained using a Digital Instruments Dimension 3100 AFM. The photovoltaic diodes were tested under vacuum (50 mbar) and illuminated with monochromatic light from a 100mW tungsten lamp.

DISCUSSION

Under monochromatic illumination, typical HBC-C_{12}:EPPTC blend photodiodes cast onto PEDOT:PSS have a maximum external quantum efficiency (EQE%) of 12.0% at 465 nm. The surface energy of the ITO/PEDOT:PSS anode was altered by functionalizing the surface with alkyl-trichlorosilanes with alkyl chains of 4-18 carbons. Photoaction spectra and open circuit voltages of HBC-C_{12}:EPPTC blend devices cast on functionalized and non-functionalized PEDOT:PSS are compared in Figure 2. Devices functionalized with a butyl chain (butyl-PEDOT:PSS) have both the highest EQE% (13.5%) and open circuit voltage (0.70 V). The open circuit voltage of the devices does not show a strong dependence on the chain length of the functionalizing group. The short circuit current, however, is dependent on the chain length of the functionalizing group on the anode. As the chain length increases, the Ratio $EQE_{365\ nm}/EQE_{465\ nm}$ – the ratio of EQE% due to the absorption of the HBC-C_{12} vs. that of EPPTC – increases (with the exception of the octadecyl functionalization). Absorption spectra (not shown) of the blended films on the same surfaces indicate no change in absorption of the materials.

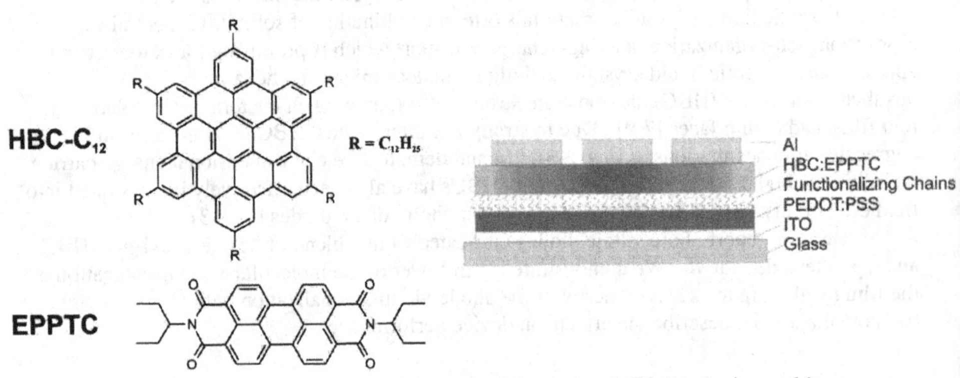

Figure 1. Left, Chemical structure of the materials used and, **Right**, device architecture.

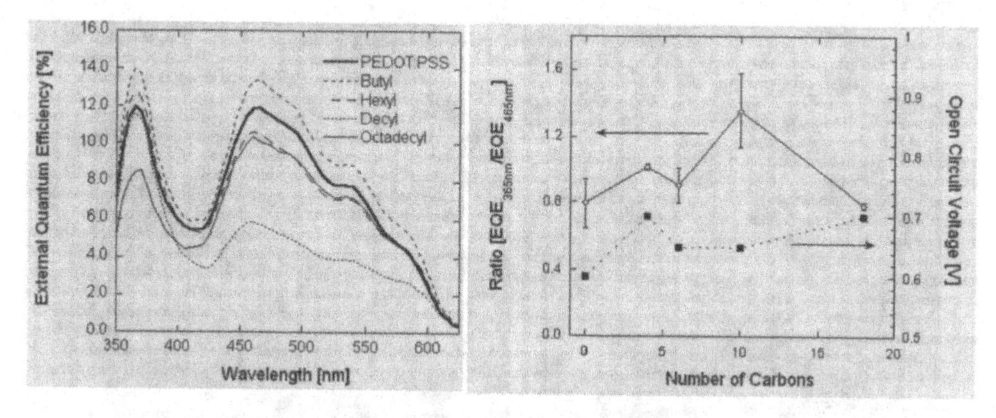

Figure 2. Left, The EQE% for photovoltaic diodes using a 40:60 blend of HBC-C_{12}:EPPTC spun-cast onto PEDOT:PSS functionalized with alkyl chains of 4-18 carbons. **Right,** The open circuit voltage for the same devices as a function of the number of carbons in the functional group on the PEDOT:PSS surface. The Ratio [EQE_{365}/EQE_{465}], the ratio of the EQE% maxima at 365 nm and 465 nm, is also presented as a function of functionalizing chain length. The Ratio [EQE_{365}/EQE_{465}] represents an average of at least 4 separate device preparations, and the error bars indicate one standard deviation.

Since the modification of the anode surface influences the relative efficiency of the two components (Ratio $EQE_{365\ nm}/EQE_{465\ nm}$), we considered whether the lower energy surface preferentially attracts one component of the blend. If the EPPTC, the electron-transporter, is preferentially attracted to the lower energy surface, the higher concentration of the electron-transporting component near the anode would result in 1) fewer charge transport pathways to either electrode and 2) an overall decrease in the efficiency of the device. A change in the Ratio $EQE_{365\ nm}/EQE_{465\ nm}$ would not be expected. Instead, a uniform decrease in the efficiency across the spectrum would be expected because each component (hole and electron transporter) would move away from their respective electrode. However, if the HBC component is preferentially attracted to the functionalized anodes, the greater concentration of hole-transporter may allow for better transport to the anode. At the same time, a higher concentration of the EPPTC electron-transporter would result near the cathode. Thus, as is seen the case of the butyl-PEDOT:PSS, a higher overall efficiency would result.

The higher concentration of the HBC-C_{12} at the functionalized anode surface may result from the attraction of the long alkyl chains of the HBC to the alkyl chains of the trichlorosilanes. In addition, the hydrophobic surfaces may induce more favorable organizations of the HBC stacks at the anode surface, which allow for better collection of charge from the HBC component. This improved organization may result in the observed increase of the Ratio $EQE_{365\ nm}/EQE_{465\ nm}$ in the alkyl-PEDOT devices, as the charges generated by the initial absorption of the HBC component are more efficiently collected than those due to EPPTC. The

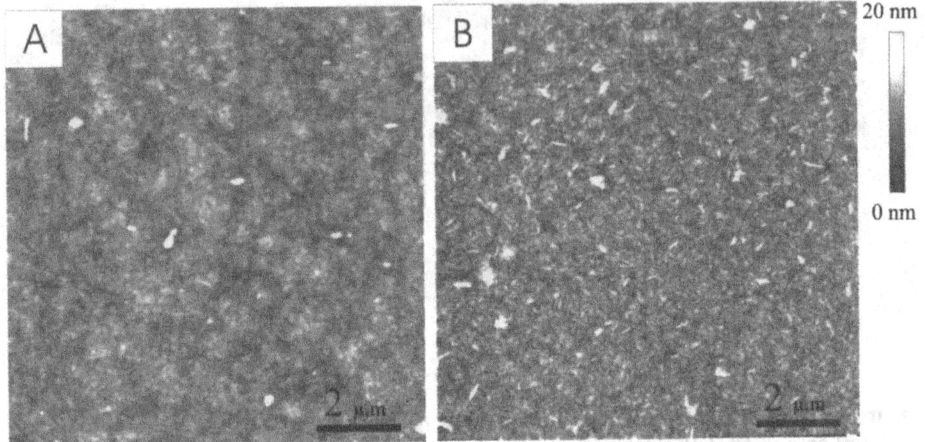

Figure 3. AFM images of the active layer surface of HBC-C_{12}:EPPTC blended photovoltaic diodes cast on PEDOT:PSS (left) and PEDOT:PSS functionalized with a butyl chain (right).

Ratio $EQE_{365\,nm}/EQE_{465\,nm}$ increases with increasing chain length, maximizing with decyl-PEDOT:PSS. Therefore, we suggest that the most favorable organization of the HBC component on the anode surface occurs on the decyl-PEDOT:PSS.). This dependence of Ratio $EQE_{365\,nm}/EQE_{465\,nm}$ upon functionalizing chain length is consistent between device preparations, despite variation in the absolute efficiency of the devices (see Ratio $EQE_{365\,nm}/EQE_{465\,nm}$ Figure 2, right).

In Fig. 3, AFM images of devices cast onto PEDOT:PSS and PEDOT:PSS functionalized with butyl-trichlorosilane are shown. The crystalline features on the surface of the blend cast onto the functionalized anode (B) are not present to such a degree when spun on the PEDOT:PSS directly (A). These morphological changes support a reorganization of the blended layer due to the anode functionalization as is indicated in the photoaction spectra.

In Figure 4, the current-voltage characteristics are shown for a series of the devices in Figure 2, left, both in the dark and under monochromatic illumination. In the dark, the devices with functionalized anodes have ~3 orders of magnitude less current than the PEDOT:PSS device. The lower dark current suggests that the functionalizing chains act as a blocking layer to hole injection into the anode. We expect that this effect will increase with increasing chain length as the distance for charge injection from the ITO/PEDOT:PSS anode to the active layer increases. As seen in Figure 4, this description is supported by the dark current data for the devices – the butyl-PEDOT:PSS has a greater dark current than the decyl-PEDOT:PSS. Under illumination, the presence of a blocking layer is evident in the decyl-PEDOT:PSS device, which has ~1 order of magnitude less current at all voltages. The butyl-PEDOT:PSS device, however, demonstrates greater current than the PEDOT:PSS device. In comparison to the large changes in dark current upon anode functionalization, the change in current under illumination is small. This effect suggests that the barriers to charge injection and charge extraction are not symmetric. Rather, the barrier to charge injection is more greatly effected by the anode functionalization than the barrier to charge extraction. Nonetheless, a long-chain functionalizing layer does form a

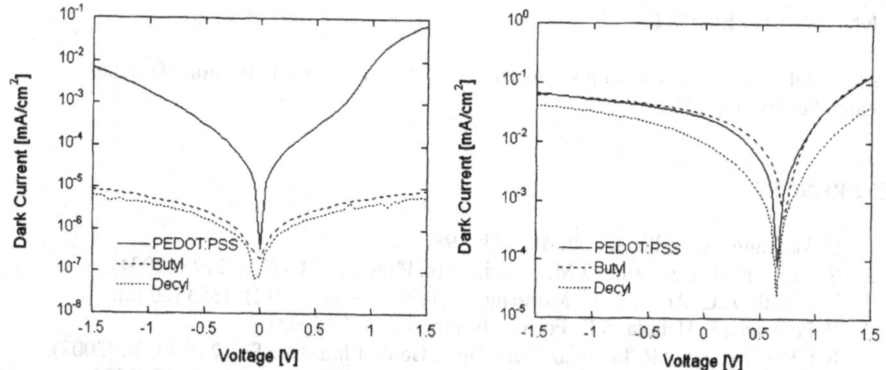

Figure 4. The current-voltage characteristics for the devices in the dark (left) and under monochromatic illumination at 550 nm are shown. For clarity, devices with PEDOT:PSS, butyl-PEDOT:PSS, and decyl-PEDOT:PSS are given as interesting representatives of the effects of anode functionalization.

significant barrier to charge extraction as well, as evidenced in the current-voltage characteristics of the decyl-PEDOT:PSS devices under illumination. In addition, the description of long functionalizing chains as a blocking layer for charge extraction correlates well with the decrease of the EQE% of the photovoltaic diodes with increasing chain length.

The implications of functionalization of the ITO/PEDOT:PSS anode therefore include two competing effects: 1) an improvement in the morphology and vertical stratification of the HBC component and 2) the formation of a blocking layer on the ITO/PEDOT:PSS anode. In the case of the butyl functional group, the short chain only causes a slight displacement of the active component relative to the anode, while the improved organization of the blend components compensates, to result in an improved overall efficiency. However, as the chain grows longer, the blocking layer becomes more efficient and thus lower device efficiencies are obtained despite the improvement in the molecular organization of the active layer.

CONCLUSIONS

The external quantum efficiency of a photovoltaic diode incorporating a hexabenzocoronene derivative was improved via modification of the anode. The efficiency of the HBC component increased rapidly with respect to the EPPTC component as the length of the functionalizing chain increased, maximizing with a decyl-chain functional group. The improvement is attributed to an improved morphology and vertical stratification of the HBC component on the lower energy surface of the functionalized anode. However, the overall efficiency of the devices functionalized with long chains was lower due to the functional groups acting as a blocking layer for hole extraction. Further work will focus on more thoroughly describing the change in morphology at the anode surface. A more complete understanding of the change in morphology at the anode surface may allow for the selection of a better surface modification that creates a minimal blocking layer while inducing an improved morphology in the active layer.

ACKNOWLEDGEMENTS

This material is based on work supported under a National Science Foundation Graduate Research Fellowship (JPS).

REFERENCES

1. C. W. Tang, Appl. Phys. Lett. **48**, 183 (1986).
2. S. Alem, R. de Bettignies, J.M. Nunzi, Appl. Phys. Lett. **84** (12), 2178 (2004).
3. H. Snaith, A.C. Arias, A. C. Morteani, et al., Nanoletters **2** (12), 1353 (2002).
4. P. Peumans, S. Uchida, S.R. Forrest, Nature **425**, 158 (2003).
5. R.J. Bushby and O.R. Lozman, Curr. Opin. Colliod Interface Sci. **7** (5-6), 12 (2002).
6. O. Bunk, M.M. Nielsen, T.I. Solling, et al., J. Am. Chem. Soc. **125**, 2252 (2003).
7. J.P. Hill, W. Jin, A. Kosaka et al., Science **304** 1481 (2004).
8. A. Tradz, J. K. Jeszka, M.D. Watson, et al., J. Am. Chem. Soc. **125**, 1682 (2003).
9. P. Samori, M. Keil, R. Friedlein, et al., J. Phys. Chem. B **105** (45) 11114 (2001).
10. A.J. Fleming, J.N.Coleman, A.B.Dalton, et al., Journal of Phys. Chem. B **107**, 37 (2003).
11. A.M. van de Craats, J.M. Warman, A Fechtenkotter, et al., Adv. Mater. **11**, 1469 (1999).
12. A.M. van de Craats, N. Stutzmann, O. Bunk, et al., Adv. Mat. **15** (6), 495 (2003).
13. L. Schmidt-Mende, A. Fehtenkotter, K. Mullen, et al., Science **293**, 1119 (2001).
14. Stabel, A., Herwig, P., Mullen, K., et al., Angew. Chem Int. Ed. **38**, 3039 (1999).
15. A.C. Arias, N. Corcoran, M. Banach, et al., Appl. Phys. Lett. **80** (10), 1695 (2002).

Poster Session

Mater. Res. Soc. Symp. Proc. Vol. 836 © 2005 Materials Research Society L5.5

Temperature Effects on Photocurrent Generation in Polymer Hetero-Junction Photovoltaic Devices

Mi Yeon Song, Kang-Jin Kim[1] and Dong Yong Kim
Optoelectronic Materials Research Center, Korea Institute of Science and Technology, P.O. Box 131 Cheongryang, Seoul 130-650, Korea
[1]Department of Chemistry and Molecular Engineering, Korea University, Seoul 136-701, Korea

ABSTRACT

In a heterojunction photovoltaic device of ITO/TiO_2/poly(3-alkylthiophene)/Au, the photocurrent was characterized with different temperature by using regio-random (P3HT), regular (RP3HT) poly(3-hexylthiophene) and regio-regular poly(3-dodecylthiophene)(RP3DT). The regio-regularity and alky chain length affected the photovoltaic characteristics due to the difference in the hole carrier transport. The drift charge mobility of those devices were analyzed by the space charge limited current (SCLC) theory using dark current versus bias relations. The photocurrent in the devices based on poly(3-alkylthiopene)s began to decrease rapidly below a temperature at which the drift charge mobility was 10^{-5} $cm^2/V{\cdot}s$.

INTRODUCTION

Recent developments in photovoltaic cell based on heterojunction between n-type and p-type materials have led to a significant increase in device performance [1]. Organic conjugated materials have been of interest as new electronic materials for active devices such as solar cells [2], LED [3], and FET [4], because these devices are flexible and wide-area, and their fabrication cost is low. Among the various conduction polymers, poly(3-alkylthiophene)s (PATs) have received much attention as electrically conducting polymers with many potential applications.

In a photovoltaic device of ITO/TiO_2/PAT/Au, the charge separation occurs at the interface between TiO_2 after generating excitons in conjugated polymer. The excited electrons under irradiation are injected to TiO_2 due to the difference in a quasi-Fermi level of TiO_2 and LUMO level of conjugated polymer. The photocurrent is delivered to a load by draining the hole carriers to back contact. The hole mobility is one of the most important parameters to achieve high performance in polymer photovoltaics because the hole carrier transport in polymers is slower than that of electrons.

Temperature-dependent photovoltaic characteristics on organic solar cell were reported in some groups using a bulk heterojunction polymer/fullerene solar cell [5,6]. In this study, three poly(3-alkylthiophene)s such as RP3HT, RP3DT and P3HT were used as hole conductors to investigate the structure and property relationships on photocurrent generation in terms of carrier mobility at different temperatures.

EXPERIMENTAL

Photovoltaic cells were fabricated with a multi-layered structure using hetero junctions of inorganic TiO_2 and organic polymer layer. The TiO_2 layer was spin-coated onto indium-tin oxide (ITO) coated glass substrates from a sol-gel solution of titanium isopropoxide in ethanol followed by annealing for 30 min at 450°C in air [7]. The regio-random poly(3-hexylthiophene)

(P3HT) which was polymerized with $FeCl_3$ oxidant [8], and the regio-regular poly(3-hexylthiophene) (RP3HT) and regio-regular poly(3-dodecylthiophene) (RP3DT) which were purchased form Aldrich and were used after re-precipitation from chloroform solution. Polymer layers were spin-coated from monochlorobenzene solution to be 500 nm. The thickness of the PAT layers was measured using a surface profiler (P-10, TENCOR. Co). The Au layer as a back contact electrode was evaporated thermally on polymer film at the pressure below 10^{-5} torr.

The typical active area of the photovoltaic device was 0.18 cm^2. Current density-voltage (J-V) characteristics were obtained with a Keithley 236 SMU placed across the ITO (negative) and Au (positive) electrodes under an illumination of a 300 W Xe arc lamp (Oriel) passed through a global AM1.5 solar simulating filter. The incident photon-to-electron conversion efficiency (IPCE) was measured using a photon counting spectrofluorometer (ISS PC1) equipped with a 350 W Xe lamp light source with a motorized monochromator. Incident light intensity was calibrated using a photodiode detector (Newport 818UV) and optical power meter (Newport 1830-C). J-V curves at low temperatures were measured in a helium-circulated cryostat (JANIS).

RESULTS AND DISCUSSION

Regio-regular and random PAT structures were represented in figure 1a. The alkyl thiophene monomeric units are connected as a head-to-tail formation in regio-regular types and randomly sequenced in regio-random polymers. The normalized absorption spectra of RP3HT, RP3DT and P3HT films were shown in figure 1b. The maximum peaks of RP3HT and RP3DT are red-shifted to longer wavelength than P3HT due to the long conjugation length which are expected in regio-regular configuration. The vibronic structures are also observed clearly in RP3HT and RP3DT.

The incident monochromatic photon to current conversion efficiency (IPCE), defined as the number of electrons generated by light in the external circuit divided by the number of incident photons, is plotted as a function of the excitation wavelength. This was derived from the photocurrents by means of equation 1.

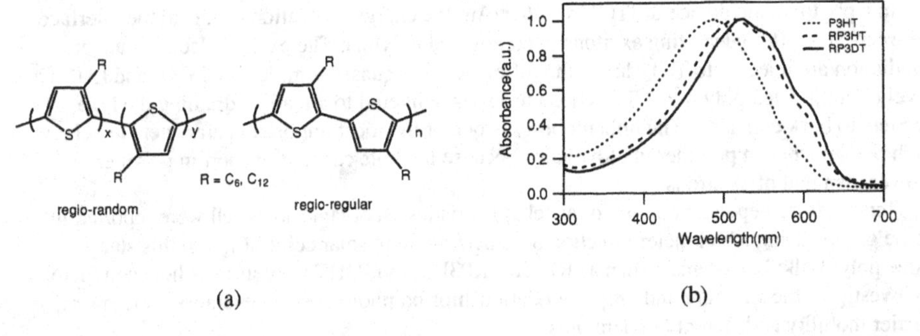

Figure 1. (a) Chemical structures of regio-random (left) and regular (right) PATs, (b) UV-Vis spectra of P3HT, RP3DT and RP3HT films.

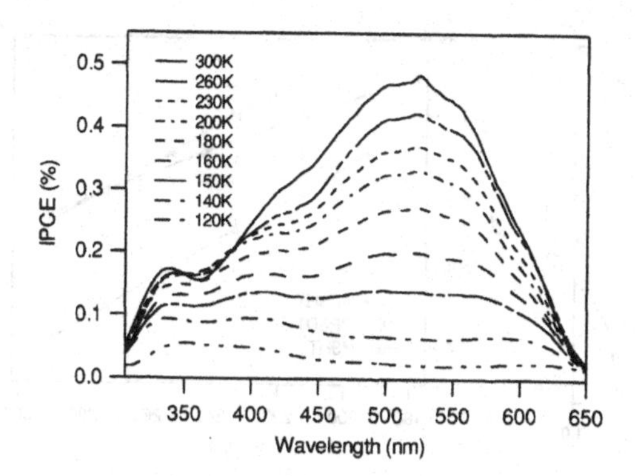

Figure 2. IPCE spectra of ITO/TiO_2/RP3DT/Au device with different temperature.

$$\mathrm{IPCE}(\%)=\frac{[(1.24\times10^{3})\times \text{photocurrent density}\,(\mu\mathrm{A/cm^{2}})]}{[\text{wavelength(nm)}\times \text{photon flux}(\mathrm{W/m^{2}})]} \quad (1)$$

The IPCE (%) spectra of ITO/TiO_2/RP3DT/Au device are shown in figure 2. The symbatic relationship in visible range between the photocurrent action spectrum and the absorption spectrum implies that the active interface for charge carrier generation in the device is the RP3DT/TiO_2 interface as a result of the formation of singlet excitons in RP3DT. TiO_2 also contributes to the photocurrents in UV range due to the energy band gap of TiO_2. In IPCE spectra, the contribution of RP3DT in visible range decreases significantly at low temperature than those of TiO_2 layer because of the difficulties in hole carrier transport.

The J-V characteristics of ITO/TiO_2/RP3HT/Au devices were measured at different temperature as shown in figure 3a. The photocurrent density decreased gradually as temperature lowered. On the other hand, the open-circuit voltage (V_{oc}) increased almost linearly decreasing the temperature range from 300 K to 180 K as shown in figure 3b. We use the model for conventional inorganic semiconductor solar cells to explain the obtained characteristics of V_{oc} as defined by equation 2 [9].

$$V_{oc}=\frac{E_g}{q}-\frac{kT}{q}\ln[(\frac{D}{\tau})^{1/2}T^{3}\frac{B}{J_{sc}}] \quad (2)$$

where Eg is band-gap energy (eV), q is electron charge (1.6×10^{-19} C), k is Boltzmann's constant, T is absolute temperature, D is diffusion coefficient ($cm^2 sec^{-1}$), and τ is minority carrier lifetime. The temperature variation of V_{oc} is often used to estimate the effective barrier height for experimental photovoltaics. The extrapolation of experimental V_{oc} versus T data to T = 0 K yields an intercept of the V_{oc} which represents the maximum theoretical V_{oc} of photovoltaics. The intercept value was about 1.0 – 1.2 V for PATs similar to the difference between the conduction band of TiO_2 (-4.2 eV) and the HOMO level of typical PATs (-5.2 eV).

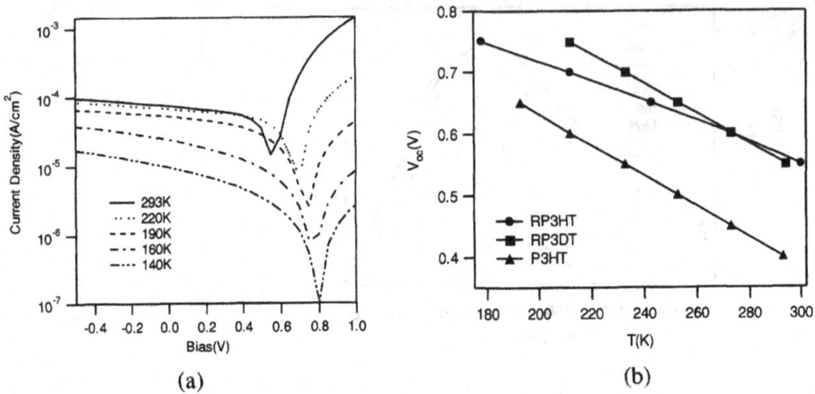

Figure 3. (a) J-V curves of ITO/TiO_2 /RP3HT/Au device with different temperature, (b) The temperature dependence of V_{oc}.

The short-circuit current (J_{sc}) of those polymers at different temperatures are shown in figure 4a under the illumination of 50 mW/cm^2. The normalized J_{sc} of RP3HT, RP3DT and P3HT are also shown in figure 4b. The J_{sc} decreased gradually with lowering temperature in PAT devices. The J_{sc} at room temperature were 0.075, 0.045 and 0.038 mA/cm^{-2} for RP3HT, RP3DT and R3HT, respectively. In the normalized J_{sc}, the curves show the sigmoidal relation versus the temperature. However, each polymer shows the different critical temperature (T_c) defined as a temperature at which the J_{sc} shows a fast decrease. In polymeric heterojunction photovoltaics, the photocurrent depends on the charge transport efficiency of hole carriers to the Au electrode.

In this study, the charge mobility was estimated according to the space charge limited current (SCLC) theory using dark J-V characteristics which were typical schottky type behaviors. One of the tools for investigating charge-carrier mobilities is to examine the SCLC through a semiconductor in the dark, since it is directly proportional to the charge-carrier mobility as equation 3 [10],

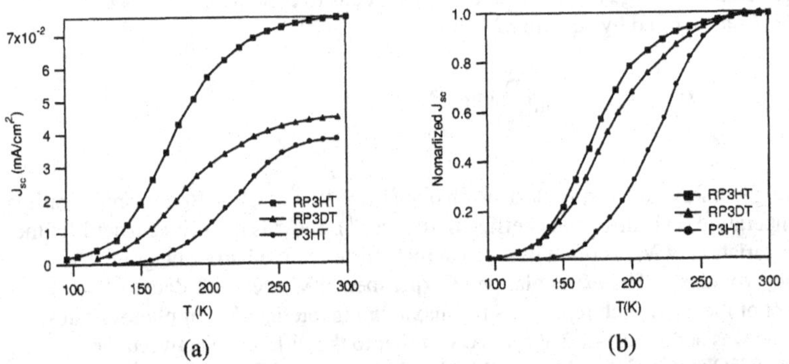

Figure 4. (a) The short-circuit current density of P3HT, RP3HT and RP3DT devices at different temperatures; (b) the normalized short-circuit current density.

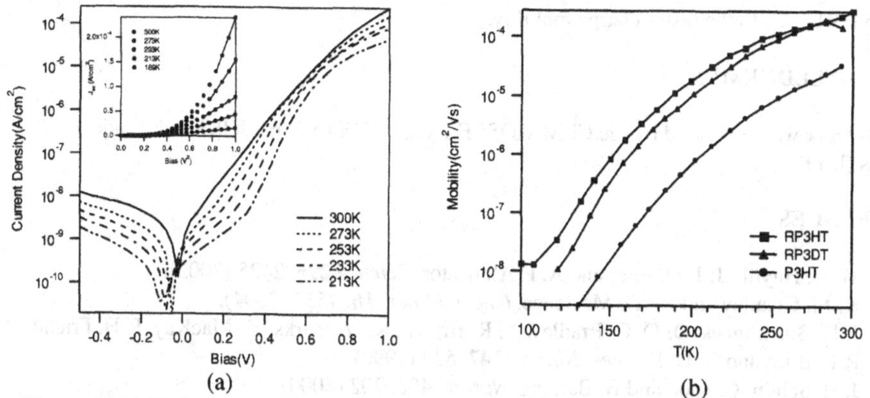

Figure 5. (a) Dark current density of ITO/TiO_2/RP3HT/Au device with temperature and slope of plot, J vs. V^2 (inset); (b) hole carrier mobility of PAT devices.

$$J = \frac{9}{8}\varepsilon_o \varepsilon_r \mu \frac{V^2}{d^3} \qquad (3)$$

where J, ε_o, ε_r, V, d and μ are dark-current density, permitivity constant, permitivity, bias voltage, device thickness and drift charge mobility, respectively.

Figure 5a shows the ohmic behavior of the J-V characteristics at low bias. But this condition breaks down at the space charge limit when the injected carrier density becomes so great that the field due to the carrier themselves dominates over that of the applied bias and then becomes space-charge limited.

In SCLC theory, the current is proportional to the square of applied bias in a trap-filled condition. The drift charge mobility is estimated from the J-V characteristics in dark condition from the slope of plot, J versus V^2. The charge mobility of RP3HT, RP3DT, and P3HT at 300 K was 2.64×10^{-4}, 1.47×10^{-4}, and 3.46×10^{-5} cm^2/V·s, respectively, and decreased also with lowering temperature (figure 5b). The T_c related to the drift charge mobility. The T_c was found to be 223, 233 and 273K for RP3HT, RP3DT and P3HT from the normalized J_{sc} under 50 mW/cm^2 illumination, respectively, and the drift mobility was 5.4×10^{-5}, 4.8×10^{-5}, and 1.8×10^{-5} cm^2/V·s. The mobility range of 10^{-5} cm^2/V·s was very critical value for the efficient photo-generated carrier transport in photovoltaic devices.

CONCLUSION

We have presented the structure and property relations in heterojunction photovoltaics based on poly(3-alkylthiopene)s with TiO_2 layer. The regio-regularity and alkyl chain length affect the photocurrent generation of the devices. Poly(3-alkylthiophene)s contributes to the photocurrent as sensitizers and hole transporting materials from the symbatic relation between the absorbance and IPCE spectra. The temperature dependence in the J_{sc} was dominated by the charge-mobility of hole carriers which was analyzed by SCLC phenomena. The regio-regular types of PAT such as RP3HT and RP3DT showed larger photocurrent and lower T_c than regio-random P3HT due to the faster charge mobility.

ACKNOWLEDGEMENTS

This work was supported by the CRM-KOSEF, Korea (KJK) and the KIST Independent Projects (DYK).

REFERENCES

1. W. U. Huynh, J. J. Dittmer and A. P. Alivisatos, *Science* **295**, 2425 (2002).
2. K. M. Coakley and M. D. McGehee, *Chem. Mater.* **16**, 4533 (2004).
3. J. H. Burroughes, D. D. C. Bradley, A. R. Brown, R. N. Marks, K. MacKay, R.H. Friend, P. L. Burn and A. B. Holmes, *Nature*, **347**, 539 (1990).
4. J. H. Schön, C. Kloc and B. Batlogg, *Nature*, **406**, 702 (2000).
5. I. Riedel, J. Parisi, V. Dyakonov, V. Dyakonov, L. Lutsen, D. Vanderzande and J. C. Hummelen, *Adv. Funct. Mater.* **14**, 38 (2004).
6. D. Chirvase, Z. Chiguvare, M. Knipper, J. Parisi, V. Dyakonov and J. C. Hummelen, *J. Appl. Phys.* **93**, 3376 (2003).
7. M. Takahashi, K. Tsukigi, T. Uchino and T. Yoko, *Thin Sol. Films* **388**, 231 (2001).
8. M. Leclerc, F. M. Diaz and G. Wegner, *Makromol. Chem.* **190**, 3105 (1989).
9. A. L. Fahrenbruch, R. H. Bube, "Fundamentals of Solar Cells", (Academic Press, 1983) chapter 6.
10. A. J. Campbel, D. D. C. Bradley and D. G. Lidzey, *J. Appl. Phys.* **82**, 6326 (1997).

Mater. Res. Soc. Symp. Proc. Vol. 836 © 2005 Materials Research Society

New Application of Electrospun TiO_2 Nanofibers as an Electrode for Dye-Sensitized Solar Cell

Mi Yeon Song, Young Rack Ahn[1,2], Seong Mu Jo[1] and Dong Yong Kim
Optoelectronic Materials Research Center, Korea Institute of Science and Technology, P.O. Box 131 Cheongryang, Seoul 130-650, Korea
[1]Hybrid Materials Research Center, Korea Institute of Science and Technology, P.O. Box 131 Cheongryang, Seoul 130-650, Korea
[2]School of Materials Science and Engineering, Seoul National University, Seoul 151-742, Korea

ABSTRACT

The electrospun TiO_2 nanofibers were employed to the quasi-solid state dye-sensitized solar cells with porous electrodes which enhanced the penetration of viscous polymer gel electrolytes. The TiO_2 fibers electrospun from poly(vinyl acetate) matrix formed the one-dimensionally aligned fibrillar morphology as a bundle of nanorods. The new TiO_2 electrodes demonstrated that the photocurrent generation with polymer gel electrolytes was over 90% of the performance in DSSC with liquid electrolytes.

INTRODUCTION

The electrospinning technique have provided a simple, cost-effective approach for producing polymeric and inorganic nanofibers within a broad range of diameters, from tens of nanometers to a few micrometers according to the selection of the processing parameters. In general, many conventional sol-gel precursors can be employed for electrospinning to provide the efficient routes to composite and inorganic nanofibers, which may be used in various applications [1,2]. In the present work, we investigated an electrospinning method to generate TiO_2 nanofiber web for an electrode in a dye-sensitized solar cell (DSSC). In a typical DSSC, the triiodide/iodide redox couples in the liquid electrolyte regenerate the photo-oxidized dyes formed after releasing electrons to the conduction band of TiO_2. However, the use of a liquid electrolyte may result in problems for the long-term stability and sealing of the DSSC. Several attempts have been suggested to replace the liquid electrolytes with polymeric gels, p-type conducting polymers or organic hole transporting materials [3-5]. But, the imperfect filling of electrolytes caused the insufficient conversion efficiencies in the quasi-solid- or solid-state solar cells [6]. We present that the electrospunTiO_2 electrodes with the one-dimensionally aligned nanofibers or nanorods enhance the penetration of gel electrolytes as well as the energy conversion efficiency in quasi-solid state DSSCs.

EXPERIMENTAL

TiO_2 fibers were electrospun from solution mixture containing polymers such as polystyrene (PS, Mw =350,000 g/mol), poly(methyl methacrylate) (PMMA, Mw =996,000 g/mol) and poly(vinyl acetate) (PVAc, Mw =850,000 g/mol), and titanium propoxide (TiP, Aldrich) with acetic acid as a catalyst for sol-gel reaction in dimethyl formamide (DMF).

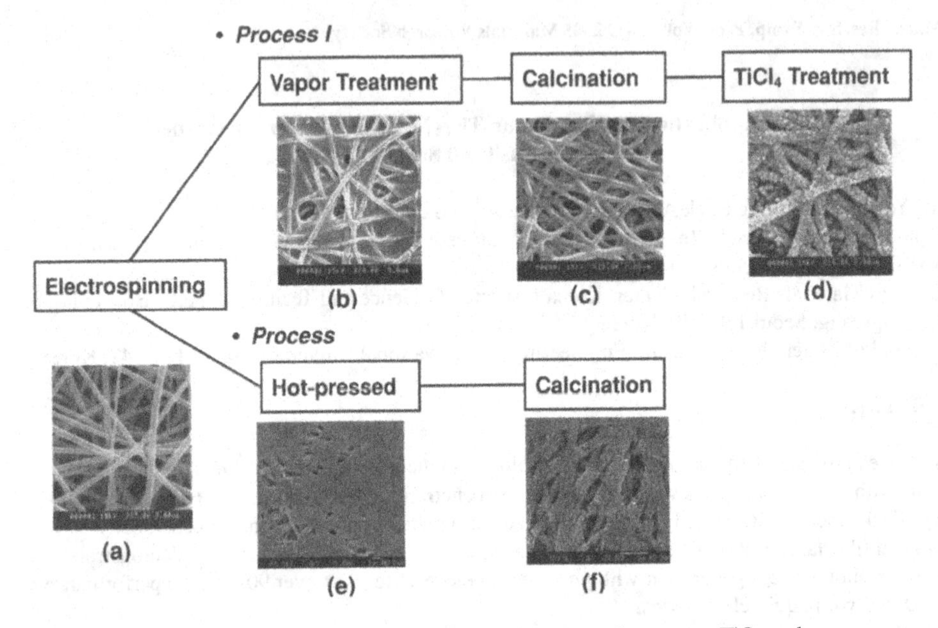

Figure 1. Electrode fabrication process for DSSCs based on electrospun TiO_2 web.

In a typical electrospinning, the precursor solution was loaded into a syringe connected to a high-voltage power supply. The electric field of 11~15 kV was applied between a metal orifice and the ground at a distance of 10 cm. TiO_2 fibers were electrospun directly onto a SnO_2:F coated glass substrate (FTO, 10 cm x 10 cm, TEC-15, Pilkington).

Electrospun TiO_2 web was stabilized for the fabrication of DSSCs with two different methods as shown in figure 1 in order to enhance the adhesion of TiO_2 web to the FTO substrate. In the process I, the electrospun TiO_2 web was treated with tetrahydrofuran (THF) vapor in a closed chamber for 1 hour prior to calcination. The calcination was carried out to remove polymers and to develop anatase TiO_2 stepwise at each temperature (duration in min) as 100 (15); 150 (15); 325 (5); 450 °C (30 min) in air. In addition, TiO_2 rutile crystal was grown on the surface of TiO_2 web electrode was immersed into a 0.1M titanium tetrachloride ($TiCl_4$) (Aldrich) aq. solution in a closed chamber for 24 hrs. In the process II, the electrospun web was pressed between pre-heated plates at 140 °C for 10 min.

For a DSSC assembly, the calcined TiO_2 web electrode was immersed overnight in ethanolic solution containing 3×10^{-4} M of ruthenium sensitizer dye, $RuL_2(NCS)_2$ (L= 2,2'-bipyridyl-4,4'-dicarboxylic acid) (N3, Solaronix). The dye-absorbed TiO_2 electrode was rinsed with ethanol and dried under the nitrogen flow. The liquid electrolyte we used consisted of 0.6 M 1-hexyl-2,3-dimethyl-imidazolium iodide (C6DMIm), 0.05 M iodine (I_2), 0.1 M lithium iodide (LiI) and 0.5 M 4-tert-butylpyridine dissolved in 3-methoxyacetonitrile. Pt-sputtered SnO_2:F glass was used as the counter electrode. A dye-sensitized solar cell with polymer gel electrolyte was also characterized using a mixture of poly(vinylidene fluoride-*co*-hexafluoropropylene) (PVDF-HFP) (Kynar 2801, 0.13g), C6DMIm (0.13 g) and I_2 (0.008 g) in propylene carbonate

(PC) (0.75 g) and ethylene carbonate (EC) (0.5 g). The typical active area of the cell was 0.16 cm^2. The J-V characteristics were measured using a Keithley SMU 2400 under the 100 mW/cm^2 irradiation of Xe lamp (Oriel, 300W) with a global AM1.5 filter for solar spectrum simulation.

RESULTS AND DISCUSSION

In the electrospinning process of inorganic fibers, the sol-gel precursors are combined with polymer solutions resulting in continuous fibers instead of particles in an electro-spray technique. The inorganic/polymer composite fibers are formed after solidification during spinning process. We can consider the fiber formation in several steps: (1) sol-gel conversion of inorganic precursor with the reaction catalyzed by ambient moisture, (2) evaporation of solvent, (3) phase separation of sol-gel precursor/polymer blends, (4) solidification into fibers. etc. In addition, the thermodynamic process is involved due to the local cooling of the fiber surface after fast evaporation of solvents. The electrospinning process of inorganic/polymer composite fiber is a complicate process to control the morphology and the final properties.

In this study, we selected three different types of polymer matrix to form TiO_2 fibers for DSSC electrode application. Polymer solution of PS, PMMA or PVAc was mixed with TiP in DMF prior to spinning as described in experimental section. The diameter of TiO_2 fibers electrospun from PS matrix (conc. PS 0.25 g/mL, TiP 0.19 g/mL in DMF) was ca. 300 nm after calcination as shown in figure 2a. Each fiber was composed of TiO_2 particles (ca. 30 nm). The selected area electron diffraction (SAED) pattern in an inset shows the clear anatase particles which are randomly oriented as observed in circular patterns for each anatase lattice plane. In the case of PMMA (conc. PMMA 0.19 g/mL, TiP 0.19 g/mL in DMF), the morphology was similar to that from PS solution as shown in figure 2b. The smaller particles were formed in PMMA matrix than in PS matrix.

The electrospun fibers from PVAc solution were found to form different morphology from PS or PMMA. We observed a sheath-and-core morphology, in which fibrillar columns were also formed as an islands-in-a-sea structure as shown in figure 2c. The SAED of fibril bundles shows the aligned patterns of anatase TiO_2 as represented in the inset of figure 2c. The diameter of a fibril was ca. 20 nm. At the early stage of electrospinning, the TiO_2 sol precursor is converted to TiO_2 gel when the electrospun fibers are exposed to moisture. The liquid-liquid

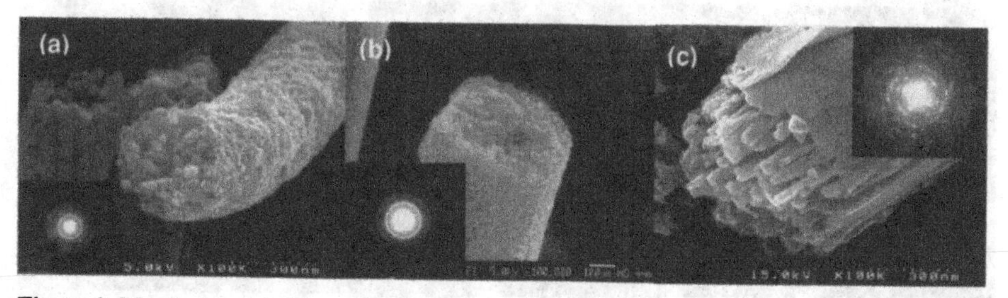

Figure 2. Morphology of electrospun TiO_2 fibers from different polymer matrix; (a) polystyrene, (b) poly(methyl metacrylate), and (c) poly(vinyl acetate).

phase separation occurs into TiO_2 rich- and PVAc rich-phase due to the concentration fluctuation after solvent evaporation. The separated phases are elongated during the spinning into fiber form. The solidification occurs faster in the sheath than in the core in which the separated phases are converted to aligned fibrillar structure in the fiber-axis direction as shown in figure 2c. The 1-D ordering of the fibrils may be also very important for charge transport in photovoltaic devices [7].

The electrospinning method is a candidate for use in the production of electronic devices composed of nanofibers. However, few electronic devices have been produced using this method thus far. One of the problems with electrospun inorganic electrodes is their poor adhesion to substrates after calcination, which is carried out to remove binder polymers. TiO_2 webs have been found to peel off substrates, so in this form cannot be used in electronic devices. In the process-I as shown in figure 1, the PVAc and TiO_2 composite fibers were treated with THF vapor in a closed chamber. Some PVAc were dissolved from the electrospun composite fibers and produced a thin film of PVAc over the fiber web, as shown in figure 1b. In figure 1a, it can be seen that each fiber as spun is stacked in a random direction, and that the untreated fibers are separate from each other. This separation results in the delamination of the untreated TiO_2 web from the substrate during calcination. The pretreated TiO_2 fibers were found to be interconnected, and after heat treatment the TiO_2 electrode was found to have maintained the porous structure of the TiO_2 web, as shown in figure 1c.

As a second method, the electrospun web was pressed directly between pre-heated plates for 10 min at 140 °C above the Tg of PVAc. The hot-pressed film shows the better interconnected fibers as shown in figure 1e before calcination than that of vapor treated one. After calcination of hot-pressed electrode, the morphology was found to be different from that in process-I (figure 1c). The original web structure in an as-spun fibers was retained as shown in figure 1f. But, interestingly, each fiber was composed of nanorods which were analyzed by the higher magnification SEM, TEM and high-resolution TEM (HRTEM) as shown in figure 3. The width and length of nanorods were about 15 nm and 60 nm, respectively. The fibrils in the island-in-a-sea structure were separated out due to the high pressure in the process-II. The regular lattice patterns in HRTEM (figure 3c) reveals that each TiO_2 nanorod is a well-ordered single crystal. Figure 4 presents the photocurrent density-voltage (J-V) curves for cells based on the electrospun TiO_2 nanofiber electrodes in combination with liquid or polymer gel electrolytes.

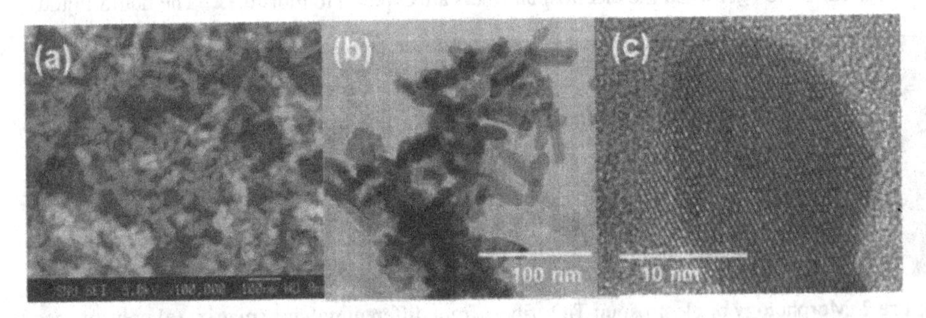

Figure 3. Morphology of electrospun TiO_2 nanorods pretreated followed by the process-II; (a) SEM, (b) TEM and (c) HRTEM images.

The photovoltaic characteristics were analyzed using equations 1 and 2,

$$FF = \frac{V_{max} \times J_{max}}{V_{oc} \times J_{sc}} \quad (1)$$

$$\eta(\%) = \frac{V_{oc} \times J_{sc} \times FF}{I_v} \times 100 \quad (2)$$

where the fill factor (FF) is the maximum power delivered to an external load, calculated from the short-circuit current density (J_{sc}) and the open-circuit voltage (V_{oc}). The overall conversion efficiency (η) is normalized by the intensity of the incident light (I_v). The device containing liquid electrolyte was found to have the following characteristics, as shown in figure 4a: V_{oc} (0.77 V), J_{sc} (8.67 mA/cm^2), FF (0.60), and η (4.01%). The device containing polymer gel electrolyte was also tested with the same electrospun TiO_2 electrode and the characteristics were summarized in table 1. For the gel electrolyte device, J_{sc} and η were found to be 8.07 mA/cm^2 and 3.80%, respectively, as shown in figure 4d. The overall η and J_{sc} obtained for the gel electrolyte system were 94.8% and 93%, respectively, of the values for the liquid electrolyte system.

The performances of the devices were found to be improved by deposition of TiO_2 crystals onto the electrospun electrodes from $TiCl_4$ aq. solution as followed by the process-I. $TiCl_4$ treatment is known to improve the injection efficiency of DSSCs because of the resulting enhancement of light scattering properties [8]. As shown in figure 2d, the diameter of the TiO_2 fibers increased slightly after $TiCl_4$ treatment, but their macroscopic porous morphology was retained. This increase results in an increase in the total volume fraction of TiO_2, and consequently in the amount of sensitizers adsorbed into the electrode. The $TiCl_4$ treatment of the electrode resulted in an increase in J_{sc} from 8.67 to 11.24 mA/cm^2 for the liquid electrolyte device, and in an increase in η from 4.01 to 5.02% (figure 4b). The η of gel electrolyte system was found to be 93% of that of the liquid electrolyte system (figure 4e).

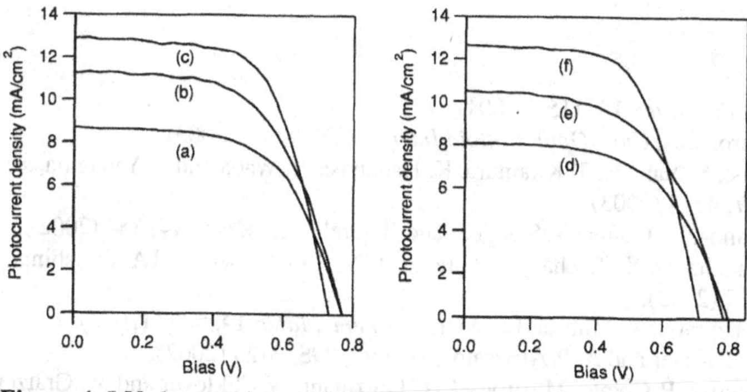

Figure 4. J-V characteristics of DSSCs with electrospun TiO_2 electrodes in liquid (a) and gel (b) electrolytes. The electrospun electrode after an additional $TiCl_4$ aq. solution treatment (process-I) in liquid (c) and gel (d) electrolytes. The TiO_2 nanorod electrodes (process-II) in liquid (e) and gel (f) electrolytes.

Table 1. Device characteristics of DSSCs with electrospun TiO_2 electrodes followed by the process-I in liquid (a), gel (b) electrolytes, and the electrospun electrode after an additional $TiCl_4$ aq. solution treatment in liquid (c) and gel (d) electrolytes. The TiO_2 nanorod electrodes by the process-II in liquid (e) and gel (f) electrolytes.

Devices	V_{oc} (V)	J_{sc} (mA/cm^2)	FF	η (%)
(a)	0.77	8.67	0.60	4.01
(b)	0.80	8.07	0.59	3.80
(c)	0.77	11.24	0.58	5.02
(d)	0.78	10.52	0.56	4.60
(e)	0.73	12.9	0.64	6.0
(f)	0.70	12.7	0.63	5.6

The TiO_2 nanorod electrode by the process-II also provided the large surface area for an efficient sensitization and the network structure enhanced the gel penetration through macro pores in the electrode. The J_{sc} is more 30% improve than the TiO_2 web electrode as shown in table 1. The overall η were 6.0% and 5.6% for the liquid and gel electrolytes, respectively.

CONCLUSION

The 1-D structured TiO_2 fibers were electrospun using the mixture of TiP and polymer solutions to control the fine morphology and properties. The TiO_2 fibers spun from PVAc matrix were composed of the nano-fibrils in a bundle and employed for the electrodes of DSSCs after the pre-treatment procedures. The performance of the gel electrolyte system with an electrospun electrode was found to be greater than 90% of an equivalent liquid electrolyte system. The electrospun TiO_2 fibers have a 1-D ordered fibril structure and the direct production of stable electrodes with the electrospinning method can give them new features and applications in electronic devices.

REFERENCES

1. D. Li and Y. Xia, *Adv. Mater.* **16**, 1151 (2004).
2. A. Frenot, I. S. Chronakis, *Curr. Opin. Colloid Interface Sci.* **8**, 64 (2003).
3. W. Kubo, S. Kambe, S. Nakade, T. Kitamura, K. Hanabusa, Y. Wada and S. Yanagida, *J. Phys. Chem. B* **107**, 4374 (2003).
4. E. Stathatos, P. Lianos, U. Lavrencic-Stanguar and B. Orel, *Adv. Mater.* **14**, 354 (2002).
5. Q.-B. Meng, K. Takahashi, X.-T. Zhang, I. Sutanto, T. N. Rao. O. Sato and A. Fujishima, *Langmuir*, **19**, 3572 (2003).
6. B. O'Regan, F. Lenzmann, R. Muis and J. Wienke, *Chem. Mater.* **14**, 5023 (2002).
7. W. U. Huynh, J. J. Dittmer and A. P. Alivisatos, *Science*, **295**, 2425 (2002).
8. C. J. Barbé, F. Arendse, P. Comte, M. Jirousek, F. Lenzmann, V. Shklover and M. Grätzel, *J. Am. Ceram. Soc.* **80**, 3157 (1997).

Mater. Res. Soc. Symp. Proc. Vol. 836 © 2005 Materials Research Society L5.8

Photocurrent of an individual ZnO nanorods synthesized by sol-gel route on a pulse laser deposited ZnO film

Seung Eon Ahn, Gyu Tae Kim, Jong Soo Lee[1], Hyunsuk kim, Sangsig Kim, Chang Hyun Bae[2], Seung Min Park[2], Jeong Sook Ha.
Department of Electrical Engineering, Korea University, 5-1 Anam-dong, Seongbuk-ku, Seoul 136-701, Republic of Korea.Electronic mail: gtkim@korea.ac.kr
[1]Dr. Jong soo Lee
Research Center of Advanced Material Development, Chonbuk National University, Duckjin-Dong, Jeonju, Jeonbuk, 561-756, South Korea.
[2]Chang Hyun Bae, Prof. Seung Min Park
Department of Chemistry, Kyunghee University, 1 Hoe-ki-dong, Dongdaemoon-gu, Seoul 130-701, South Korea.

ABSTRACT

ZnO nanorods were grown on SiO_2/Si substrates and ZnO thin film by a sol-gel method at low temperatures of around T = 95 °C. The diameters and the lengths of ZnO nanorods increased at high concentrations of zinc nitrate hexahydrate and methenamine solution. And the grain size of ZnO film is strongly correlated with the shape and the diameters of synthesized ZnO nanorods. Current–voltage characteristics of the ZnO nanorod followed a typical nonlinear behavior with significant photoresponse below λ=400 nm in air, and the conductance was enhanced in vacuum with the decrease of the absolute magnitudes of the photocurrent. In photoluminescence (PL) and photocurrent (PC) spectra, the PL peak (λ_{peak}=380 nm and 3.26 eV) did not match the PC edge (λ_{edge}=400 nm and 3.1 eV), indicating the nondirect band-gap transition in photocurrent. The origin of the photocurrent was discussed from the point of the influence of desorption of adsorbed water molecules on the surface or inside the ZnO nanorods.

INTRODUCTION

ZnO nanostructures of nanowires, nanorods or nanotubes have drawn much attention due to their interesting multi-functionality for the applications as opto-electronic, optical devices. ZnO is known to be a direct bandgap semiconductor of 3.37 eV at room temperature with a large exciton binding energy of 60 meV. A large binding energy enables a stable formation of the exciton, enhancing the efficiency of the luminescence and the sensitivity of photo-response

prerequisite for the nanoscale optoelectronic devices. ZnO nanostructures have been synthesized by chemical vapor deposition,[1] physical vapor deposition,[2] vapor-liquid-solid method,[3] catalyst-free method,[4] metal organic chemical vapor deposition (MOCVD),[5] thermal evaporation process,[6] and sol-gel process. [7] Among them, sol-gel process has been recently reported as a simple way to grow ZnO nanostructures at low temperatures with the several advantages such as low cost facilities, non-strict synthesis condition, the simplicity of the process and the possibility of the mass production. [8] However, sol-gel-processed ZnO nanorods were normally thicker than 50 nm with a high density of defects. Recently, the usage of a seed layer of 4 nm sized ZnO nanoparticles deposited on Si/SiO_2 substrate was demonstrated for reducing the diameter of ZnO nanorods, but was not so efficient because it yielded a rather thicker ZnO nanorods with a diameter above 100 nm than the size of seed ZnO nanoparticles. [9]
In our experiments, we developed a simple but controllable synthetic route for producing the narrow widths of single crystalline ZnO nanorods below 50 nm in a large quantity and measured the photoresponse of an individual ZnO nanorod.

EXPERIMENTAL DETAIL

A ZnO thin film was deposited on a Si substrate by pulsed laser deposition (PLD). A ZnO target was irradiated by a focused Nd:YAG laser ($\lambda = 355$ nm) with a spot size of 1.0 mm. The Si substrate was placed at the position 23 mm apart from the target and the temperature of the substrate was maintained at T=600□. Three substrates deposited by ZnO film were prepared at different oxygen pressures of 1, 0.1 and 10^{-4} Torr, respectively. For the sol-gel process of ZnO nanorods on the prepared substrates with Si/SiO_2 substrate, an aqueous solution of 0.006 M zinc nitrate hexahydrate ($Zn(NO_3)_2.6H_2O$) and 0.006 M methenamine ($C_6H_{12}N_4$) was prepared. Each substrate was immersed in such prepared solution and kept in an oven at T = 95 °C for several hours. The structures of synthesized ZnO nanorods were investigated by field-emission scanning electron microscopy (FE-SEM), transmission electron microscopy (TEM), and energy-dispersive X-ray spectroscopy (EDX). The optical property and electrical conductivity of ZnO nanorods were confirmed by photoluminescence (PL) spectrum and by the photocurrent measurement on a single ZnO nanorod.

DISCUSSION

Figures 1(a), 1(b) and 1(c) show the FE-SEM images of ZnO nanorods synthesized on three different PLD grown ZnO seed layers deposited under the different ambient oxygen pressure of 1 Torr, 100mTorr and 0.1 mTorr respectively. Left insets of Fig. 1(a), 1(b) and 1(c) show the different roughness of the surface of the PLD grown ZnO films. The grain size of the ZnO thin film grown under 1 Torr of the oxygen pressure is larger than that under 0.1 mTorr of the oxygen pressure. When the oxygen pressure was 0.1 mTorr, the ZnO film was too smooth to estimate the grain size from the SEM images. As clearly distinguished in the SEM images, the same conditions of the sol-gel process produced the totally different morphologies of ZnO nanorods depending on the grain sizes of ZnO seed layers: the larger aspect ratio of thin ZnO nanorods with the diameters of 33 nm on the rough ZnO film (1 Torr) and the smaller aspect ratio of thicker ZnO nanorods with the diameters of 75 nm on the smooth ZnO film (0.1 mTorr). This result shows that the surface morphology of the PLD grown ZnO seed layers plays an important role in controlling the morphologies of the crystalline ZnO nanorods. The grain size of ZnO film is strongly correlated with the shape and the diameters of synthesized ZnO nanorods as plotted in Figure 1(d). The average diameters of ZnO nanorods synthesized on the PLD grown ZnO seed layers were estimated to be 33 nm, 23 nm, and 75 nm under the ambient O2 pressure of 1 Torr, 100 mTorr, and 0.1 mTorr, respectively, which were quite thinner than the previous reports for the case of the sol-gel process on bare Si substrates. Right inset exhibits the relationship between the concentrations of the solution and the diameter of ZnO nanorods synthesized on Si/SiO_2 substrate of the experiment we previously reported. [10]

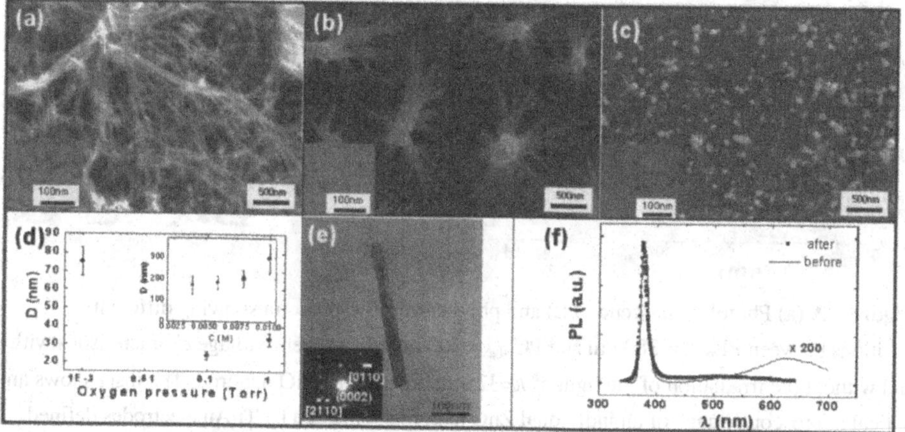

Figure 1. (a), (b) and (c). Field-emission scanning electron microscope (FE-SEM) images. (d). The relationship between the oxygen pressure for the first ZnO layer and the diameter of nanorods in the ZnO layer estimated from SEM images. Right inset: a concentration dependence of the diameter of ZnO nanorods synthesized on Si/ SiO2 substrate. (e).TEM image of a single

ZnO nanorod. Left inset: selected-area electron diffraction pattern from ZnO nanorods.
(f) Photoluminescence spectra of ZnO thin films deposited on Si substrate by pulsed laser ablation at 600□ under ambient O_2 pressure of 1 Torr before (solid line) and after (filled squares) the synthesis of ZnO nanorods.

Figure 1(e) shows a TEM image of the synthesized ZnO nanorod. The average diameter of the individual nanorods is estimated to be around 30 nm. A selected-area electron diffraction (SAED) pattern in the left inset confirms a good crystallinity with the preferential growth of ZnO nanorods along the [0001] direction. Figure 1(f) shows photoluminescence (PL) spectra taken from PLD grown ZnO seed layers before and after the sol-gel process for synthesizing ZnO nanorods, showing the strong peak at 380nm in the same way. The ZnO thin film was grown at the substrate temperature of 600 °C under the ambient O2 pressure of 1 Torr. Considering that the sharp PL peak at 380 nm originates from the single crystalline structure and the broad PL band around ~ 650 nm originates from the oxygen defects, a strong enhancement of the PL peak at 380nm with a noticeable decrease of the PL peak in the broad green-yellow band at ~ 650 nm may be attributed to a good stoichiometric composition and crystallinity of the sol-gel-synthesized ZnO nanorods. With the structural and chemical analysis as in Figure 1, a systematic sol-gel synthetic route for the thinner ZnO nanorods could be confirmed to be reproducible on the PLD deposited ZnO seed layers below 50 nm with a high quality of the crystallinity, which was impossible by a normal sol-gel process.

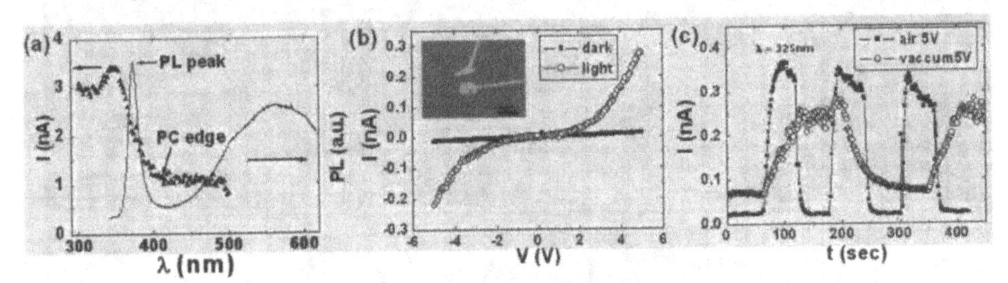

Figure 2. (a) Photoluminescence(PL) and photocurrent(PC) spectra showing different positions between PL_{peak} = 380 nm and PC_{edge} = 400 nm. (b) Current-voltage characteristics with and without the irradiation of the light of λ=325nm on a single ZnO nanorod. The inset shows an optical microscope image of an individual ZnO nanorod contacted by Ti/Au electrodes defined by electron beam lithography. (c) Time dependence of the photocurrent by irradiating the light periodically in vacuum or in air.

Figure 2(a) compares the photoluminescence (PL) and photocurrent (PC) spectra of ZnO nanorods. The PC edge at around λ=400 nm (3.1eV) did not fit to the PL peak of λ=380 nm (3.26eV), indicating the photocurrent did not originate from the direct band-gap transition. Figure 2(b) shows the current-voltage characteristics of a single individual ZnO nanorod. The inset shows a two-probe configuration of the electrodes on a single ZnO nanorods defined by conventional electron beam lithography. The scale bar represents 6 μm. A very low dark current indicates the dominant contact barriers at the interface between Ti/Au electrodes and ZnO nanorods. Under the illumination of the light (λ=325nm from He-Cd laser), a significant current increase was observed, indicating the photo-generated carriers. Figure 2(c) represents the overall tendencies of the current depending on the existence of the light or the vacuum with the same bias voltage of 5V. The evacuation induced the increase of the dark current but the decrease of the absolute magnitudes of the photocurrent. The relaxation time of the photo-generated current was much longer in vacuum than in air. A gradual decrease of the photocurrent was noticeable at the ambient condition under the steady illumination of the light, but no corresponding decrease of the photocurrent in vacuum.

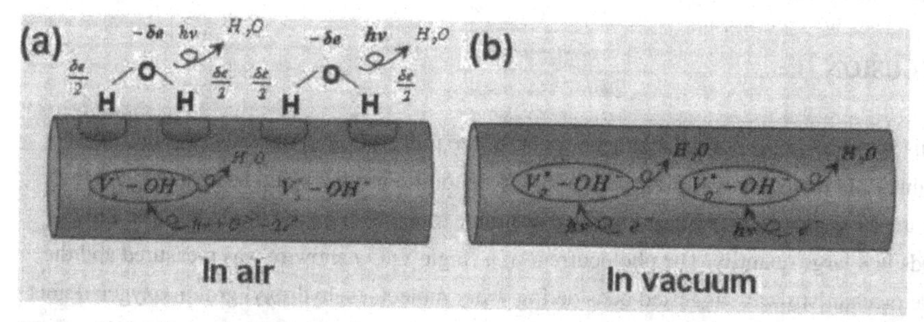

Figure 3. A schematic diagram of charge status surrounding ZnO nanowire including water molecules, oxygen defects, and hydrocyl group. The wound arrows indicate the case under the illumination of the light.

Figure 3 illustrates a possible scenario of the photocurrent observed in our ZnO nanowires. As shown in Fig. 3(a), water molecules can adsorb physically on the surface of nanowires with the hydrogen side to the surface. Considering the hydrogen side of a water molecule can be regarded to be positively charged by δe/2, one adsorbed water molecule on the surface of nanowire can capture the electrons by e/2, reducing the free electrons in ZnO nanorods. Because of the n-type nature of ZnO nanowires, the reduction of available electrons in air can reduce the dark current. Although oxygen defects should be doubly charged by +2e$^-$ as V_O^{++}, the hydroxyl group (OH^-) can join a singly ionized oxygen defect (V_O^+), forming a V_O^+-OH^- pair as shown in Fig. 3(a) and 3(b) with an annihilation of a free carrier by e$^-$. By the irradiation of the light, the excitation of electron and hole pair will be generated an decomposed, contributing the free carrier under the bias voltage by moving in opposite direction ($h\nu \rightarrow e^- + h^+$). In air, the hole carrier might encounter the binding electron at defect sites (V_O^+-OH^-) and recombine, converting the defect states to the oxygen filled sites (O_O^x) by introducing the oxygen molecular. The conversion of the defect sites to the filled oxygen will consume the available electrons by $1/2O_2 \rightarrow O_2^- - 2e^-$ to the doubly charged defect sites (V_O^{++}), explaining the gradual decrease of the photo-current in the steady state of the illumination. The noticeable photocurrent should accompany the generation of the oxygen defect sites by the relation of $O_O^x + h\nu \rightarrow 1/2O_2 + V_O^{++} + 2e^- + pe^- + ph^+$. Practically the relative ratio between the recombination of the defect sites with hydroxyl group (OH^-) and the generated photo-excited electron (pe^-) will decide the absolute magnitude of the observed photocurrent. In vacuum, the deficiencies of water molecules and oxygen gas molecules may indicate the electronic filling of the defect sites by ($V_O^{++} + 2e^- \rightarrow V_O^X$), reducing the absolute magnitudes of the photocurrent with a repetition of the photocurrent generation. The photogenerated electron-hole pair will contribute the total current because of the opposite direction of the flow.

CONCLUSION

In this study, a simple sol-gel synthetic route for the preparation of sub-50 nm single crystalline ZnO nanorods was proposed. A low-temperature growth could be achieved via the help of the PLD grown ZnO film, offering a desirable route for the growth of very thin ZnO nanorods in a large quantity. The photocurrent of a single ZnO nanowire was measured and the possible mechanism was suggested considering water molecules, hydroxyl group, oxygen defect and the photo-generated electron-hole pairs.

---

[1]. J.-Jen Wu and S.-Chang Liu, *Adv. Mater.* **14**, 215 (2002)

[2]. Y. C. Kong, D. P. Yu, B. Zhang, W. Fang, and S. Q. Feng, *Appl. Phys. Lett.* **78**, 407 (2001)

[3]. M. H. Huang, S. Mao, H. Feick, H. Yan, Y. Wu, H. Kind, E. Weber, R. Russo and P. Yang, *Science.* **292**, 1897 (2001)

[4]. J. J. Wu and S. C. Liu, J. Phys. Chem. B 106, 9546 (2002); *Adv. Mater.* **14**, 215 (2000)

[5]. W. I. Park, D. H. Kim, S. W. Jung, and G. Yi, *Appl. Phys. Lett.* **80**, 4232. (2002)

[6]. K. Park, J. Lee, M. Sung and S. Kim, J. *Appl. Phys.* **41**, 7317. (2002)

[7]. L. Vayssieres, K.Keis, S.-E. Lindquist, A. Hagfeldt, *J. Phys. Chem. B*, **105**, 3350. (2001)

[8]. L. Vayssieres, *Adv. Mater.* **15**. 464. (2003)

[9] Jin-Ho Choy, Eue-Soon Jang, Jung-hee Won, Jae-Hun Chung, Du-Jeon Jang, and Young-Woon kim, *Adv. Mater.* **15**. 1991. (2003)

[10] .Seung-Eon Ahn, Jong Soo Lee, Hyun Suk Kim, Sang Sig Kim, Byung-Hyun Kang, Kang-Hyun Kim, Gyu-Tae Kim, *Appl. Phys. Lett.* **84**, 5022. (2004)

Synthesis and Characterization of $Cu_x(In,Ga)_ySe_z$ Nanoparticles by Colloidal Route

Ki-Hyun Kim[1,2], Young-Gab Chun[1], Byung-Ok Park[2] and Kyung-Hoon Yoon[*1]
[1]Solar Cells Research Korea Institute of Energy Research,
71-2 Jang-dong, Yusong-gu, Daejeon, 305-343, Republic of Korea
[*]Tel: +82-42-860-3191/ Fax: +82-42-860-3739/ E-mail: y-kh@kier.re.kr
[2]Dept. of Inorganic Materials Engineering, Kyungpook National University,
1370 Sankyuk-dong, Puk-ku, Daegu, 702-701, Republic of Korea

ABSTRACT

The quarternary CIGS nanoparticles for absorber layer of solar cells have been synthesized with various mole ratios by colloidal route. The CIGS nanoparticles were prepared by reacting CuI, InI_3, GaI_3 in pyridine with Na_2Se in methanol at 0℃ under inert atmosphere. For $Cu_{0.9}In_{0.8}Ga_{0.3}Se_2$ and $Cu_{0.9}In_{0.7}Ga_{0.4}Se_2$ stoichiometric ratios, tube-type nanofibers were obtained with the widths in the range of 20-50 nm and lengths of 0.1-3 μm from reaction at 0℃ for 20 min. For $Cu_{1.1}In_{0.68}Ga_{0.23}Se_{1.91}$, and $Cu_{0.9}In_{0.68}Ga_{0.23}Se_{1.91}$ ratios, spherical nanoparticles were obtained from the same reaction condition. As compared to particles from $Cu_{0.9}In_{0.68}Ga_{0.23}Se_{1.91}$ ratio, more uniform and smaller nanoparticles with diameter in the range of 5-20 nm were obtained from the $Cu_{1.1}In_{0.68}Ga_{0.23}Se_{1.91}$ stoichiometric ratio. The morphology change of the CIGS particles seems to be closely related to the ratio of Cu/(In,Ga).

INTRODUCTION

The synthesis of nanostructure semiconductors has been of interest to material scientist and chemists because of their potential applications in recent years [1-5]. Compared with microscale material, the nanoscale materials have potential application in both mesoscopic research and development of nanodevices [6-7]. The I-III-VI_2 nanoscale materials, ternary and quaternary semiconductor compounds such as $CuInSe_2$ (CIS) and $CuInGaSe_2$ (CIGS) with the chalcopyrite structure have shown to be useful as nonlinear optical and direct energy gap materials [8-9]. The CIGS is known to be a very prominent absorber layer for high efficiency thin film solar cell devices. Most of the research groups developing CIGS solar cells have used physical vapor deposition (PVD) techniques such as evaporation [10] or sputtering [11] for depositing the absorber layer of CIGS. However, it is expensive to manufacture photovoltaic (PV) device by PVD method and it is difficult with almost of these methods to obtain satisfactory stoichiometric compound. In order to overcome such problem, it has been tried to use nanoparticles for solar cells device by non-vacuum and low temperature processes. The colloidal route of non-vacuum process allows to prepare the CIGS semiconductor compounds for absorber layer without high-pressure Se vapor treatment and H_2Se treatment of Cu-In-Ga alloys.

In this work, a novel approach has been tried to fabricate CIGS solar cells in which a non-vacuum process has been employed for depositing the CIGS absorber layer. We have synthesized CIGS nanoparticles by colloidal route and address some characteristic features on the obtained CIGS nanoparticles.

EXPERIMENTAL

Compared to the solvothermal routes [12] by which we have recently reported the synthesis of spherical CIGS nanoparticles at reaction temperatures over 230℃, the synthesis of the CIGS nanoparticles was carried out at lower temperature such as 0℃ by colloidal method[13].

The $Cu_x(In,Ga)_ySe_z$ (w=0.9, 1.1; y=0.9, 0.91; z=1.91, 2) nanoparticles were prepared by reacting CuI, InI_3, GaI_3 in pyridine with Na_2Se in methanol at 0℃ in a glove box under inert atmosphere. The pyridine and methanol were distilled and deoxygenated before use. The metal iodines of CuI, InI_3, and GaI_3 and Na_2Se were dissolved in the pyridine and methanol solvents, respectively. The pyridine solution was mixed with the methanol solution. The solution mixture was reacted with stirring for the reaction times in the range from 10 to 3600 seconds in ice bath at 0℃.

The reaction mixture was allowed to settle down for 24 hours and then, the reaction product of dark red color supernatant was decanted and discarded. Thereafter, the remaining reaction products were transferred to 50 ml centrifuge tube, which was added with distilled and deoxygenated methanol to fill it. Then, it was subjected to sonication for 10 minutes. The products of yellow color were centrifuged for 15 minutes at 4000 rpm.

At the completion of centrifugation, the reaction products of dark red and yellow colors were turned out to be NaI and CIGS nanoparticles, respectively, by inductively coupled plasma atomic emission spectrometer (ICP-AES: Ultima-C, Jobin Yyon, France) and energy dispersive spectrometry (EDS: Phoenix, EDAX Inc., USA) analyses. The particle size and morphology of the products were investigated by high-resolution scanning electron microscopy (HRSEM: XL30SFEG, Philips Co., Holland at 10 kV) and transmission electron microscopy (TEM: EM912Ω, Carl Zeiss Co., Germany). The phase and the crystallographic structure of the products were identified by X-ray diffraction (XRD: D/max-A, Rigaku, Japan, Cu Kα: λ=1.54178Å) and TEM.

RESULTS AND DISCUSSIONS

In order to obtain CIGS colloid, the colloidal route reactions were carried out in ice-bath as function of reaction times and Cu/(In+Ga)/Se ratios of raw material. First of all, in order to obtain the information about the chemical composition of the nanoparticles, every product was analyzed with EDS (Fig. 1). The average atomic composition of the surface of the nanoparticles was determined to be approximately a ratio of 1.1: 0.9: 1.91 for Cu: In+Ga: Se. The chemical composition of the reaction product of dark red color supernatant obtained from the reactions in ice-bath for 20 minutes was determined with ICP-AES and given in Table I.

Table I: Chemical compositions of the reaction product of dark red color supernatant

Element	Na	I	Cu
Atom.%	30.14	55.83	14.83

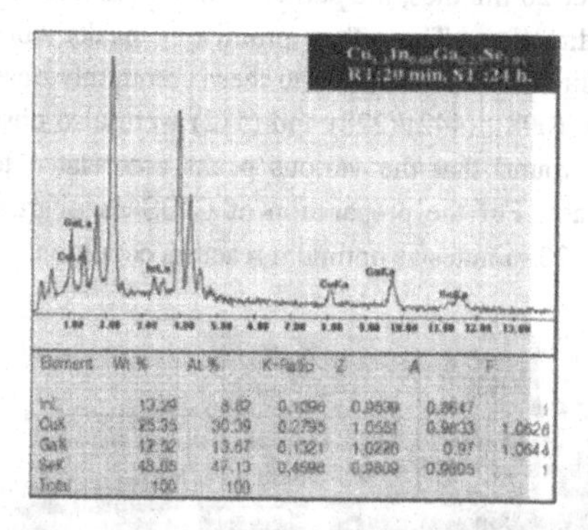

Fig. 1 Energy dispersive spectra of the CIGS products obtained from the reaction in ice-bath for 20 minutes.

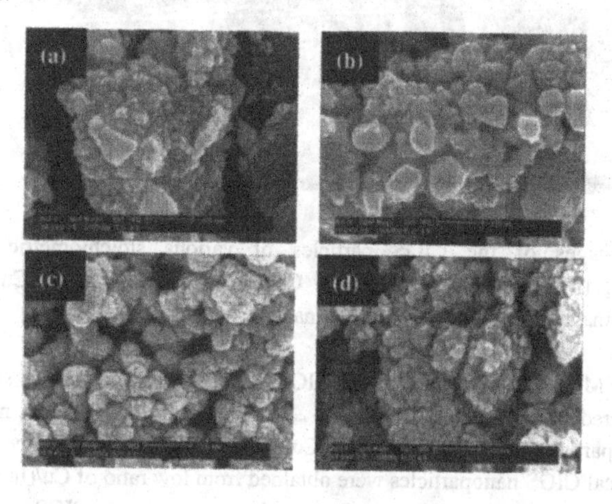

Fig. 2 SEM images of the $Cu_{1.1}In_{0.68}Ga_{0.23}Se_{1.91}$ nanoparticles obtained from the reactions in ice bath with stirring for various reaction times: (a), 10 sec; (b), 1 minute; (c), 20 minutes; (d), 60 minutes.

Fig. 2 shows SEM images of the $Cu_{1.1}In_{0.68}Ga_{0.23}Se_{1.91}$ nanoparticles obtained from the reaction with stirring in ice bath at 0℃ for 10 sec - 60 minutes. The reaction time of 20 minutes led to the most spherical nanoparticles which were characterized as being fine and uniform nanoparticles with diameter below 20 nm. For the reaction times of 10 second and 1 minute, mixtures of plate-type and spherical particles were obtained with diameter below sub-micron meter as shown in fig. 2(a) and (b). The reactions for forming these particles were not completely proceeded because the XRD patterns did not show a typical chalcopyrite structure of the CIGS particles. The $Cu_{1.1}In_{0.68}Ga_{0.23}Se_{1.91}$ particles were bigger and irregular for the reaction time of 60 minutes rather than of 20 minutes.

For the reaction time of over 20 minutes, the particles showed an intense peak at $2\theta=26.6°$ oriented along the (112) direction. The other prominent peaks corresponded to the (220)/(204) and (312)/(116) directions. In addition to these commonly observed orientations, the weak peaks such as (400)/(008), (442)/(228) and (512) were also observed in the XRD patterns. In addition, it was found that the various peaks are related to the presence of $InGaSe_2$, In_2Se_3 and InSe phase. For the preparation of CIGS nanoparticles, therefore, we have used the reaction time of 20 minutes as optimum reaction condition.

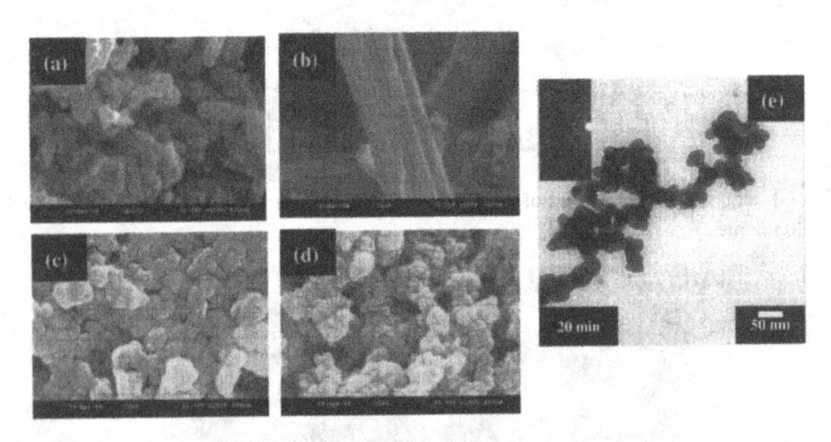

Fig. 3 SEM images of the CIGS particles of various stoichiometric ratios: (a), $Cu_{0.9}In_{0.7}Ga_{0.4}Se_2$; (b), $Cu_{0.9}In_{0.8}Ga_{0.3}Se_2$; (c), $Cu_{0.9}In_{0.68}Ga_{0.32}Se_{1.91}$; (d), $Cu_{1.1}In_{0.68}Ga_{0.32}Se_{1.91}$; and TEM image of $Cu_{1.1}In_{0.68}Ga_{0.23}Se_{1.91}$ nanoparticles.

Fig. 3 shows SEM and TEM images of the CIGS particles with the various stoichiometric ratios. As compared to the particles of $Cu_{0.9}In_{0.68}Ga_{0.23}Se_{1.91}$ ratio (Fig. 3(c)), more uniform and smaller nanoparticles were obtained of the $Cu_{1.1}In_{0.68}Ga_{0.23}Se_{1.91}$ stoichiometric ratio (Fig. 3(d)). The spherical CIGS nanoparticles were obtained from low ratio of Cu/(In+Ga), such as $Cu_{1.1}In_{0.68}Ga_{0.23}Se_{1.91}$, and $Cu_{0.9}In_{0.68}Ga_{0.23}Se_{1.91}$, whereas tube-type CIGS nanofiber with length of several ㎛ and width in the range of 20-50 ㎚ were obtained from $Cu_{0.9}In_{0.7}Ga_{0.4}Se_2$, and $Cu_{0.9}In_{0.8}Ga_{0.3}Se_2$ (Figs.3(a) and (b)). The spherical $Cu_{1.1}In_{0.68}Ga_{0.23}Se_{1.91}$ nanoparticles obtained from the reaction in ice-bath for 20 minutes were confirmed to be consisted of small agglomerate with diameter in the range of 5-20 ㎚ by TEM. The electron diffraction patterns corresponded to the prominent peaks, (112), (204)/(220), (116)/(312) of the tetragonal CIGS phase.

Fig. 4 shows XRD diffraction patterns of chalcopyrite structure of the CIGS nanoparticles with various stoichiometric ratios. The CIGS nanoparticles showed a typical X-ray diffraction pattern for chalcopyrite structure with an intense peak at $2\theta=26.6°$ oriented along the (112) direction. The other prominent peaks corresponded to the (220)/(204) and (312)/(116) directions. In addition to these commonly observed orientations, the weak peaks such as (400)/(008), (442)/(228) and (512) were also observed in XRD patterns.

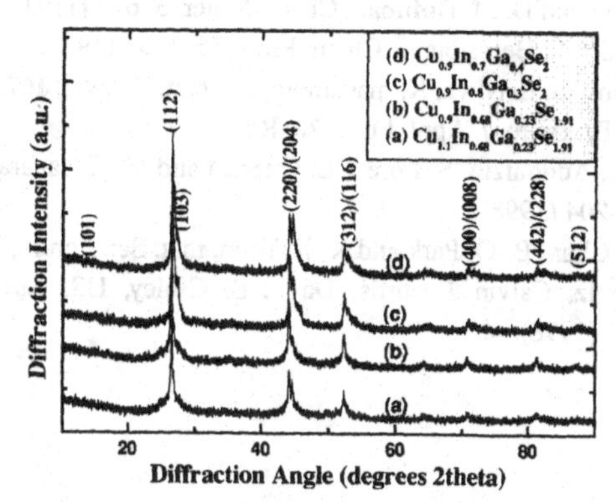

Fig. 4 XRD patterns of the CIGS particles of various stoichiometric ratios: (a), $Cu_{1.1}In_{0.68}Ga_{0.32}Se_{1.91}$; (b), $Cu_{0.9}In_{0.68}Ga_{0.32}Se_{1.91}$; (c), $Cu_{0.9}In_{0.8}Ga_{0.3}Se_2$; (d), $Cu_{0.9}In_{0.7}Ga_{0.4}Se_2$.

CONCLUSION

We synthesized CIGS nanoparticles for the CIGS absorber layer of solar cells by colloidal route. Morphological change of the nanoparticles was observed from the various compound ratios. With the higher value of Cu / (In+Ga), the shape of CIGS particles changed from tube-like nanofiber to spherical nanoparticle. For $Cu_{0.9}In_{0.8}Ga_{0.3}Se_2$ and $Cu_{0.9}In_{0.7}Ga_{0.4}Se_2$ stoichiometric ratios, tube-type CIGS nanofibers with widths in the range of 20-50 nm and lengths of 0.1-3 μm were obtained from reaction at 0℃ for 20 min. For $Cu_{1.1}In_{0.68}Ga_{0.23}Se_{1.91}$ and $Cu_{0.9}In_{0.68}Ga_{0.23}Se_{1.91}$ ratios, spherical nanoparticles were obtained with diameter of 20 nm from the same reaction condition. As compared to particles from $Cu_{1.1}In_{0.68}Ga_{0.23}Se_{1.911}$ ratio, more uniform and smaller nanoparticles with diameter in the range of 5-20 nm were obtained from the $Cu_{0.9}In_{0.68}Ga_{0.32}Se_{1.9}$ stoichiometric ratio.

ACKNOWLEDGEMENT

The present work was supported by Korea Ministry of Science and Technology through National R&D Project for Nano Science and Technology.

REFERENCE

1. J. Xiao, Y. Xie, Y. Xiong, R. Tang and Y. Qian, J. Meter Chem., **11**, 1417-1420 (2001).
2. M. A. Malik, P. O'Brein and N. Revaprasadu, Adv. Mater., **11**, No. 17, 1441-1444 (1999).

3. Y. Jiang, Y. Wu, S. Yuan and B. Xie, J. Mater. Res., Vol.16, No. **10**, 2805-2808 (2001).
4. C. Eberspacher, C. Fredric, K. Pauls and J. Serra, thin solid films, **387**, 18-22 (2001).
5. W. Wang, Y. Geng, Y. Qian, M. Ji and X. Liu, Adv. Mater., **10**, No. 17, 1479-1481 (2001).
6. R. Fix, R. G. Gordon and D. M. Hoffman, Chem. Meter, **3**, 1138 (1191).
7. R. Fix, R. G. Gordon and D. M. Hoffman, Chem. Meter, **5**, 614 (1193).
8. S. C. Abrahams and J. L. Bernstein, J. Chem. Phys. **59**, 1695 (1973).
9. I. Yonenaga, K. Sumino, E. Niwa, K. masumoto, J. Cryst. Growth, **167**, 616 (1996).
10. A. Rockett, R.W. Birkmire, J. Appl. Phys. **70**, R81 (1991).
11. A. Parretta, M.. L. Addonizio, S. Loreti, L. Quercia and M. K. Jauaraj, Journal of Crystal Growth **183**, 196-204 (1998).
12. K.-H. Kim, Y.-G. Chun, B.-O. Park and K.-h. Yoon, mat. Sci. Forum, **273**, 449-452 (2004).
13. Douglas L. Schulz, Calvin J. curtis, David S. Ginley, US. partent, patent number, **6**,126,740, jan. 27 (1998).

Mater. Res. Soc. Symp. Proc. Vol. 836 © 2005 Materials Research Society L5.10

Preparation of $CuInGaSe_2$ Absorber Layer by Nanoparticles-Based Spray Deposition

Ki-Hyun Kim[1,2], Young-Gab Chun[1], Byung-Ok Park[2] and Kyung-Hoon Yoon[*1]
[1]Solar Cells Research Center, Korea Institute of Energy Research,
71-2 Jang-dong, Yusong-gu, Daejeon, 305-343, Republic of Korea
[*]Tel: +82-42-860-3191/ Fax: +82-42-860-3739/ E-mail: y-kh@kier.re.kr
[2]Dept. of Inorganic Materials Engineering, Kyungpook National University,
1370 Sankyuk-dong, Puk-ku, Daegu, 702-701, Republic of Korea

ABSTRACT

CIGS nanoparticles for the CIGS absorber layer have been synthesized by low temperature colloidal routes. The CIGS absorber layers for solar cells have been prepared by spray deposition of CIGS nanoparticle precursors (~20 nm) in glove box under inert atmosphere. An automatic air atomizing nozzle spray system with computer controlled X-Y step motor system was used to spray. The nanoparticle precursor CIGS film was deposited onto molybdenum-coated soda-lime glass substrates (2.5 ㎝ X 5.0 ㎝) heated to 160℃. The film thickness in the range of 2 ㎛ ± 0.3 ㎛ was attained by spraying of 3 mM colloidal over an area of 12.5 ㎠. The coalescence between particles was observed in the CIGS absorber layer under post-treatment of over 550℃. This is related to the reactive sintering among the nanoparticles to reduce surface energy of the particles. The Cu_xSe thin film, formed on Mo film by evaporation, improved adhesion between CIGS and Mo layers and enhanced the coalescence of the particles in the CIGS layer. These are closely related to the fluxing of Cu_2Se phase which has relatively low melting temperature. The CdS buffer layer was deposited on the CIGS/Mo/soda-lime glass substrate by chemical bath deposition. The CIGS nanoparticles-based absorber layers were characterized by using energy dispersive spectroscopy (EDS), x-ray diffraction (XRD) and high-resolution scanning electron microscopy (HRSEM).

INTRODUCTION

Compound solar cell devices consist of electrode/transparent conducting oxides (TCO)/buffer layer/absorber layer/back contact/soda-lime glass. In general, chalcopyrite material of $CuInGaSe_2$ (CIGS) is known to be a very prominent absorber layer for high efficiency thin film solar cell devices. Recently, improved efficiencies for CdS/$Cu(InGa)Se_2$ (Eg~1.1 eV) solar cells have been reported with a value of 19.2% [1,2]. While this CIGS device possesses the highest thin film polycrystalline solar cell efficiency to date, the scaling of this technology toward commercialization presents many challenges. Amongst various physical and chemical deposition techniques of preparing thin semiconducting film, the spray deposition is a simple, easy, and economical technique with the unique feature to control the preparative parameters, especially by controlling the substrate temperature.

In this work, we have tried a novel approach to fabricate CIGS solar cells in which a non-vacuum process is used for depositing the CIGS absorber layer. With the CIGS nanoparticles (~20 nm) synthesized by colloidal route, we prepared and analyzed the CIGS absorber layer of solar cells by spray deposition.

EXPERIMENTAL

In order to obtain CIGS nanoparticles for spray-deposition, the CIGS nanoparticles were synthesized by colloidal routes. The CIGS nanoparticles were prepared by reacting CuI, InI_3, GaI_3 in pyridine with Na_2Se in methanol at 0℃ in a glove box under inert atmosphere. The pyridine and methanol were distilled and deoxygenated before use. The solution mixture was reacted with stirring for the reaction times in the range between 10 sec - 60 minutes in ice bath at 0℃. The reaction mixture was allowed to settle down for 24 hours and then, the reaction product was washed to remove NaI. Consequently, the CIGS nanoparticles were obtained with diameter in the range of 5-20 nm.

The CIGS absorber layers for solar cells were prepared by spray-deposition of the CIGS nanoparticle precursors in glove box under inert atmosphere. An automatic air atomizing nozzle spray system with computer controlled X-Y step motor system was used to spray. The nanoparticle precursor CIGS film was deposited onto molybdenum-coated soda-lime glass substrates (2.5 ㎝ X 5.0 ㎝) heated to 160℃. The as-deposit CIGS films were subjected to the reduction and the selenization processes for stoichiometric ratio control. Another step for solar cells was proceeded by deposition of CdS buffer layer by chemical bath deposition at 68℃.

The morphology and thickness of the CIGS absorber layer were investigated by high-resolution scanning electron microscopy (HRSEM: XL30SFEG, Philips Co., Holland at 10 kV). The phase and the crystallographic structure of the products were identified by X-ray diffraction (XRD: D/max-A, Rigaku, Japan, Cu Kα: λ=1.54178 Å).

RESULTS AND DISCUSSIONS

In order to obtain CIGS absorber layer, CIGS nanoparticles were synthesized by colloidal routes for spray-deposition. With the CIGS nanoparticles (~20 nm) synthesized by colloidal route, the CIGS absorber layer of solar cells was prepared by spray deposition in glove box under inert atmosphere. The spherical $Cu_{1.1}In_{0.68}Ga_{0.23}Se_{1.91}$ nanoparticles for spray-deposition were confirmed by TEM to be small agglomerate with diameter in the range of 10-20 nm (Fig. 1(a)).

The precursor CIGS films were deposited onto molybdenum-coated soda-lime glass substrate (2.5 ㎝ X 5.0 ㎝) heated to 160℃ which was controlled by home-made heating system with a thermocouple feedback loop. The as-sprayed CIGS film with thickness of 2.5 ± 0.2 ㎛ was attained by spraying CIGS solution of ~20 g of 3 X 10^{-3} M with an automatic air atomizing nozzle spray system (Fig. 1(c)). Various parameters, such as solution concentration (1, 3, 5 X 10^{-3} M), spraying time (5~20 sec), spray frequency (1-7 times) and sprayer-to-substrate distance (1-13 cm) were tried to obtain optimum spraying condition giving good CIGS precursor films. The solution concentration, spraying rate, spray frequency and sprayer-to-substrate distance were optimized to be 3 X 10^{-3} M, 2.5 ㎤/18 sec, 5 times and 5 cm, respectively.

Figs. 1(b) and (c) show SEM images of surface and cross-section of the CIGS precursor films deposited with optimized spray conditions, respectively. The thickness of as-sprayed CIGS film was about 2.5 ㎛. In order to obtain the information about the chemical composition of the CIGS absorber layer, all deposited CIGS films were analyzed with EDS. The average atomic composition of the surface of CIGS films deposited onto Mo substrate after selenization treatment was determined to be approximately a ratio of 0.9: 0.9: 1.91 for Cu: In+Ga: Se.

Fig. 1 TEM image of $Cu_{1.1}In_{0.68}Ga_{0.23}Se_{1.91}$ nanoparticles for spray deposition which is obtained from the reaction in ice-bath for 20 minutes (a) and SEM images of CIGS precursor film: (b), surface morphology; (c), cross-section.

The reduction treatment was carried out at 500℃ for 30 min under inert atmosphere with in-situ RTA system to remove the oxygen in the CIGS absorber layer. The EDS analysis indicates that oxygen concentration was decreased below 1 atom% by reduction treatment. The reduction process did not influence the morphology of the CIGS precursor films. Selenization causes cracking of the CIGS absorber layer in pieces of tens of microns independent of the used selenization technique. However, this cracking mainly depends on the deposition technique and the thickness of the CIGS absorber layer. The cracks were absent in thinner regions of the CIGS absorber layer and were suppressed by low increasing rate of temperature and keeping the CIGS layer at 80℃ for long time. Some evidences of crack onset could be already found when the solvent from the deposition paste was evaporated. The coalescence between particles was appeared at selenization temperature up to 575℃. This is due to the reactive sintering among nanoparticles to reduce surface energy of the particles.

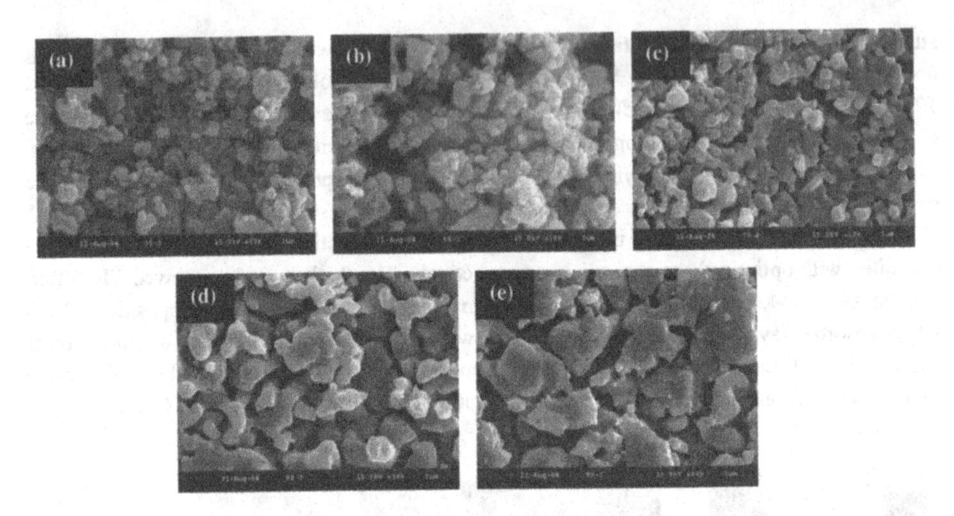

Fig. 2 SEM images of the $Cu_{0.9}In_{0.68}Ga_{0.32}Se_{1.91}$ absorber layer subjected to various selenization temperatures after the reduction treatment at 500℃ for 30 min: (a), as-sprayed CIGS film; (b), 450; (c), 500; (d), 550; (e), 575℃.

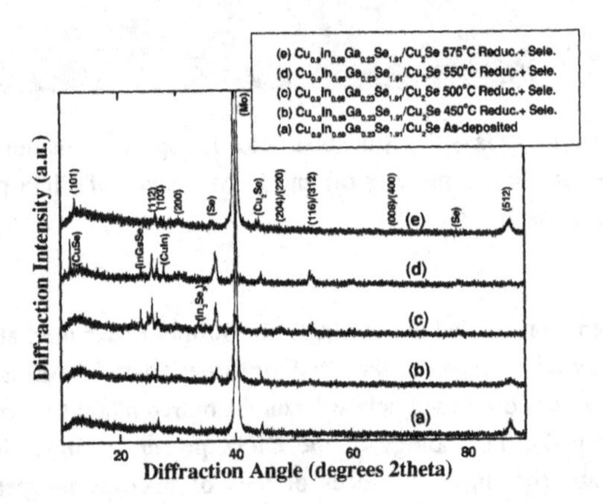

Fig. 3 XRD diffraction patterns of the $Cu_{0.9}In_{0.68}Ga_{0.32}Se_{1.91}$ absorber layer subjected to various selenization temperatures for 30 min: (a), as-sprayed CIGS absorber layer; (b), 450; (c), 500; (d), 550; (e), 575℃.

Fig. 3 shows XRD diffraction patterns of chalcopyrite structure of the $Cu_{0.9}In_{0.68}Ga_{0.32}Se_{1.91}$ absorber layer subjected to various selenization temperatures for 30 min. The reduction temperature was maintained at 500℃ for 30 min. For the selenization temperature of over 550℃, XRD patterns of the CIGS absorber layer showed a typical chalcopyrite structure with an intense peak at 2θ=26.6° oriented along the (112) direction. The other prominent peaks corresponded to the (220)/(204) and (312)/(116) directions. In addition to these commonly observed orientations, the weak peaks such as (400)/(008), (442)/(228) and (512) were also observed in XRD patterns. In addition, the various peaks related to the presence of CuIn, CuSe, Cu_2Se, InGaSe, In_2Se_3, Mo and Se phase were observed.

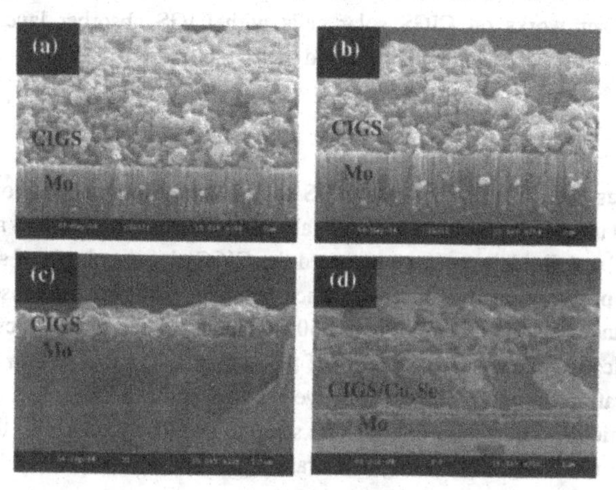

Fig. 4 SEM images for cross-section of CIGS absorber layer: (a), as-sprayed; (b), after reduction treatment at 500℃; (c), after reduction at 500℃ and selenization treatment at 550℃; (d), after reduction and selenization treatment of CIGS layer deposited on Cu_xSe/Mo films.

Fig. 5 Photos of the experimental equipments used by nanoparticle-based non-vacuum processing for the CIGS absorber layer deposition: (a), automatic spray system with air atomizing nozzle; (b), RTA system for reduction and selenization processes; (c), CBD bath for CdS film formation.

The structure of the sintered film is relatively uniform throughout the thickness of the film. Provided the CIGS films do not actually melt during sintering, lateral material movement is modest, hence, coverage and surface morphology of the final CIGS film resembles that of the precursor layer (fig. 4(b)). It was confirmed that bulk grain growth appeared in the CIGS cross-section at selenization temperature over 550℃ (fig. 4(c)). The as-sprayed CIGS films with thickness of 2.5 ± 0.2 μm can be obtained by spray-deposition of uniform and fine CIGS nanoparticles (Fig. 4(a)). The Cu_xSe thin film, formed on Mo film by evaporation, improved adhesion between CIGS and Mo layers and enhanced the coalescence of the particles in the CIGS layer(fig. 4(d)). This is due to the fluxing of Cu_2Se phase of which the melting temperature is about 525 ℃. Further works on CIGS solar cells with CIGS absorber layer formed by particle-based, non-vacuum processing will be carried out in the future

CONCLUSION

A novel approach has been tried to fabricate CIGS solar cells by spray deposition in which a non-vacuum process is used for depositing the CIGS absorber layer. With the CIGS nanoparticles (~20 nm) synthesized by colloidal route, we prepared the CIGS absorber layer of solar cells by spray deposition and post-treatment. The coalescence between particles was observed in the CIGS absorber layer under post-treatment of over 550℃. This is related to the reactive sintering among the nanoparticles to reduce surface energy of the particles. The Cu_xSe thin film, formed on Mo film by evaporation, improved adhesion between CIGS and Mo layers and enhanced the coalescence of the particles in the CIGS layer. These seem to be closely related to the fluxing of Cu_2Se phase which has relatively low melting temperature.

ACKNOWLEDGEMENT

The present work was supported by Korea Ministry of Science and Technology through National R&D Project for Nano Science and Technology.

REFERENCE

1. A. Rockett, R.W. Birkmire, J. Appl. Phys. 70, **R81** (1991).
2. K. Ramanathan, M.A. Contreras, C.L. Perkins, S. Asher, F.S. Hasoon, J. Keane, D. Young, M. Romero, W. Metzger, R.Noufi, J. Ward, A. Duda, Prog. Photovolt: Res. Appl. **11**, 225 (2003).
3. A. Parretta, M.. L. Addonizio, S. Loreti, L. Quercia and M. K. Jauaraj, J. Cryst. Growth, **183**, 196-204 (1998).
4. K.-H. Kim, Y.-G. Chun, B.-O. Park and K.-h. Yoon, mat. Sci. Forum, **273**, 449-452 (2004).
5. C. Eberspacher, C. Fredric, K. Pauls and J. Serra, thin solid films, **387**, 18-22 (2001).

Dye Sensitized Solar Cells Using Nanostructured Thin Films of Titanium Dioxide

Douglas A. Gish, Gregory K. Kiema, Martin O. Jensen, and Michael J. Brett
Department of Electrical and Computer Engineering, University of Alberta,
Edmonton, Alberta T6G 2V4, Canada

ABSTRACT

Dye sensitized solar cells (DSSCs) were fabricated using porous thin films of TiO_2. These films were deposited by electron beam evaporation and an advanced substrate motion technique called PhiSweep. PhiSweep, an extension of glancing angle deposition (GLAD), allows for greater control over the surface area of nanostructured thin films than is possible with traditional GLAD. The as-deposited films were amorphous, so the films were annealed to improve their crystal structure. The films were sensitized with a photoactive dye and implemented into a DSSC configuration as the electron collecting electrode. It was expected that the higher surface area of the films produced using the PhiSweep method would improve the cell performance compared with cells made using traditional GLAD films of TiO_2. However, the performance of the cells prepared using PhiSweep films was likely hindered by higher internal resistance of the films compared to the films prepared by traditional GLAD. The highest photoelectric conversion efficiency of the dye sensitized solar cells produced using the PhiSweep method was 1.5%.

INTRODUCTION

Recently there has been increased interest in the development of dye sensitized solar cells (DSSCs), which were first proposed by O'Regan and Grätzel [1]. This is due to their high conversion efficiencies, low fabrication cost, and simple cell technology. DSSCs are based on photoactive dye molecules adsorbed onto the surface of a wide band-gap semiconductor. The semiconductor that is typically used is TiO_2 because its conduction band energetics matches the excited state of many types of dyes and it has useful surface chemistry and material properties [2]. The dye sensitized TiO_2 electrode is placed in contact with an electrolyte and completed by an inert counter electrode.

To achieve high photoelectric conversion efficiencies, a TiO_2 electrode with a high surface area is desirable. This allows more dye molecules to adsorb to the electrode surface, and should result in increased absorption of the incoming light. This has led to the use of a porous network of TiO_2 nanoparticles, prepared by sol-gel methods, as the electron collecting electrode, rather than a flat unstructured film [1]. Very recently, Kiema *et al.* implemented an alternate method of fabricating the porous TiO_2 electrode based on glancing angle deposition (GLAD) [3].

GLAD is a single-step deposition process utilizing physical vapour deposition at an oblique angle combined with substrate rotation [4-6]. The morphology of thin films can be tailored on nanometre size scales by exploiting the self-shadowing effects which arise when vapour flux arrives at highly oblique angles to the substrate surface. GLAD films, which have nominal porosities of 70%, have engineered microstructures such as slanted columns, polygonal spirals, zigzags, or vertical posts. The high surface area of these films makes them suited for application in DSSCs.

Very recently, an extension of the GLAD process called PhiSweep GLAD has been developed, which utilizes advanced substrate motion to provide even greater control over the porosity of nanostructured thin films [7]. In particular, it allows for control over the quantity and size scale of voids in the film independent of the film's density or helical structure. This capability allows for the fabrication of highly porous thin films consisting of fibres that are much smaller in diameter than those attainable with traditional GLAD. For example, a 4 μm thick Si film with fibre diameters of 20 nm to 30 nm and a constant average film density of 30%-40% bulk density was recently demonstrated by Jensen and Brett [7]. Also, these fibres maintain a more consistent diameter throughout the entire thickness of the film, whereas traditional GLAD films typically broaden considerably as the film thickness increases. Although PhiSweep films have the same volume and mass density as traditional GLAD films, PhiSweep films have a greater number of smaller pores than are present in traditional GLAD films. Consequently, thin films fabricated using the PhiSweep method have an estimated one order of magnitude higher surface area than traditional GLAD films [7], which already have a greatly enhanced surface area compared to conventional thin films.

This paper reports our initial efforts in fabricating and characterizing DSSCs incorporating TiO_2 films, prepared using the PhiSweep method, as the dye sensitized electrode. The motivation behind using the PhiSweep method is to maximize the surface area compared to traditional GLAD. The photoelectric conversion efficiency results are compared with DSSCs based on traditional GLAD films.

EXPERIMENTAL DETAILS

The substrates used in the solar cell experiments were 2.5 cm square fluorinated tin (IV) oxide coated 2.2 mm thick sodalime glass with an as-delivered resistivity of 5.6 Ω/□ and optical transmission of ~ 80%. Bare silicon substrates were also mounted beside the conducting glass substrates as witness samples which were imaged by scanning electron microscopy (SEM). The substrates were loaded on a motor controlled substrate holder which is 42 cm above the electron beam crucible. The substrate orientation is described by the angle between the crucible normal and the substrate normal, α, and the angle of rotation about the substrate normal, φ. The vapour source material used in this study was rutile TiO_2 from Cerac with a purity of 99.9%. The chamber was evacuated to base pressures below 5×10^{-6} Pa by cryogenic pumping. Once base pressure was achieved, the chamber pressure was increased to 7×10^{-3} Pa by the addition of oxygen near the substrate surface. The background oxygen pressure was found to be essential to achieving the desired film stoichiometry [8]. Deposition of the TiO_2 was performed by electron beam evaporation at an oblique angle, α, of 84°. The deposition rate was measured by a crystal thickness monitor and was kept at ~15-20 Å/s.

PhiSweep method

In the PhiSweep method [7], the incident flux and the film growth directions are decoupled. This is accomplished by sweeping the φ angle about a central axis defining the direction of growth. One half of the angle covered by each sweep is called the sweep angle, which is typically 45°. At each outer extreme of the sweep curve, the substrate is paused until a certain thickness of film is deposited. The thickness that is deposited between each sweep is called the sweep pitch. After every phi-sweep, the competition process between neighbouring fibres, which leads to extinction of some fibres and broadening of others in traditional GLAD, is effectively reset. If the sweep pitch is sufficiently small, such that very little competition has occurred before the process is reset, the original fibre diameter can be maintained throughout the thickness of the film. However, the sweep pitch must not be too small or the vapour flux will not have time to grow into a new fibre segment before the flux direction changes. Rather, the flux will accumulate onto either side of the nucleated fibres, which causes the fibres to broaden. Therefore, there exists an optimum sweep pitch, which is in general material dependent.

Two different PhiSweep film morphologies were used for the solar cell experiments. The first film was a PhiSweep square spiral with a spiral pitch (the height of one 360° rotation) of 1.8 μm, a sweep pitch of 50 nm and a sweep angle of 45°. The second film was a PhiSweep vertical post film with a sweep pitch of 40 nm. In the case of the vertical post, the film was not swept from side to side since that would result in a slanted post segment. Instead, the film was always swept in the same direction by a 90° angle. This motion is identical to a traditional GLAD square spiral [9], but with a very small spiral pitch of 160 nm.

Solar cell assembly

Three different solar cells were prepared. Cell A was prepared with an as-deposited PhiSweep square spiral film. Cell B used a PhiSweep square spiral film deposited simultaneously with the film used in Cell A, but it was annealed for 3 h at 500°C. Cell C was made from a PhiSweep vertical post film which was annealed for 6 h at 550°C. The annealing was carried out in air with a Thermolyne 48000 muffle furnace. Previous work has shown that as-deposited TiO_2 films are amorphous, and annealing the films produces polycrystalline anatase TiO_2, which is nearly stoichiometric [3]. The annealing temperature was chosen to be above the transition point from amorphous to anatase titanium dioxide (350°C) and below the glass transition point of standard glass (~550-600°C) [3].

Dye sensitization of both as-deposited and annealed films was accomplished by first heating the films to ~100°C for 1 h to drive off any water adsorbed to the films' surface. The films were then immersed in the dye solution when their temperature had cooled to ~80°C. The dye solution used in Cell A and Cell B was 0.3 mM cis-bis(isothiocyanato)bis(2,2'-bipyridyl-4,4'-dicarboxylato)-ruthenium(II)bis-tetrabutylammonium in ethanol. The dye used for Cell C was a 0.3 mM solution of tris(isothiocyanato)-ruthenium(II)-2,2':6',2''-terpyridine-4,4',4'-tricarboxylic acid in ethanol. The dyes were supplied by Solaronix SA, Switzerland, and were sold under the names Ru 535-TBA and Ru 620-1H3TBA, respectively. Immersion took place over night and the excess dye was removed by rinsing with ethanol. The dye-coated film was then soaked in 4-tert-butylpyridine for 2 minutes and dried under nitrogen stream.

The DSSCs were constructed using two transparent conducting glass substrates (SnO_2:F). One substrate was coated with the dye-sensitized TiO_2 film, and the other substrate acted as the inert counter electrode. The counter electrode was prepared by immersing the substrate in a transparent platinum catalyst solution (Pt-Catalyst T/SP, Solaronix) and then baking it for 30 minutes at 450°C. The two electrodes were separated by paper spacers which were ~40μm thick, and a redox electrolyte from Solaronix purchased under the name of Iodolyte TG 50 was introduced into the inter-electrode space by capillary action. A clip was used to hold the sandwiched electrodes together. The assembled cells were illuminated with an Oriel Instruments solar simulator (Model 91191), which contains a 1000 W Xenon arc lamp. Photocurrent versus voltage measurements were taken with air mass 1.5 and 10% transmittance neutral density filters. The measured solar intensity was ~65 mW/cm^2.

DISCUSSION

The SEM images of the films used to make Cells A, B, and C, as well as a traditional GLAD slanted post film deposited at an oblique angle, α, of 85° are shown from the side view in figure 1 and from top-down view in figure 2. The thickness of the film used for Cell A and B was measured to be 9.2 μm, while the film used in Cell C was 6.9 μm thick. These images clearly illustrate that the PhiSweep method reduced the size of the individual fibres compared to the traditional GLAD film. Since the PhiSweep films and the traditional GLAD films were deposited at nearly the same angle of α (84° for the PhiSweep films and 85° for the traditional GLAD film), they have mass and volume densities that are nearly the same. However, since the PhiSweep films comprise a larger number of smaller fibres compared to the traditional GLAD films, the PhiSweep films have a greatly enhanced surface area, allowing more dye to be adsorbed. SEM images were taken before and after annealing, and the morphology of the films appears unchanged at the resolution attainable with the SEM. However, Kiema *et al.* reported that the crystal structure of TiO_2 films deposited by GLAD changes from amorphous to polycrystalline anatase after annealing, as shown by X-ray diffraction (XRD) measurements [3].

The photoelectric measurements of the constructed DSSCs are shown in figure 3, and table I provides a summary of the solar conversion performance in terms of the open circuit voltage (V_{oc}), short circuit current density (J_{sc}), and overall conversion efficiency. The short circuit current density was found to be significantly higher for the cells incorporating the two annealed films, Cell B and C, than Cell A which was made from the as-deposited film. This is likely due to the differences in film crystal structure. Heat treatment increases the ordering of the films, which likely improves the transport of injected electrons. This results in reduced recombination with the electrolyte and higher quantum efficiencies. The maximum voltage obtainable from this type of solar cell, V_{oc}, is dependent on the relative position of the conduction band edge, V_{CB}, of the TiO_2 electrode and the redox potential of the electrolyte, V_{red}, as given by

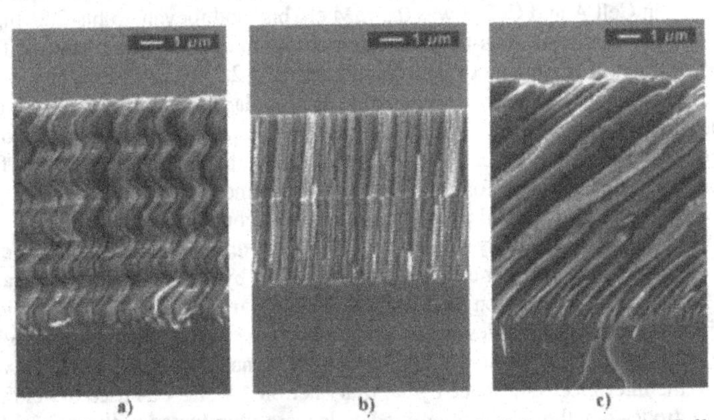

Figure 1. SEM side views of the films used to make a) Cells A and B, b) Cell C, and c) a traditional GLAD slanted post TiO_2 film deposited with α=85°. The discontinuity in b) is an artefact of the deposition process.

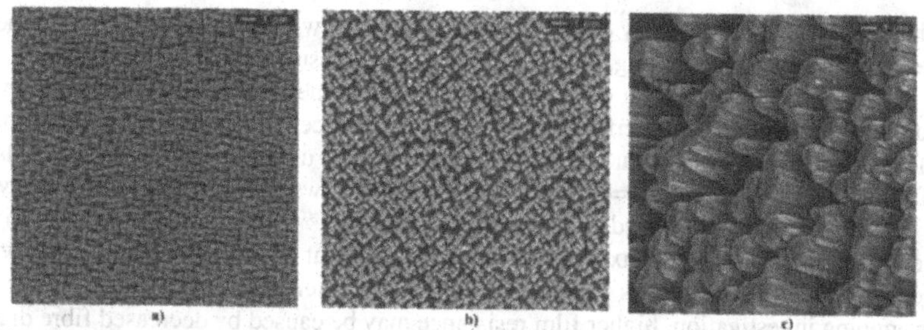

Figure 2. SEM top-down views of the films used to make a) Cells A and B, b) Cell C, and c) a traditional GLAD slanted post TiO_2 film deposited with α=85°.

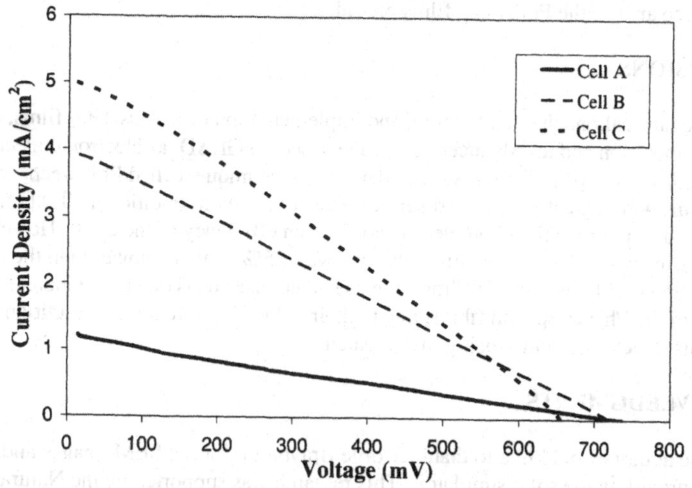

Figure 3. Current-voltage (*I-V*) curves for DSSCs incorporating PhiSweep deposited TiO_2 films.

Table I. Results of photoelectric measurements on DSSCs incorporating PhiSweep deposited TiO_2 films.

Cell	V_{oc} (mV)	J_{sc} (mA/cm^2)	Fill Factor	Efficiency (%)
A	737	1.4	0.20	0.3
B	721	3.9	0.26	1.2
C	655	5.1	0.28	1.5

equation 1.

$$V_{oc}=V_{CB}-V_{red} \quad (1)$$

The open circuit voltage was similar for Cell A and B, and about 70 mV lower for Cell C. The overall efficiency and fill factor were higher for the annealed films, and the highest efficiency was that of Cell C. This is likely due to the fact that Cell C was annealed at a 50°C higher temperature and for twice as long as Cell B, which would result in better crystallinity.

It was expected that solar cells incorporating the higher surface area of PhiSweep films would have increased the cell performance relative to solar cells prepared using traditional GLAD films. However, the solar cell made by Kiema *et al.* using a traditional GLAD slanted post film with α=85°, near the α=84° used here for the PhiSweep films, had an efficiency of 1.8% [3]. In addition, cells based on traditional GLAD films with α between 60° and 75° had efficiencies above 3% with a maximum efficiency of 4.1% at α=60°. The comparatively poor efficiency of the PhiSweep films is likely due to a higher internal film resistance. While it is still under ongoing investigation, higher film resistance may be caused by decreased fibre diameter. The smaller diameter fibres may have restricted the size of crystal grains of anatase TiO_2 that formed. The increased number of grain boundaries would impede the transport of injected electrons. It is possible that this effect may dominate over any increase in efficiency that the larger surface area of the PhiSweep films provide.

CONCLUSIONS

The fabrication, characterization, and implementation of porous TiO_2 films, deposited by reactive evaporation and an advanced technique based on GLAD, as electron collecting layers in DSSCs has been reported. This advanced deposition technique, called PhiSweep, was used to provide films with a greatly increased surface area compared to traditional GLAD films, which was expected to improve the photoelectric conversion efficiency of the cells. However, the highest efficiency obtained using this technique was 1.5%, which is lower than the efficiencies obtained using traditional GLAD films. We speculate that this is due to the increased internal resistance of the PhiSweep film fibres, due to their reduced size relative to tradition GLAD film fibres. This effect is currently being investigated.

ACKNOWLEDGMENTS

The authors would like to thank George Braybrook for the SEM images and Dr. Joel Haber for providing the solar simulator. This research was supported by the Natural Sciences and Engineering Research Council of Canada (NSERC), the Informatics Circle of Research Excellence (iCORE), Micralyne Inc., and Alberta Ingenuity.

REFERENCES

[1] B. O'Regan and M. Grätzel, Nature **353**, 737 (1991).
[2] A. Hagfeldt and M. Grätzel, Acc. Chem. Res. **33**, 269-277 (2000).
[3] G. K. Kiema, M. J. Colgan, and M. J. Brett, Sol. Energy Mater. & Sol. Cells, In press (2004).
[4] K. Robbie and M. J. Brett, U.S. Patent No. 5 866 204, (1999).
[5] K. Robbie and M. J. Brett, J. Vac. Sci. Technol., **A15**, 1460-1465 (1997).
[6] K. Robbie, M. J. Brett, and A. Lakhtakia, Nature, **384**, 616 (1996).
[7] M. O. Jensen and M. J. Brett, Appl. Phys. A, In press (2004).
[8] M. J. Colgan, B. Djurfors, D. G. Ivey, and M. J. Brett, Thin Solid Films, **466**, 92-96 (2004).
[9] S. R. Kennedy and M. J. Brett, Nano Letters, **2** (1), 59-62 (2002).

Mater. Res. Soc. Symp. Proc. Vol. 836 © 2005 Materials Research Society L5.14

Novel Strategies for the Preparation of TiO_2 Nanofibers

Kenneth J. Balkus, Jr., Chunrong Xiong and Minedys Macias-Guzman
Department of Chemistry and the UTD NanoTech Institute, University of Texas at Dallas
Richardson, TX 75083-0688, U.S.A.

ABSTRACT

The various polymorphs of titanium oxide (TiO_2) are probably the most widely studied materials for photovoltaic applications. It is well known that composition and particles size can have a profound effect on photoactivity. It is thought that high surface to volume nanostructures such as nanofibers and nanorods may possess the optimal properties. We have developed technology for the preparation of TiO_2 nanofibers in a variety of forms including core shell structures, papers and films. The formation of mesoporous TiO_2 fibers as spider webs and papers has been achieved by electrostatic deposition. The mesoporous fibers and other forms such as shaped particles and films have been employed as templates to grow TiO_2 nanofibers in various configurations. The proposed research below will address the synthesis and characterization of the TiO_2 fibers developed by these techniques

INTRODUCTION

Titanium dioxide (TiO_2) is widely studied for its potential application in high performance photovoltaics. Dimensionality becomes a crucial factor in determining the properties of nanomaterials including surface area. Therefore, efforts have been made to decrease the particle size of porous or dense forms of titania [1-4]. In addition to nanoparticles, there may be advantages to preparing TiO_2 as high surface area fibers. The probability of electron-hole recombination at surface trap sites increases with nanoparticles. In contrast, nanofibers or nanorods the electron-hole recombination may be reduced because of improved delocalization of carriers. TiO_2 nanorods have been prepared by controlled growth using an organic capping technique [5]. Well aligned rutile and anatase TiO_2 nanorods have also been prepared by MOCVD [6]. Titania nanofibers have also been electrospun from titanium isopropoxide mixed with acetic acid and high molecular weight polyvinylpyrrolidone (PVP) to form amorphous titania nanofibers [7]. Amorphous titania films with micrometer sized pores has also been formed by electrostatic sol-spray deposition (ESSD) using titanium isopropoxide [8]. Although there has been progress in fabricating dense TiO_2 in various configurations there has been little reported regarding the morphogenesis of mesoporous TiO_2 molecular sieves. Mesoporous TiO_2 has been prepared as bulk powders and films [9-16]. For example, Grätzel and co-workers achieved 7% efficiency by employing a ruthenium dye complex adsorbed onto a mesoporous TiO_2 nanocrystalline film [17]. Subsequent studies have shown that several characteristics of the TiO_2 film including the size and degree of crystallinity, affect the overall efficiency of the cell [18-19]. Yanagida and co-workers concluded that the diffusion coefficient of electrons increased with an increase in TiO_2 particles size while electron recombination lifetimes decreased [18]. Recently, we reported the first example of mesoporous TiO_2 fibers prepared by electrospinning [20]. The TiO_2 fibers have pores 3.7nm in diameter with crystalline pore walls composed of anatase. More recently the e-spun TiO_2 fibers have been prepared as free standing papers. Additionally, the TiO_2 have been co-spun with donor polymers such as MEH-PPV to form a donor-acceptor mesh. One of the key issues that remain with this technology is the fiber diameter. We have discovered that the mesopores in the electrospun fibers or other configurations for these molecular sieves can be exploited as templates for the growth of small (<15nm) TiO_2 nanofibers. In this case, template free mesoporous particles are allowed to adsorb a TiO_2 precursor such as $TiCl_4$ under moderate temperatures (<100C) and pressures in an autoclave. Then controlled exposure to moisture results in growth of TiO_2 nanofibers from the pores of mesoporous template. There have been some examples of semiconductor growth in porous media. For example, the growth of TiO_2 nanorods in the pores of anodic alumina has been demonstrated [21-23]. The resulting rods maybe nanosized (<100nm) but more typically they are several hundred nanometers in diameter and the alumina template must then be dissolved away to recover the TiO_2 nanorods. TiO_2 nanorods 125-200 nm in diameter have also been grown by sol-gel electrophoresis in track etched polycarbonate with 200 nm pores [24]. One of the distinct features of the molecular sieve templated growth of TiO_2 nanofibers is the ability to configure the molecular sieves as particles, fibers and films. Additionally, the molecular sieve could function in a synergistic fashion by supplying adsorption sites for sensitizers. The dissolution of the molecular sieve template leaves behind a core shell structure, such as hollow spheres or fibers. Therefore, the TiO_2 nanofibers may be configured in a variety of hierarchal forms such as hollow spheres and fibers as well as films which could then be incorporated into a solar cell.

EXPERIMENTAL

The formation of TiO_2 fibers in the form of spider webs was previously reported [20]. The precursor gel in this case, had a molar ratio of 0.0037 P-123 (Pluronic): 0.016 12M HCl: 0.13 EtOH : 0.01 $Ti(OCH_4H_9)_4$. TiO_2 yarns were prepared by twisting the spider web fibers. The freestanding TiO_2 paper resulted from a gel having a molar ratio of 0.0003 P-123: 0.0137 12M HCl: 0.0868 EtOH: 0.01 $Ti(OCH_4H_9)_4$. This gel was prepared by adding a solution of 0.001 mol $Ti(OCH_4H_9)_4$ and 0.55 mL 12 M HCl to a 50 wt % P-123 solution in 5.07 mL of ethanol. The gel was aged for 22 hours and electrospun at 20 kV to obtain a paper of mesoporous TiO_2 fibers.

The TiO_2 nanorods were templated from mesoporous materials in the form of shaped particles films or papers (fibers). The spherical core shell structures were prepared from DAM-1 (Dallas Amorphous Material) and SBA-15 (Santa Barbara) spheres, DAM-1 and SBA-15 spheres were prepared according to literature [25,26]. The mesoporous templates were placed in a small cup made of stainless steel screen which was held in a Teflon lined autoclave 4 cm above $TiCl_4$. Filling of the mesopores by $TiCl_4$ was conducted under autogeneous pressure in a 23 ml Parr autoclave at 90 °C. The TiO_2 nanorods were then formed by hydrolysis in air or controlled air flow. The mesoporous Ta_2O_5 papers employed as templates, were prepared as previously described [27]. A precursor gel having a molar ratio of 0.01 $TaCl_5$: 0.08 1-propanol: 0.01 12 M HCl: 0.09×10^{-2} Brij 76 was electrospun at 18 kV and calcined at 350 °C. The formation of TiO_2 nanorods follows the same procedure used for the particles.

RESULTS AND DISCUSSION

Electrospinning involves the application of a high voltage to a viscous precursor gel which then traverses some distance to an oppositely charged conductive surface. As the resulting jet travels towards the grounded target, fibers form and deposit on the substrate of choice. Only recently have we discovered how to apply this technique to the production of mesoporous metal oxide fibers. For example, the preparation of mesoporous TiO_2 fibers by electrospinning was recently reported [20]. In this case, a spider web like fibrous mesh of TiO_2 fibers was formed as

Figure 1. Mesoporous TiO_2 Fibers

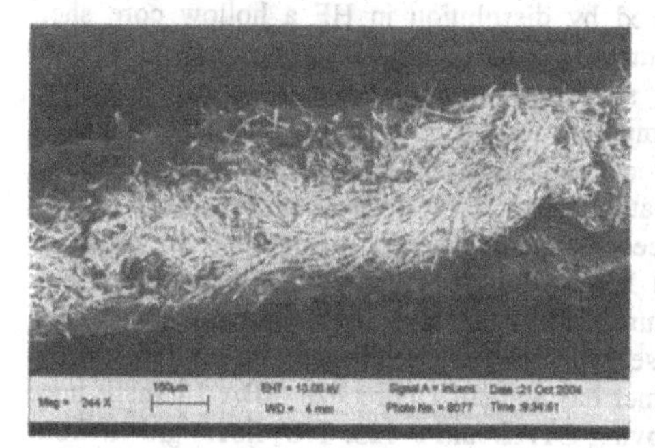

Figure 2. Mesoporous TiO_2 yarn.

shown in Figure 1. The pores having a wormhole structure were 3.7nm in diameter with a surface area of $200m^2/gm$. The pore walls are composed of crystalline anatase as determined by XRD and Raman spectroscopy. The fibers are flexible and retain their shape even after heating up to 1000°C. This is somewhat surprising since the TiO_2 undergoes a phase change to rutile above 700°C with a loss in porosity. The spider web like configuration makes these fibers difficult to handle. However, this can be exploited in the preparation of TiO_2 yarns or rope as shown in Figure 2. The molecular sieve yarn can be prepared by simply twisting the spider web fibers. Mechanical testing to determine if the TiO_2 yarn is stronger than the individual fibers is in progress. The high surface to volume ratio of TiO_2 molecular sieve fibers may also be beneficial in photovoltaic applications. The configuration of the mesoporous TiO_2 fibers in textiles and papers may represent new opportunities for solar cells. In fact by simply changing the precursor gel composition a free standing paper can be formed as shown in Figure 3. The fiber diameter is dependent on several parameters including the gel concentration, voltage, target distance, needle size etc. In general, the smallest mesoporous TiO_2 fibers obtained by electrospinning so far are 100 -200nm in diameter. Smaller fibers can be achieved by acid etching The mesoporous TiO_2 fibers in the paper are ~2 μm in diameter which is larger than those prepared as spider webs (<1μm). An interesting strategy for employing the TiO_2 fibers in a solar cell would be to co-electrospin a donor polymer to form an intimately entwined mesh of donor acceptor fibers. We have recently reported the preparation of an MEH-PPV (poly[2-methoxy-5-(2-ethyl-hexyloxy)-1,4-phenylenevinylene]) composite fiber mesh with the mesoporous silica, SBA-15 [28]. An interesting consequence of the electrospinning process and the intimate contact with the SBA-15 fibers is the lack of polymer aggregation. This manifests itself in large shifts (>60nm) in the emission spectra. There are many examples of conducting polymers employed in photovoltaics and an important issue is controlling the degree of aggregation in the film. Additionally, the fiber mesh should increase the surface contact between the donor and acceptor relative to layered films. In a preliminary experiment, a TiO_2 precursor gel was electrospun together with the fluorescent donor polymer MEH-PPV by using a dual syringe method to obtain a donor-acceptor intertwined composite fibers as shown in Figure 4. The composite fibers exhibited a quenching of fluorescence in the donor polymer. Construction of a solar cell based on this mesh is in progress.

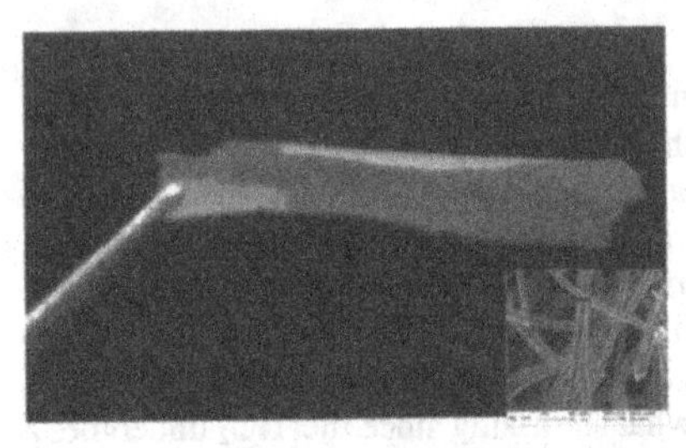

Figure 3. Mesoporous TiO_2 paper

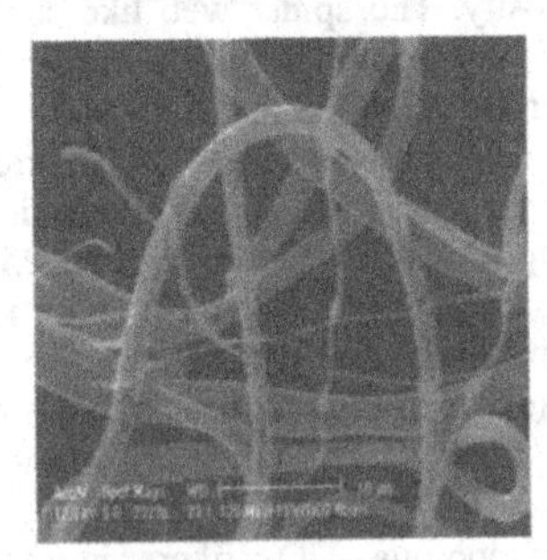

Figure 4. TiO_2/MEH-PPV fiber mesh

The mesoporosity in the metal oxide fibers can be exploited for the adsorption of small molecules or dyes that could be employed in a solar cell. The uniform pores could also be used to template the growth of nanorods of TiO_2. We have found that the adsorption of a TiO_2 precursor such as $TiCl_4$ under high pressure followed by controlled hydrolysis results in nanorods of anatase. The size of the nanorods depends on the size of the mesopores and if the mesopores are blocked then no TiO_2 nanorods are formed. Figure 5 shows an electrospun mesoporous Ta_2O_5 paper before and after templated growth of TiO_2 nanorods. If thc molecular sieve fiber is removed by dissolution in HF a hollow core shell structure is formed. In addition to the electrospun fibers, shaped molecular sieve particles can also serve as templates for the growth of TiO_2 nanofibers or nanorods. When DAM-1 spheres were used as templates, many TiO_2 fibers can be found on the surface of spheres as seen in Figure 6. Removal of DAM-1 spheres resulted in formation of core shell structure, and fibers range 30~50 nm in diameter. The anatase phase was confirmed by XRD. When comparing SBA-15 spheres to DAM-1, the former has a smaller pore size. SBA-15 small pore size arises from a dual template synthesis employing CTAB and P-123. TiO_2 fibers grown from SBA-15 exhibited smaller diameters of 14 to 20nm. In the same way, The application of these templated TiO_2 nanorods in solar cells by dispersion and coating with donor polymers is in

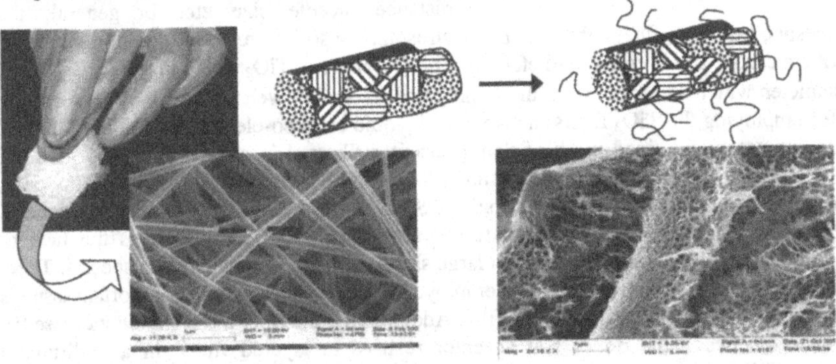

Figure 5. TiO_2 nanorods grown from electrospun mesoporous Ta_2O_5

Molecular Sieves Are filled with $TiCl_4$ Under Pressure → **Filled Particles are Exposed to Air** → **Surface TiO_2 Nanorods Form on the Particles**

progress. *The support of the Air Force Office of Research and the R. A. Welch Foundation is gratefully acknowledged.*

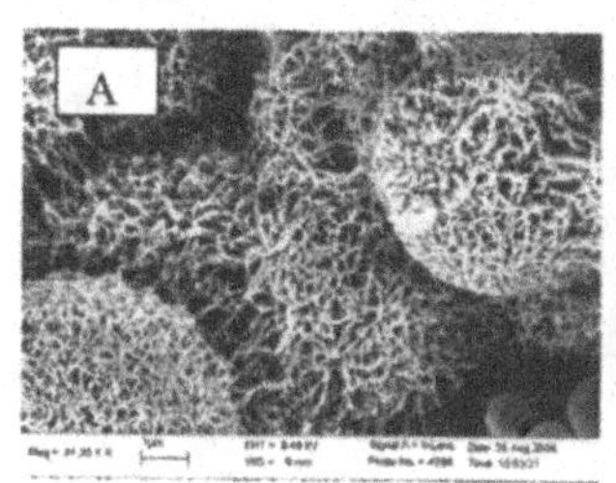

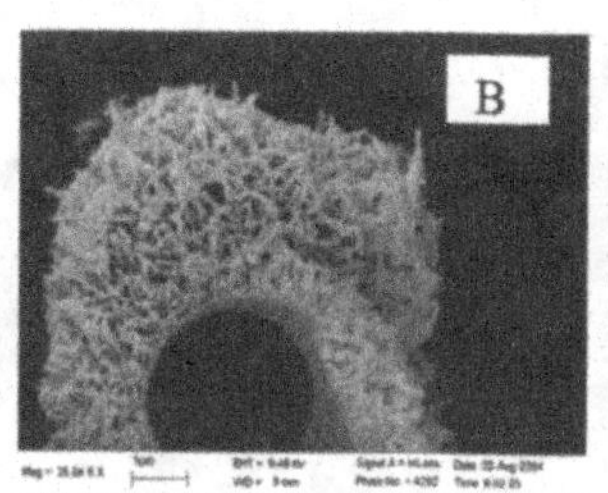

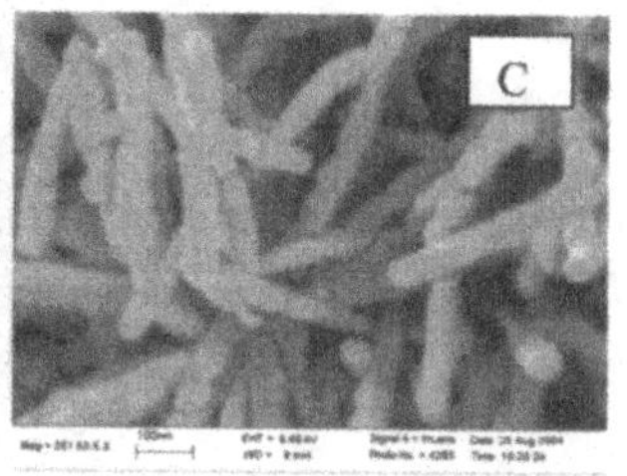

Figure 6 TiO_2 fibers on Dam-1. A: as-synthesized, B and C: after removal of Dam-1.

REFERENCES

1) X.F. Zhou, D.B. Chu, S.W. Wang, C.J. Lin, Z.Q. Tian, *Mater. Res. Bull.* **37,** 1851 (2002).
2) Z. Yuan, J. Colomer, B. Su, *Chem. Phys. Lett.* **363**, 362 (2002).
3) Z. Wang, J. Chen, X. Hu, *Mater. Lett.* **43,** 87 (2000).
4) S. Jeon, P. V. Braun, *Chem. Mater.* **15**, 1256 (2003).
5) P.D. Cozzoli, A. Kornowski, W. Weller, *J. Am. Chem. Soc.*, **125,** 14539 (2003).
6) J. Wu, C. Yu, *J. Phys. Chem. B*, **108** 3377 (2004).
7) D. Li, Y. Xia, *Nanolett.*, **3,** 555 (2003).
8) M. Nomura, B. Meester, J. Schoonman, F Kapteijn, J Moulijn *Chem. Mater.* **15**, 1283 (2003).
9) P. Alberius, K. Frindell, R.Hayward, E. Kramer, G. Stucky, B. Chmelka, *Chem. Mater.***14**, 3284 (2002).
10) P. Yang, D. Zhao, D. Margolese, B.F. Chmelka, G.D. Stucky, *Chem. Mater.* **11**, 2813 (1999).
11) Y. Yue, Z. Gao, *Chem. Commun.* 1755 (2000).
12) Y.K. Hwang, K.C. Lee, Y.U. Kwon, *Chem. Commun.*1738 (2001).
13) H. Yoshitake, T. Sugihara, T. Tatsumi, *Chem. Mater.* **14,** 1023 (2002).
14) K. Kim, K.D. Benkstein, J.V. Lagemaat, A.J. Frank, *Chem. Mater.* **12**, 1042 (2002).
15) D.S. Lee, T.K. Liu, J. *Sol-gel. Sci. Technol.* **25**, 121 (2002).
16) G.J. de A.A. Soler-Illia, D. Grasso, E.L. Crepaldi, F. Cagnol, C. Sanchez, *Mat. Res. Soc. Symp. Proc.* **726**, Q7.3.1 (2002).
17) O'Regan, B. and Grätzel, M. *Nature* **353**, 737 (2003)
18) S. Nakade, W. Kubo, T. Kitamura, Y. Wada, S. Yanagida *J. Phys. Chem. B* **107**, 8607 (2003)
19) Wang, Z-S.; Kawauchi, H.; Kashima,T. Arakawa, H. *Coord Chem. Rev.* **248**, 1381 (2004)
20) S.Madhugiri, B. Sun, P. Smirniotis, J.Ferraris' K. Balkus, Jr. *Micropor. Mesopor. Mater.* **69**, 77 (2004).
21) M. Zhang, Y. Bando, K. Wada, *J. Mater. Sci. Lett.*, **20**, 167 (2001).
22) L. Miao, S. Tanemura, S. Toh, K. Kanecko, M. Tanemura, *Appl. Surf. Sci.* **238,** 175 (2004).
23) B.B. Lakshmi, P.K.Dorhout, C.R. Martin, *Chem. Mater.* **9,** 857 (1997).
24) S. Limmer, T.L. Hubler, G. Cao, *J. Sol-Gel Sci Tech.*, **26,** 577 (2003).
25) D. Coutinho, R. Orozio, R. Reidy, K. Balkus Jr, *Micropor. Mesopor. Mater.* **54**, 229 (2002).
26) D.Zhao, J.Sun, Q.Li, G. D.Stucky, *Chem. Mater.* **12,** 275 (2000)
27) M. Macias, A. Chacko, K.J. Balkus, Jr. Submitted.
28) S. Madhugiri, A Dalton, J. Ferraris, K.J.Balkus, Jr., *J. Am. Chem. Soc.*, **125**, 14531 (2003).

Mater. Res. Soc. Symp. Proc. Vol. 836 © 2005 Materials Research Society L5.16

Bulk Heterojunction Photovoltaic Cells from Polymer Mixtures with Soluble Oxadiazole and Quinoline Polymers as Electron Acceptors

Solon Economopoulos[1,2], Christos L. Chochos[1,2], Giannis K. Govaris[1,3], Panagiotis Yiannoulis[3], Joannis K. Kallitsis[1,2] and Vasilis G. Gregoriou[1]*
[1] Foundation for Research and Technology Hellas Institute of Chemical Engineering and High Temperature Chemical Processes, (FORTH-ICEHT), P.O. Box 1414, Patras 26504, Greece.
[2] Department of Chemistry, University of Patras, Patras 26500 Greece.
[3] Department of Physics, University of Patras, Patras 26500 Greece.
* Corresponding author. Phone: (+30)2610-965205. Fax: (+30)2610-965223. E-mail: gregoriou@iceht.forth.gr

ABSTRACT

The interplay between phase separation in polymer blends consisting of the electron donating poly(3-hexylthiophene-2,5-diyl) (P3HT) or poly[2-methoxy-5-(2'-ethylhexyloxy)-,4-phenylenevinylene] (MEH-PPV), with either polyquinoline or a fluorene-oxadiazole copolymer as the electron accepting polymers is presented herein. The bulk heterojunction photovoltaic performance has been examined for these blends along with a new copoly(aryl ether) containing substituted anthracene and fluorene segments coupled with aromatic 1,3,4-oxadiazole moieties. Photoluminescence quenching in the MEH-PPV blends as well as a 100-fold photocurrent enhancement of the bulk heterojunction device P3HT/SDPQ photovoltaic device were observed compared to the single layer P3HT device. Finally, the structure and morphology of these films was investigated using atomic force microscopy and scanning electron microscopy in an attempt to correlate the role of morphology to photovoltaic performance.

INTRODUCTION

Since the discovery of electroluminescence in conjugated polymers [1], this class of materials has been used to build efficient light emitting diodes, field effect transistors, photodetectors and photovoltaic diodes. The discovery of photoinduced electron transfer in composites of conjugated polymers and buckminsterfullerene, C_{60} [2] provided an efficient molecular approach for improved photovoltaic conversion (up to 3%). In general, increased conversion efficiencies can be accomplished through improvement of the charge carrier generation, selective charges transportation and collection and light harvesting. Although polymeric photovoltaic devices have been mainly based on the C_{60} as the electron acceptor[3], the use of polymers as both the electron donor and acceptor offers many advantages. The energy gaps and ionization potentials of conjugated polymers can be finely tuned through chemical modification of the polymeric structure. Indeed, purely polymeric photovoltaic cells with high efficiencies have been reported, in which distributed heterojunctions on a nanometer scale provided effective dissociation of excitons.

Conjugated polymers containing quinoline or oxadiazole moieties are characterized by high thermal stability, outstanding mechanical properties, and can form optically clear films. Their electroluminescent and photoconductive properties have been extensively studied while their potential application in solar cells has been also demonstrated[4].

In this present study aiming at all polymeric solar cells, two widely used polymers have been employed as electron donors; regioregular P3HT (poly(3-hexylthiophene-2,5-diyl) and poly[2-methoxy-5-(2'-ethylhexyloxy)-,4-phenylenevinylene] (MEH-PPV) (Scheme 1). In addition, poly(6-(4-vinyl-phenyl)-2,4-diphenyl-quinoline) (SDPQ) (Scheme 1) and an oxadiazole-fluorene containing copolymer (PFOXD) (Scheme 1) were synthesized and employed as the electron accepting polymeric parts. The interconnection between morphology and charge transfer processes in these bulk heterojunction polymeric blends was extensively studied [5]. Furthermore, a copolymer incorporating both electron and hole transporting moieties (PAFOXD) (Scheme 1) has also been investigated aiming towards single layered photovoltaic architectures.

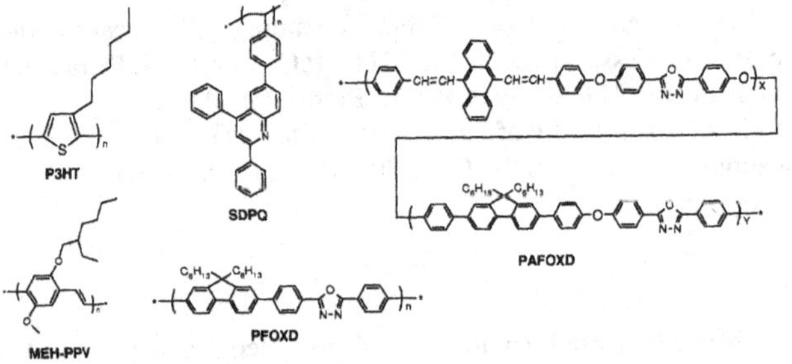

P3HT

SDPQ

PAFOXD

PFOXD

MEH-PPV

Scheme 1. Chemical structures of P3HT, MEH-PPV, SDPQ, PFOXD and the PAFOXD copolymer.

EXPERIMENTAL DETAILS

Films of pristine regioregular P3HT, SDPQ or PFOXD and blends of 1:2 weight ratio for the two materials were spin-coated from chloroform solution of 3.75 g/L concentration onto glass substrates for UV-Vis and PL measurements as well as for the morphological study of the polymers. Optical absorption spectra of the individual polymers and polymer blends were obtained by using a Hewlett Packard 8452A UV-vis spectrometer and corrected for substrate influences. Photoluminescence emission (PL) spectra of films on glass substrates were obtained by using a Perkin-Elmer LS45 spectrofluorimeter. Tapping-mode AFM was carried out using a Topometrix Explorer SPM Microscope and SEM images were obtained using a Jeol Scanning Microscope J.S.M. 6300.

P3HT and MEH-PPV were purchased from Aldrich and were used without further purification. The synthetic routes for SDPQ [6], PFOXD [7] and PAFOXD [8] can be found in the literature.

DISCUSSION

Optical properties

The solid state absorption and emission spectra for thin films of the polymers are depicted in Table I. The optical band gaps, *Eg*, of P3HT and SDPQ are 1,9 eV and 3,2 eV as estimated from the absorption onset wavelengths of the UV-Vis spectrum of the polymer film. Evidence of photoinduced charge transfer in polymer blends or heterojunction devices can be provided by PL quenching because once the singlet exciton has been dissociated it can no longer decay radiatively to the ground state. After photoexcitation, the exciton may undergo either charge transfer, which is the desired process in photovoltaic cells, or energy transfer as it reaches an interface between the two materials. The MEH-PPV blend exhibited partial quenching of about 15% at room temperature (Fig. 1), suggesting that some photogenerated excitons are dissociated at the polymer/polymer interface, leaving the electron on the SDPQ (electron acceptor) and the hole on the MEH-PPV (electron donor).

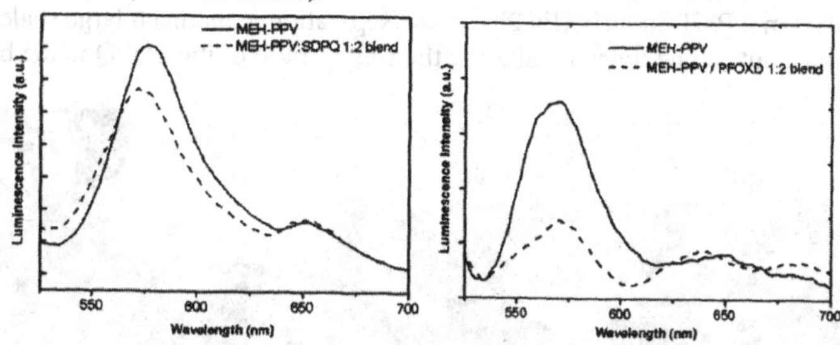

Figure 1 : Photoluminescence spectra of the MEH-PPV:SDPQ 1:2 blend (left) and of the MEH-PPV:PFOXD 1:2 blend (right).

The absorption spectra of the polymers PFOXD and PAFOXD are depicted in Table I. The poly-anthracene-fluorene-oxadiazole copolymer exhibits two well resolved peaks attributed to the π-π^* electronic transitions of the oxadiazole and the anthracene moieties respectively. From the optical spectra of the PFOXD and PAFOXD copolymers their energy gaps of 3.0 and 2.5 eV respectively were calculated.

Blends of PFOXD with both electron donating polymers were prepared and examined with regards to the photoluminescence quenching. Based on Figure 1, the MEH-PPV / PFOXD blend exhibits a significant but incomplete quenching of the PL intensity similar to the MEH-PPV / SDPQ, but in higher percentage. In the case of P3HT / PFOXD and P3HT / SDPQ blends, no quenching of the PL spectra were observed.

Table I. The spectral maxima of the materials in thin films, spin coated from chloroform solution.

	P3HT	MEH-PPV	SDPQ	PFOXD	PAFOXD (ODCB)
UV-Vis absorption (nm)	540, 605	500	295, 330	362	314, 422
Photoluminescence (nm)	680	575	415	428, 460, 507	568, 600

Morphological studies

It is well known that morphology plays an increasingly important role in the performance of the final device. Therefore, its understanding and subsequent control of it allows for improvements in the photovoltaic performance. Tapping-mode AFM was used in order to investigate the morphology of the blend films.

Figure 2 shows the atomic force microscopy AFM (2a) image of the surface of P3HT:SDPQ 1:2 (w/w) spin coated using chloroform solution. Two distinct phases were observed in the morphology of the blend. Spherical domains of average thickness 500 nm within a continuous matrix are visible, which is an indication of phase separation. In order to clarify the nature of the spherical particles the following SEM experiment was conducted.

A blend of P3HT:SDPQ 1:2 (w/w) was prepared and spin coated onto two glass substrates from chloroform solution. From selective dissolvation of SDPQ polymer using formic acid, the white spheres in the SEM image were identified as SDPQ rich regions with average diameter of 500nm dispersed in a P3HT matrix (fig 2b). Phase separation is on much large scale than optimum (10nm), but can be tuned by altering the weigth ratio of the SDPQ in the blend.

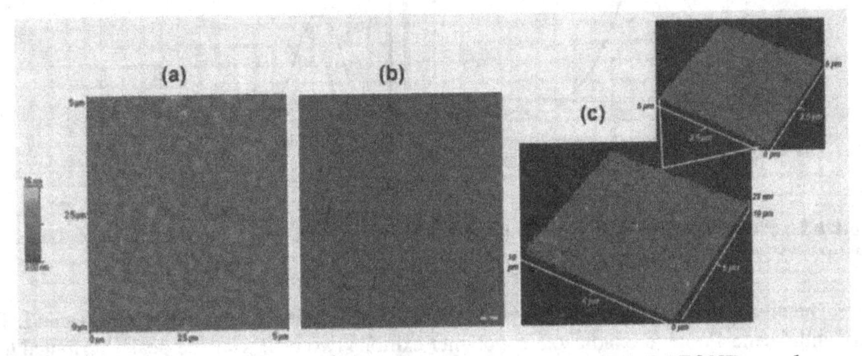

Figure 2. AFM images of the P3HT:SDPQ 1:2 blend 5x5μm area (2a), PAFOXD copolymer 10x10μm area (2c), and SEM image of the P3HT:SDPQ 1:2 blend (2b). The bar in the SEM image indicates a 500nm scale and the white spherical particles are SDPQ-rich regions in a P3HT matrix.

PAFOXD copolymers' morphological study as an active layer in an organic optoelectronic device revealed the formation of continuous "pathways" as depicted in Figure 2c. This morphology resembles a bicontinuous network, but no definitive region size can be determined from such a morphology.

<u>Photovoltaic devices.</u>

Polymer solar cells were fabricated by using either SDPQ or PFOXD as electron acceptors. The photovoltaic cell was constructed by sandwiching the photoactive mixture, constiting of P3HT or MEH-PPV and SDPQ or PFOXD in 1:2 weight ratio. The composite layer was spin coating from a chloroform solution of polymer blend on the surface of the anode. We have used high-work-function electrodes of indium tin oxide/poly(3,4-ethylene-dioxythiophene)-poly(styrene sulfonate) (ITO/PEDOT-PSS) to collect the holes and Al as low-work-function electrode for electron collection.

The current-voltage (*I-V*) characteristics of ITO/PEDOT-PSS/P3HT-SDPQ(1:2)/Al device with an active area of 0.04 cm^2 in the dark and under an irradiation with white light of 50 mW cm^{-2} are shown in figure 3a. A clear photovoltaic behavior can be seen, when illuminating the cell from the ITO side. The short circuit current (I_{sc}) is 28.51 μA cm^{-2} whereas the open circuit voltage (V_{oc}) is 0.44 V. The energy conversion efficiency is n=0.0044%. The P3HT:SDPQ bulk heterojunction solar cell has a 100-fold enhancement in energy conversion efficiency compared to the ITO/PEDOT-PSS/P3HT/Al device, which was tested under the same conditions.

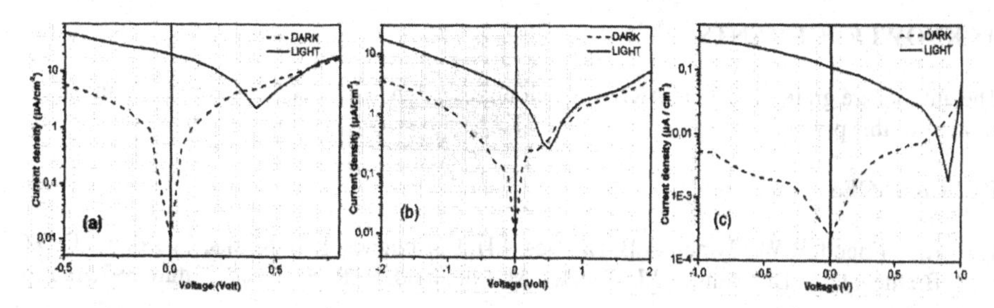

Figure 3. Current–voltage characteristics of P3HT:SDPQ (a), MEH-PPV:PFOXD (b) blends and the PAFOXD copolymer (c) devices in the dark (dash line) and under illumination (AM1.5, 50 $mWcm^{-2}$, solid line) in a semilogarithmic plot.

The current-voltage characteristics of ITO/PEDOT-PSS/PAFOXD/Al are presented in figure 3c. Clear photovoltaic behavior can be observed with a Isc = 0.11 $\mu A/cm^{-2}$ and an open cirquit voltage Voc = 0.89 V. Despite the low power conversion efficiency measured for the PAFOXD-based device, the high open circuit voltage (close to the theoretical upper limit of the device based on the MIM model) makes this material very promising candidate. Inefficient photon collection from the chromophore is the main reason for the low short circuit current observed. The ITO/PEDOT-PSS/MEH-PPV-PFOXD(1:2)/Al device was examined as well. The I-V curves plotted in a semilogarithmic scale can be seen in Figure 3b.

We note that processing of the photoactive layer and the I-V characterization of the devices were done in normal lab atmosphere. Since this may negatively affect the performance of the device, we believe that the energy conversion efficiency using these materials can be further enhanced.

CONCLUSIONS

In conclusion, we have prepared binary blends of the p-type conjugated polymers P3HT or MEH-PPV with n-type polymers SDPQ and PFOXD (an oxadiazole-fluorene containing polymer), and observed quenching of photoluminescence as an indication of photoinduced electron transfer from MEH-PPV to the electron acceptor in the blend. These blends along with a new copolymer, incorporating both electron donating and electron accepting units in the polymeric backbone (PAFOXD), have been examined towards their morphology and photovoltaic performance. Atomic force microscopy characterisation showed that the phase separated blend consisted of SDPQ rich spherical domains in a continuous P3HT matrix. The bulk heterojunction solar cell based on the above polymer blend shows a 100-fold enhancement of power conversion efficiency compared with the single P3HT device, but recombination of excitons on the conjugated polymer represents the primary mechanism by which efficiency is lost. The device based on the PAFOXD copolymer gave very interesting device characteristics reaching an open circuit voltage of 0.89V and providing great potential for optimization towards obtaining higher power conversion efficiencies. Further work is in progress to optimize the morphology of thin-film polymer composites for photovoltaic applications.

ACKNOWLEDGEMENTS

The authors are grateful to KONARKA technologies of Lowell, MA, USA for the financial support of this project.

REFERENCES

1. R.H. Friend, R.W. Gymer, A.B. Holmes, J.H. Burroughes, R.N. Marks, C. Taliani, D.D.C. Bradley, D.A. Dos Santos, J.L. Brédas, M. Lögdlund, W.R. Salaneck, Nature **397**, 121 (1999).
2. N.S. Sariciftci, L. Smilowitz, A. J. Heeger, F. Wudl, Science **258**, 1474 (1992)
3. M.M.Wienk, J.M. Kroon, W.J.H. Verhees, J. Knol, J.C. Hummelen, P.A. van Hal, R.A.J. Janssen, Angew. Chem. Int. Ed. **42**, 3371 (2003).
4. M. M. Alam, S. A. Jenekhe, Chem. Mater. **16**, 4647 (2004).
5. S.P. Economopoulos, G.K. Govaris, C.L. Chochos, N.P.Tzanetos, A.K. Andreopoulou, J.K. Kallitsis, P. Yianoulis, V.G. Gregoriou, Macromolecular Symposia **205**, 19 (2004)
6. S.P. Economopoulos, A.K. Andreopoulou, V.G. Gregoriou, J.K. Kallitsis, submitted to Chem. Mater.
7. X. Zhan, Y. Liu, X. Wu, S. Wang, D. Zhu, Macromolecules **35**, 2529 (2002).
8. C.L. Chochos, G.K. Govaris, F. Kakali, P. Yiannoulis, J.K. Kallitsis, V.G. Gregoriou, submitted to Polymer.

Mater. Res. Soc. Symp. Proc. Vol. 836 © 2005 Materials Research Society L5.19

Efficient Bulk Heterojunction Photovoltaic Cells using Sublimable Transition Metal Complex as Photosensitizers

Hei Ling Wong[1], Kitty Ka Yan Man[1], Wai Kin Chan[1]*, Chung Yin Kwong[2], Aleksandra B. Djurišić[3]
[1]Department of Chemistry, The University of Hong Kong, Pokfulam Road, Hong Kong
[2]Department of Electrical and Electronic Engineering, The University of Hong Kong, Pokfulam Road, Hong Kong
[3]Department of Physics, The University of Hong Kong, Pokfulam Road, Hong Kong

ABSTRACT

Multilayer bulk heterojunction photovoltaic cells using chlorotricarbonyl rhenium (I) bis(phenylimino)acenaphthene (Re-DIAN) complex as photosensitizer were studied. The complex is sublimable, has lower band gap, good thermal stability and good processibility. It has broad absorption in UV-visible region. Therefore, Re-DIAN exhibits good photosensitising property for photovoltaic cells. Multilayer bulk heterojunction photovoltaic cells with simple structures were fabricated base on Re-DIAN complex. The active layer consists of a blend of Re-DIAN and fullerene that were co-deposited in the same layer by vacuum deposition. The photovoltaic properties of the devices were studied by irradiaton under AM1.5 simulated solar light. The effects of changing the co-deposition film thickness, amount of Re-DIAN photosensitizer, and hole transporting materials were studied.

INTRODUCTION

Many scientists are increasingly interested in high power and efficient solar cells based on both small organic molecules and conjugated polymers. It is because of the advantages of organic materials such as low fabrication costs, lightweight, tunability of the electronic band gap by chemical synthesis, and simple fabrication into flexible and thin film panels. However, many studies showed that organic solar cells in general exhibit much lower efficiencies of energy conversion compared to current inorganic silicon solar cells. One of the major barriers to fabricate efficient organic photovoltaic cells is the insufficient exciton dissociation at low bias and the low charge carrier mobility. In order to design an efficient organic photovoltaic cell, the efficiency for dissociation of excitons and the charge transport process should be improved.

A typical organic photovoltaic device consists of multilayer organic thin films. The photovoltaic performances in multilayer bulk heterojunction devices can be greatly improved compared to that in multilayer heterojunction devices [1]. The bulk heterojunction consists of a blend of two materials, which form interpenetrating networks that one of the materials serve as the electron donor and the other one serve as the electron acceptor. After Re-DIAN is excited, it transfers an electron to C_{60} immediately so that a pair of free hole and electron is generated. Then, these charges are transported to the corresponding electrodes, which results in the generation of photocurrent.

EXPERIMENTAL DETAILS

Synthesis of photosensitizer

Chlorotricarbonyl rhenium (I) bis(phenylimino)acenaphthene (Re-DIAN) complex [2] was used as the photosensitizer to fabricate photovoltaic cells [3]. DIAN was synthesized according to literature procedure [4]. The synthesis of Re-DIAN was carried out in the following procedure. A mixture of DIAN (1 g, 1 mmol) and rhenium pentacarbonyl chloride (0.92 g, 1 mmol) in 40 ml toluene was refluxed at 140 °C under nitrogen atmosphere for overnight. The mixture was cooled to room temperature and filtered off. The solid was washed with toluene and dried in vacuum. The structure of the Re-DIAN is shown in Figure 1. The complex can be conveniently purified by sublimation under high vacuum (280 °C at 10^{-5} mbar) [1].

Fabrication of photovoltaic devices

Photovoltaic devices with multilayer bulk heterojunction structure ITO/CuPc/Re-DIAN:C_{60}/C_{60}/Al were fabricated by vacuum sublimation. Copper phthalocyanine (CuPc) and fullerene (C_{60}) were used as the hole and electron transport materials, respectively. The thickness of both CuPc and C_{60} layers were kept at 10 nm. The active layer was fabricated by co-deposition of Re-DIAN and C_{60}. C_{60} can capture electrons more efficiently once excitons are formed. The thickness of Re-DIAN:C_{60} mixed layer was varied from 25 to 100 nm. Aluminium (thickness = 40 nm) was used as the cathode. The photovoltaic properties of the devices were studied by irradation under AM 1.5 simulated solar light (100 mW/cm^2) .The device active area is 0.13 cm^2.

RESULTS AND DISCUSSION

Photovoltaic performances with different thickness of co-deposition film and amount of Re-DIAN photosensitizer in mixed layer in the device structure ITO/CuPc(10 nm)/Re-DIAN:C_{60} [1:1] (25, 50, 75, 100 nm)/C_{60}(10 nm)/Al(40 nm) were studied. Their current-voltage (*I-V*) characteristics are summarized in Table I. When the thickness of the mixed layer was 25 nm, highest power conversion efficiency (η_p) was observed in the device with 30% photosensitizer in mixed layer. For devices with active layer thickness 50 nm, 75 nm and 100 nm, maxima η_p were observed when the amount of photosensitizer were 50%. These results clearly show that the amount of photosensitizer and electron transport molecules may strongly affect the balance between the photon absorption, exciton formation, dissociation, and charge transport processes. In order to improve the devices performance, determining exciton diffusion length and optimising the thickness of each layer should be achieved.

Table I. Effect of different thickness of mixed layer and amount of photosensitizer in mixed layer in device structure ITO/CuPc(10 nm)/Re-DIAN:C_{60}(25, 50, 75, 100 nm)/C_{60}(10 nm)/Al(40 nm) irradiated with AM1.5 simulated solar light.

Thickness of mixed layer (nm)	Amount of photosensitizer in mixed layer (%)	I_{sc} [a] (mA/cm^2)	V_{oc} [b] (V)	FF [c]	η_p [d] (%)
25	10	4.13	0.37	0.49	0.74
	30	4.03	0.42	0.45	0.76
	50	3.02	0.47	0.33	0.46
50	10	2.67	0.37	0.40	0.40
	30	2.72	0.44	0.52	0.62
	50	5.07	0.51	0.51	1.29
75	10	1.48	0.30	0.39	0.17
	30	2.93	0.46	0.43	0.58
	50	4.34	0.42	0.42	0.76
100	10	2.29	0.30	0.46	0.31
	30	0.03	0.30	0.12	0.001
	50	3.23	0.46	0.32	0.47

[a] Short circuit current
[b] Open circuit voltage
[c] Fill factor = $(I_{max} \times V_{max})/(I_{sc} \times V_{oc})$, where I_{max} and V_{max} corresponding to the point of maximum power output
[d] Power conversion efficiency = $(I_{sc} \times V_{oc} \times FF)/P$, where P is the incident light power.

The effect of changing the hole transporting materials (HTMs) is summaried in Table II. The structure of multilayer bulk heterojunction device is ITO/HTM(10 nm)/Re-DIAN:C_{60}[1:1](50 nm)/C_{60}(10 nm)/Al(60 nm). When (N,N'-di(naphthalene-1-yl)-N,N'-diphenyl-benzidine) (NPB), 1,1'-bis[(di-4-tolylamino)phenyl]cyclohexane (TAPC), N,N'-diphenyl-N,N'-bis(3-methylphenyl)(1,1'-biphenyl)-4,4'-diamine (TPD) and 4,4',N,N'-dicarbazolyo biphenyl (CBP) were used as HTMs, the η_p of these devices were 0.08%, 0.05%, 0.03%, and 0.001%, respectively. It is well-known that the oxidation potential or ionization potential of HTM can be varied by modifying the chemical structures. The ionization potentials of HTMs reported in the literature are: NPB (5.1-5.4 eV) [5], TAPC (5.4 eV) [5], TPD (5.3-5.5 eV) [5], CBP (6.0-6.3 eV) [6-7], CuPc(5.0 eV) [8], ZnPc(4.7 eV) [8]. Thus, different HTMs will yield different energy barriers at ITO/HTM interface [5]. Also, the hole carrier mobility may strongly affect the photovoltaic performance. When copper phthalocyanine (CuPc) and zinc phthalocyanine (ZnPc) were used as HTMs, the η_p were 1.29% and 0.59%, respectively. These two photovoltaic cells show greatly improved performance compared to the devices mentioned above. The metallophthalocyanines may strongly affect the photon absorption and exciton diffusion in photovoltaic devices.

The external quantum efficiency (η_q) is given by the ratio of the number of electrons flowing through external circuit to the number of incident photons. The η_q is given by $\eta_q = (1240 \times J_\lambda[\mu A/cm^2]) / (\lambda[nm] \times P_\lambda[W/m^2])$. This shows that the quantum efficiency can be calculated from a measurement of the short circuit current at a given wavelength with input radiant power

P_{λ} [9]. Figures 1 and 2 show the absorbance and η_q of Device 3 and 5. Figure 1 shows the photocurrent response of device structure ITO / TPD(10 nm)/ Re-DIAN:C_{60}[1:1](50 nm) / C_{60}(10 nm) / Al(60 nm) at different wavelength. The absorption spectrum of the device is shown for comparison. This device has the maximum η_q (1.9%) at 510 nm. The photocurrent action spectrum agrees well with the corresponding Re-DIAN absorption spectrum. The UV-visible absorption spectrum of Re-DIAN thin film shows broad absorption band that covers most of the visible region in the range between 430 and 600 nm. The maximum absorption peak at around 500 nm is assigned to be the MLCT [d_{π}(Re) to π*(DIAN)] electronic transition. Therefore, Re-DIAN may be a promising candidate as photosensitizer in photovoltaic cells. Figure 2 shows the photocurrent response of the device structure ITO / CuPc(10 nm) / Re-DIAN:C_{60}[1:1] (50 nm) / C_{60}(10 nm) / Al(60 nm) at different wavelength. This device has the maximum η_q (25%) at 620 nm. The photocurrent action spectrum agrees well with the corresponding CuPc absorption spectrum. This result clearly shows that the CuPc:C_{60} junction also contributes to the photovoltaic process.

Table II. Effect of different hole transporting materials (HTMs) in device structure ITO/HTM(10 nm)/Re-DIAN:C_{60}[1:1](50 nm)/C_{60}(10 nm)/Al(60 nm) irradiated with AM1.5 simulated solar light.

Device	HTM	I_{sc} (uA/cm^2)	V_{oc} (V)	*FF*	η_p (%)	η_q at 510nm (%)
1	NPB	384	0.72	0.29	0.08	3.50
2	TAPC	297	0.58	0.31	0.05	1.52
3	TPD	203	0.48	0.36	0.03	1.89
4	CBP	14	0.34	0.26	0.001	0.45
5	CuPc	5067	0.51	0.51	1.29	5.42
6	ZnPc	2412	0.48	0.51	0.59	4.31

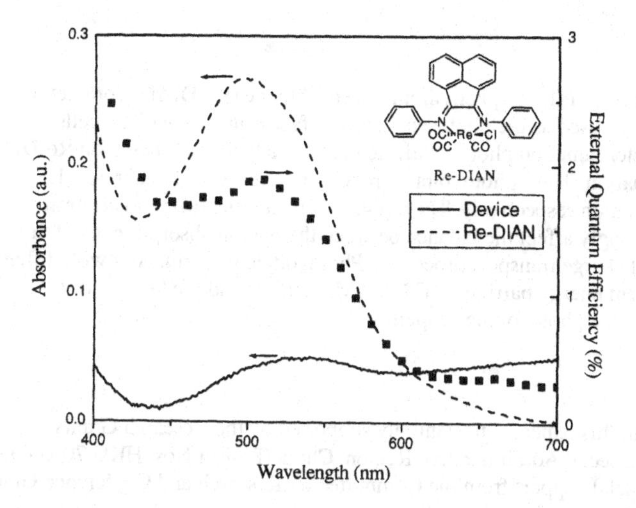

Figure 1. Structure of Re-DIAN and plot of external quantum efficiency and the absorbance of the device ITO/TPD(10 nm)/Re-DIAN:C_{60}[1:1](50 nm)/C_{60}(10 nm)/Al(60 nm) and Re-DIAN thin film at different wavelength.

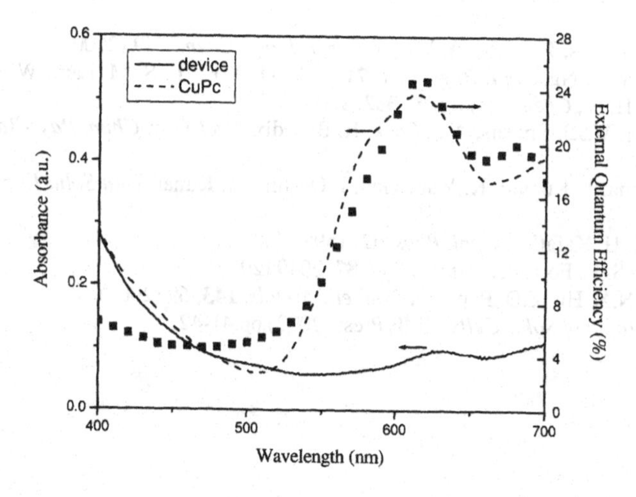

Figure 2. Plot of external quantum efficiency and the absorbance of the device ITO/CuPc(10 nm)/Re-DIAN:C_{60}[1:1](50 nm)/C_{60}(10 nm)/Al(60 nm) and CuPc thin film at different wavelength.

CONCLUSIONS

Chlorotricarbonyl rhenium (I) bis(phenylimino)acenaphthene (Re-DIAN) complex was synthesized. The complex was used as the photosensitizers to fabricate photovoltaic cells. Efficient multilayer bulk heterojunction photovoltaic devices were fabricated based on Re-DIAN complex by vacuum sublimation. Their photovoltaic properties were studied under AM 1.5 simulated solar light illumination respectively. The amount of photosensitizer and electron transport molecules may strongly affect the balance between the photon absorption, exciton formation, dissociation, and charge transport processes. Photovoltaic performances with different HTMs were studied. Different energy barriers at ITO/HTM interface and the hole carrier mobility may strongly affect the photovoltaic properties.

ACKNOWLEDGMENTS

The work reported in this paper is substantially supported by the Research Grants Council of the Hong Kong Special Administrative Region, China (Project Nos. HKU 7009/03P, and 7008/04P). Partial financial support from the Committee on Research and Conference Grants (University of Hong Kong) is also acknowledged.

REFERENCES

1. H.L. Wong, L.S.M. Lam, K.W. Cheng, K.Y.K. Man, W.K. Chan, *Appl. Phys. Lett.* **84**, 2557 (2004).
2. G. Knör, M. Leirer, T. E. Keyes, J. G. Vos, A. Vogler, *Eur. J. Inorg. Chem.* 749 (2000).
3. W. K. Chan, X. Gong, W. Y. Ng, *Appl. Phys. Lett.* **71**, 2919 (1997); L. S. M. Lam, W. K. Chan, A. B. Djurisic, E. H. Li, *Chem. Phys. Lett.* **362**, 130 (2002).
4. R.V. Asselt, C.J. Elsevier, W.J.J. Smeets, A.L. Spek, R. Benedix, *Recl Trav. Chim. Pays-Bas.* **113**, 88 (1994).
5. L.Zhu, H. Tang, Y. Harima, Y. Kunugi, K. Yamashita, J. Ohshita, A. Kunai, *Thin Solid Films.* **396**, 213 (2001).
6. A. J. Makinen, I. G. Hill, H. Kafafi, *J. Appl. Phys.* **92**, 1598 (2002).
7. C. Adachi, M. A. Baldo, S. R. Forrest, *J. Appl. Phys.* **87**, 8049 (2000).
8. G. Vamvounis, H. Aziz, N.X. Hu, Z.D. Popovic, *Synthetic Metals*, **143**, 69 (2004).
9. G.P. Smestad, *Optoelectronic of Solar Cells* (SPIE Press, 2002) pp.41-42.

Mater. Res. Soc. Symp. Proc. Vol. 836 © 2005 Materials Research Society L5.34

Dendritic CuIn Films Grown by Electroless Deposition

David W Lane, Jonathan D Painter, Keith D Rogers, Ian Forbes*, Robert W Miles*, Kathleen M Hynes*
Centre for Materials Science and Engineering, Cranfield University at Shrivenham, Swindon, Wiltshire, SN6 8LA, UK
* Northumbria Photovoltaics Applications Centre, Northumbria University, Newcastle upon Tyne, NE1 8ST, UK

ABSTRACT

$CuInS_2$ films are often grown in a two-step process with the deposition of a Cu-In alloy followed by high temperature sulphurisation in either S vapour or H_2S. Numerous techniques exist for the deposition of Cu-In. In this work Cu-In films have been deposited on tin oxide coated glass using the electroless deposition technique, a low cost, low temperature approach. The films were found to consist of a two layered structure. Initially, Cu islands grew forming a layer on the substrate. The deposition process subsequently produced dendritic Cu-In alloys from the tops of the Cu islands. Various CuIn phases were observed, namely Cu_9In_4 and CuIn, the presence of which was related to the solution pH. None of the films were found to contain metallic In, indicating that this process may be well suited to subsequent high temperature sulphurisation.

INTRODUCTION

The drive for low cost photovoltaic devices continues and Si, CdTe and $CuInSe_2$ are among the materials actively used in commercial photovoltaic devices. There are however, environmental concerns regarding the production and use of these materials and consequently alternatives are continually being investigated. Relatively recently $CuInS_2$ has re-emerged as a possible absorber material for photovoltaic solar cells, as its band gap of 1.52 eV is well matched to the solar spectrum. Thin film $CuInS_2$ can be grown by several different processes, these include its direct growth using techniques such as spray pyrolysis and its growth by a two stage process in which a precursor layer of $CuIn_X$ alloy, grown by processes such as sputtering or co-evaporation, is converted to $CuInS_2$ by a high temperature reaction with either S vapour or H_2S. In this paper we present some results for precursor layers grown by the electroless method. This is a low-cost, convenient deposition method that has been previously reported to successfully deposit $CuIn_X$ films [1,2,3].

The conditions for electroless deposition of $CuIn_X$ require a conducting substrate (e.g. SnO_2 coated glass) that is electrically connected to an easily oxidizable redox component (e.g. Al). Both electrodes are then immersed in the chemical solution containing Cu^{2+} and In^{3+} ions and an electrochemical potential difference is generated between the two electrodes. These ions are then reduced at the conducting substrate surface to form the Cu-In alloy film. The electroless deposition of specific phases of Cu-In alloy requires careful complexing of the ions because of the large difference between the standard potentials for the reduction of Cu^{2+} and In^{3+}.

In our work we have specifically aimed to use the electroless technique to produce precursor layers with a highly textured surface, for subsequent conversion into $CuInS_2$ layers that retain this texture to efficiently trap light. In addition, we have employed chemical solutions with a relatively low concentration in order to provide better utilization of materials and consequently have less environmental impact. This paper concentrates on the growth of such textured $CuIn_x$ layers and their characterization by X-ray diffraction (XRD) and scanning electron microscopy (SEM) with energy dispersive X-ray analysis (EDX).

EXPERIMENTAL

A range of aqueous solutions was prepared using $CuCl_2$ (4 mM), $InCl_3$ (4 mM), 0.75 volume % NH_4OH aqueous solution[♦] and 1.0 volume % triethanolamine, which was included as a complexing agent. The pH of each solution was set by the addition of HCl. SnO_2 coated glass substrates were cleaned by soaking in a 5% Decon 90 solution, followed by thorough rinsing in distilled water followed by acetone. They were then blown dry with nitrogen. During deposition the Cu-In solution was held in a Pyrex jacketed beaker and was maintained at 45 °C by a circulating water bath. A cleaned substrate was then placed in the solution with its SnO_2 coated side facing a length of Al wire that had also been cleaned in 5% Decon 90 solution and rinsed thoroughly. This was used as the redox electrode and was electrically connected to the substrate to initiate deposition, which lasted typically 2 hours.

The deposited layers were analyzed by XRD and SEM/EDX. X-ray powder diffraction spectra were collected on a Philips 1820 diffractometer using CuKα radiation and a diffracted beam monochromator. The samples were continually rotated as data was acquired over a 2θ range of 20-80° in 0.02° steps. A JEOL scanning electron microscope was used to study the structure of the films, with energy dispersive X-ray analysis used to aid the identification of specific elements and phases.

RESULTS AND DISCUSSION

Film growth generally occurred quickly, with a pink deposit clearly visible on the substrates within a few seconds. This was identified as a fine-grained, pure Cu powder using SEM/EDX, and was noted to be present on all films, see figures 1a and 1c. Continued growth resulted in the formation of a layer with a gray/black appearance, which SEM examination revealed to be constructed from complex dendritic structures that appeared to grow out from individual Cu grains, see figures 1b and 1d. Analysis of these structures by EDX indicated the growth of a Cu-In alloy. Whilst these layers appear to have a very similar structure, there does appear to be some statistically significant[♣] differences in their feature size, see table I. The average size of the Cu grains in the initial layer of the pH 1.15 film appears to be smaller than that of the pH 1.3 film, and the resulting dendritic growths appear to be longer and thinner for the film deposited from the pH 1.15 solution compared to that for the pH 1.3 film, although it was difficult to measure the feature size of the dendritic growths due to their overlapping complex structure.

[♦] Standard 30 volume % aqueous solution.

[♣] As determined by Student's T-test, $p < 0.1\%$.

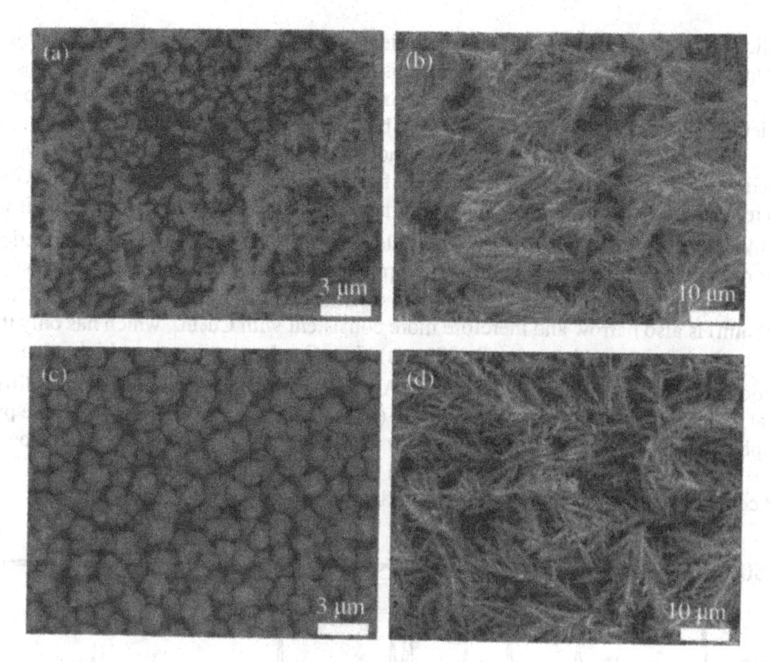

Figure 1. SEM micrographs of film growth. Solution of pH 1.15: (a) initial layer of pink growth, (b) gray dendritic Cu-In top layer. Solution pH 1.3: (c) initial layer of pink growth, (d) gray dendritic Cu-In top layer.

Table I. Average feature sizes for the films shown in Figure 1

	Solution pH	
	1.15	**1.3**
Cu grain diameter (µm)	0.6 (±0.1)	1.3 (±0.3)
Width of dendrite (µm)	0.28 (±0.05)	0.35 (±0.05)
Dendrite length (µm)	16 (±4)	10 (±4)

Figure 2 compares the XRD patterns from dendritic layers grown from solutions with different pH. These exhibit similar features, although those grown at a higher pH appeared to have additional lines indicating the presence of at least one extra phase. These XRD patterns were analyzed by comparison to the ICDDS files for low temperature Cu-In phases. These were Cu (4-836), In (5-642), Cu_9In_4 (42-1476), $Cu_{11}In_9$ (41-0883) and CuIn (35-1150), and $CuIn_2$ from the lattice constants and symmetry group of Keppner et al [4]. SnO_2 (21-1250) was also included, as it was known to be present in the substrate. Table II lists the observed peaks for the dendritic layer grown at a pH of 1.3. The low temperature region of the Cu-In phase diagram indicates the possibility of several different phases, however the analysis of the XRD patterns is complicated by the possibility of preferred orientation and the phases having crystallographic structures that result in very similar d-spacings. A general comparison of the peak positions for all four specimens indicated the presence of SnO_2 from the substrate, Cu possibly from the initial pink deposit and at least one more phase. We believe that all four specimens mainly contained Cu_9In_4. This phase exhibits its strongest diffraction peak from the (411) reflection, which has a d-spacing of 0.2144 nm, accounting for the diffraction peak at a 2θ angle of 42.2°. Whilst the (511) reflection of $Cu_{11}In_9$ (d-spacing 0.2145 nm) would also account for this peak, it would most likely possess strong diffraction peaks close-by from its $(11\bar{2})$ and (313) reflections at angles of 41.5° and 42.3° respectively. The limited width of the observed line suggests that these peaks are not present. Similarly, the observed diffraction peak at a 2θ angle of 29.5° (d-spacing 0.3025 nm) is also narrow and therefore more consistent with Cu_9In_4, which has only the (300) reflection in this region (d-spacing 0.3036 nm), than $Cu_{11}In_9$ that has three high intensity reflections in this area ((311), $(11\bar{1})$, (402)). All four spectra also exhibited a small diffraction peak at 77.0°, which is only accounted for by (522) refection of Cu_9In_4. However the presence of other phases cannot be ruled out, as the minor peak observed at 55.8° could result from either $Cu_{11}In_9$ or $CuIn_2$ and the slight broadening on the bottom of the diffraction peak at a 2θ angle of 42.2° could be caused by the three $Cu_{11}In_9$ reflections in this area.

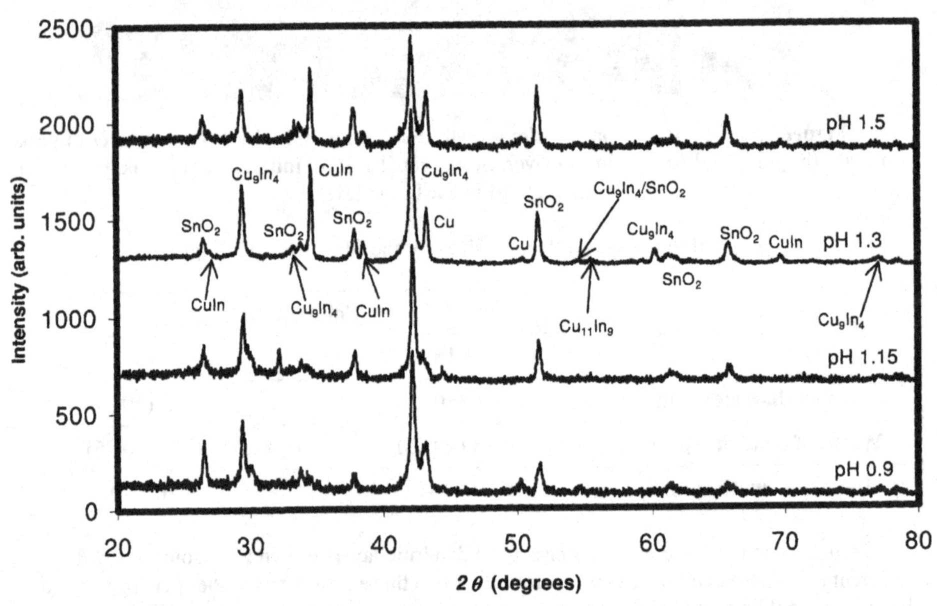

Figure 2. XRD spectra for films grown from different pH solutions.

Table II includes five lines that are only present on the two higher pH specimens. These are consistent with the presence of CuIn, showing that there is a significant change in chemistry between pHs of 1.15 and 1.3. The film deposited at pH 1.15 also appears to have less Cu than the other layers; possibly because deposition began with a thinner Cu layer, see Figure 1a. In was not detected in any of the layers, in contrast to the previously published work on a similar electroless Cu-In film grown by Calixto et al [2]. Their Cu-In layer was deposited from a solution with $CuCl_2$ and $InCl_3$ concentrations of 25 mM and a pH of 1.5 and was interpreted to be largely $Cu_{11}In_9$. However, we believe there are too many peaks absent from their diffraction pattern for this phase to be reliably identified.

Table II. Observed diffraction peaks and d-spacings for the film deposited at pH 1.3 compared to major reflections (intensity > 10%) for expected compounds (* - peak not detected for lower pH films, a – close but minor reflections with total intensity ≥ 10%).

Observed peaks		d-spacing of expected compounds (nm)					
2θ (degrees)	d-spacing (nm)	SnO_2	Cu	Cu_9In_4	$Cu_{11}In_9$	CuIn	$CuIn_2$
26.53	0.3358	0.3351					
27.03*	0.3297					0.3312	0.3323
29.50	0.3025			0.3036	0.3016 0.3034 0.3052		
33.26	0.2692				0.2716		0.2688
33.81	0.2649	0.2644		0.2624			
34.62*	0.2589				0.2607	0.2596	0.2601
37.91	0.2372			0.2434	0.2381		
38.57*	0.2333	0.2369				0.2347	0.2333 0.2349
42.22	0.2139			0.2144	0.2136 0.2145 0.2177		0.2101
43.31	0.2087		0.2088	1.9411	0.2001	0.2099	0.2090
50.45	0.1807		0.1808	0.1857			
51.64	0.1769	0.1765		0.1786	0.1765		0.1769 0.1743
54.57	0.1680	0.1675		0.1680			
55.77	0.1647				0.1646		0.1661 0.1656
59.23	0.1559			0.1517		0.1563[a] 0.1561[a]	
60.25*	0.1535				0.1525	0.1542	
61.43	0.1508	0.1498			0.1509		
65.76	0.1419	0.1415					
69.71*	0.1348			0.1341 0.1314 0.1286	0.1350 0.1372	0.1370[a] 0.1367[a]	
77.00	0.1237			0.1238			
78.63	0.1216	0.1215					

Electroless deposition is a low cost production technique with a low energy requirement. Environmental hazard assessment has been performed on this deposition technique and the four main materials involved have hazardous properties that must be controlled. Care must be taken as $CuCl_2$ is classified as an irritant and as harmful by the EU, and solutions may be corrosive. There is little information available on $InCl_3$, however, toxicity data and occupational exposure limits exist which would suggest toxic properties. NH_4OH has corrosive properties and the disposal of waste material is licensed, as it is highly toxic in the aquatic environment. When triethanolamine is heated it gives off a vapour that is an irritant to the respiratory tract, eyes and skin and may cause dermatitis. As the process is running at 45°C, care must be taken with aerosols produced above the solution and adequate ventilation is essential. These hazardous properties, however, will be adequately controlled using regular health and safety procedures.

CONCLUSIONS

We have successfully grown highly textured Cu-In alloy films using an electroless process from chemical solutions with a relatively low concentration. Film growth was found to commence with the formation of a thin layer of Cu particles with an average diameter of 0.6 to 1.3 μm depending on solution pH. These appeared to 'seed' the subsequent growth of Cu-In alloy dendrites. The dimensions of the dendrites appeared to be dependant on the pH of the chemical solution, and XRD indicated that they were constructed primarily from Cu_9In_4. This is in contrast to other published work that appeared to show that $Cu_{11}In_9$ was the dominant phase, possibly indicating that solution concentration may affect the structure of the films. This was found when varying the solution pH, with Cu and Cu_9In_4 produced at pH 0.9 and 1.15, and Cu, Cu_9In_4 and CuIn formed at pH 1.3 and 1.5, with the general structure of layers unaffected by the presence of the additional phase. In contrast to other published work none of the films appeared to contain metallic In. This may indicate that films deposited from more dilute solutions are better suited to subsequent conversion to $CuInS_2$ by high temperature sulphurisation as they are more likely to retain their texture, beneficial for trapping light in photovoltaic devices. In addition, an electroless process with low concentration solutions would provide better utilization of materials, which is particularly important with In-based compounds where resources are limited. Adherence to correct working procedures would ensure that the environmental impact would be negligible taking into consideration the hazardous nature of some processing materials.

ACKNOWLEDGEMENTS

We gratefully acknowledge the financial support of the EPSRC through grants GR/N03709/01 and GR/N04645/01.

REFERENCES

1. A. Gupta, A. S. N. Murthy, *J. Mater. Sci. Let.* **8** 559 (1989).
2. M. E. Calixto, P. J. Sebastian, *J. Mat. Sci.* **33** 339 (1998).
3. A. Gupta, A. S. N. Murthy, *J. Mat. Chem.* **16** 929 (1991).
4. W. Keppner, R. Wesche, T. Klas, J. Voigt, G. Schatz, *Thin Solid Films* **143** 201 (1986).

Mater. Res. Soc. Symp. Proc. Vol. 836 © 2005 Materials Research Society L5.35

Bath pH Dependence on the Structural and Optical Properties of Chemical Bath Deposited CdS Thin Films

Udaya S. Ketipearachchi, David W. Lane, Keith D. Rogers, Jonathan D. Painter, Michael A. Cousins
Center for Materials and Engineering, Cranfield University, RMCS, Shrivenham, Swindon, SN6 6LA, UK

ABSTRACT

High quality CdS thin films made with ultrasonic agitation during chemical bath deposition were investigated. The change in band gap and variation of atomic stacking during the film growth was observed as a function of bath pH. The band gap of as-deposited thin films of CdS was found to be between 2.26 eV and 2.50 eV, the highest being observed when the pH was 9.0. The transition from cubic (β-CdS) to hexagonal (α-CdS) was observed with decreasing pH. The lattice parameters of $CdCl_2$ treated CdS were found to be a = 0.414 nm and c = 0.672 nm. Calculations based on the Sherrer formula showed significant grain growth after annealing with $CdCl_2$. No apparent effect of ultrasonication on crystalline structure of CdS was seen in this method, although ultrasonication was noted to produce films with a higher quality optical surface. A maximum bath temperature was fixed at 70°C in order to suppress the rate of homogeneous reaction and minimise the evaporation of ammonia from the chemical bath.

INTRODUCTION

The use of polycrystalline thin films in the solar cell industry has become an attractive alternative to the conventional silicon solar cell process due to its low production cost. Absorber layers based on $CuInSe_2$ and CdTe have been reported to have the highest efficiencies of 18.8%[1] and 16.8 %[2] respectively. This has focused more research on this area over past few years. In these devices, an n-type window layer having a good transparency for visible light is used to make a junction with a p-type optical absorber layer. Cadmium sulphide was found to be a promising window layer for the fabrication of high efficiency thin film solar cells with $CuInSe_2$ [1], CdTe [2] and Cu_2S [3]. In addition to applications for thin films of CdS in photovoltaic devices, their application for photoconductive sensors and many other optical devices is also documented. There are a number of different approaches for the fabrication of thin films of CdS such as chemical bath deposition (CBD), electrodeposition, closed spaced sublimation (CSS), spray pyrolysis, successive ionic layer adsorption and reaction (SILAR), molecular beam epitaxy, etc. However, CBD CdS is attractive due to its low cost, large area and low temperature fabrication process. The role of pH on the structural and optical properties of CBD CdS thin films is not clear due to its variation with the concentration of species in the solution and also the lack of information in the literature in this regard. Therefore, the work in this communication highlights the effect of bath pH on some important aspects of the film. In comparison to published data [4], it has also been seen that the role of pH on CBD CdS is strongly dependent on solution chemistry and deposition conditions.

Difficulties have been reported in growing thick films of CdS by CBD [5]. Film peeling and the formation of a powdery CBD CdS film surface have been observed [6]. Film peeling could be due to the increase of stress with the increase of film thickness. The intrinsic stress level in the film can affect not only the mechanical properties of the films but also other bulk properties such as electrical and optical characteristics. In addition to this, film discontinuities and formation of pinholes are also reported [7]. Nevertheless, good quality thin films of CBD CdS have been reported, but using Cd^{2+} ion concentration as high as 1M [4].

Cd is a highly toxic compound in the laboratory with a human exposure limit of 0.025 mg m^{-3}. Therefore it is always advisable to minimise its use in the laboratory and industrial plants. This minimises not only the hazard but also the production cost. However, the range of $CdCl_2$ concentrations used in CBD CdS is large and in the range of 0.001-1 M [4,6,8,9].

Moreover, it is known that the homogeneous process of CBD CdS deposition yields a powdery and non-adherent film [9]. Therefore, suppressing the homogeneous process by lowering the temperature and reaction rate may yield good quality films. However, high quality films have been obtained while carrying out the reaction in a specially designed ultrasonicator [10], and this method would be advantageous if the ultrasonicator is chemically inert. In our investigation, we have successfully developed a method to produce high quality thin solid films of CBD CdS using a commercially available ultrasonic cleaner while keeping the $CdCl_2$ concentration low as 0.0024M. In this method the reactions are carried out in a glass vessel and the chemical components have no contact with ultrasonic cleaner.

EXPERIMENTAL

Thin films of CBD CdS were grown on both the conducting and non-conducting sides of indium tin oxide (ITO) glass substrates. The substrates were thoroughly cleaned before CdS deposition by rinsing sequentially with Decon 90, de-ionized water (resistivity=15MΩ), acetone and de-ionized water. A solution containing cadmium chloride (0.0024M), ammonium chloride (0.028M), thiourea (0.057M) and de-ionized water was transferred into a glass jar having a double walled water jacket, which was placed in the centre of an ultrasonic cleaner. The bath pH was adjusted by the addition of ammonia solution (22%). Analytical grade chemicals with purity over 99.9% were used. The substrates were mounted on a glass holder and were immersed in the solution. The solution was continuously stirred at a constant speed during the film deposition. A constant bath temperature of 70 (±2) °C was maintained using a thermostat. This temperature was maintained in order to minimise the evaporation of ammonia from the chemical bath. The experiments were carried out in the pH range of 8.3 to 11.3 with and without ultrasonic agitation. The CdS films prepared without agitation were cleaned ultrasonically after the deposition. The annealing process of thin film CBD CdS was carried out in air, with and without $CdCl_2$. X-ray diffraction (XRD) analysis was carried using a Philips powder diffractometer with CuK_α radiation. The incident angle was kept fixed at 3° in order to minimise the strong reflections from the substrate underneath. The data was collected over 2-theta range of 20-70°. In this investigation a comparative study was also undertaken with some residue that was found in the chemical bath after the film deposition.

Scanning electron microscope and XRD were employed to estimate the grain size of CdS. In the case of XRD, the "grain size" was estimated using Scherrer formula, $t = \frac{0.9\lambda}{B\cos\theta_B}$ Where t is the grain size, B is the full width at the half-maximum (FWHM) corrected for instrumental broadening , λ is the wavelength of the incident X-rays, θ_B is the Bragg angle. This assumes there is negligible microstrain present. The grain size measurements were performed using a incident angle (3°) applying to Scherrer equation to the (002) peak.

In common with a number of polycrystalline materials, CdS exhibits sub-gains within the radius of the larger grains visible by, for example, SEM. The Scherrer formula may give relative the size of these sub-grains.

RESULTS AND DISCUSSION

Structural characterisation

The literature contains a large amount of information for different phases of CdS. Among them, there are a number of claims either for cubic (β-CdS) [4,5,11,14] or hexagonal (α-CdS) [6,8,12]. However, a few communications also appear concerning the polytype or mixed phase structure of CdS [13,15].

For the films, a change from cubic CdS to hexagonal was seen with decreasing bath pH (see figure 1). However, it was found that the (200) peak associated with fig. 1(c) was always absent. The XRD analysis of the residue collected from the bath at pH=10.2 gave a shoulder instead of a sharp (200) maxima (see figure 2). This data suggests that preferred orientation is not responsible for the absence of the (200). This pattern is also similar to the simulation pattern that was obtained by Gibson et al [15] for polytype (predominantly cubic) CdS. Thus, the observed results suggest the formation of polytype CdS during the stacking changing from cubic to hexagonal with decreasing pH. However, the hexagonal phase with all the characteristic peaks including the (102) reflection and cubic phase including (200) reflection appear when the bath pH is 8.5 and 10.5 respectively.

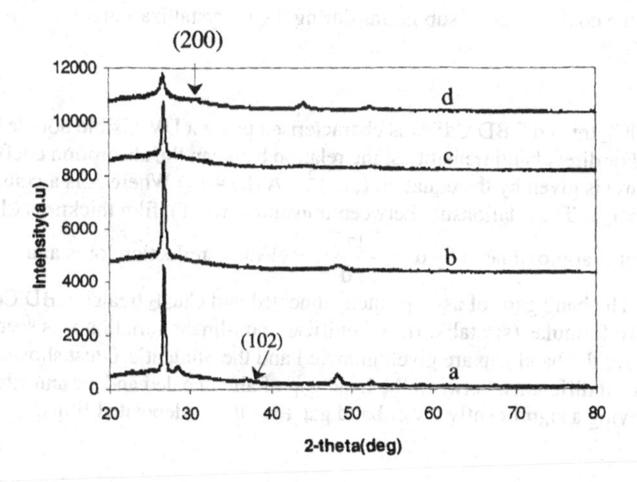

Figure 1. XRD pattern of as deposited CBD CdS on glass substrates (a) pH=8.5, (b) pH 9.0, (c) pH=10.2, (d) pH=10.5.

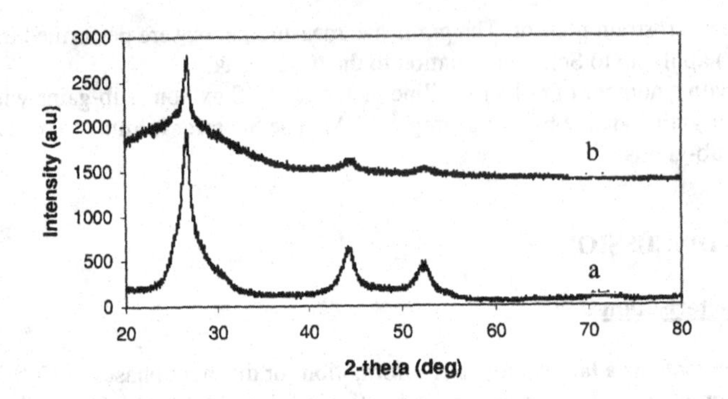

Figure 2. (a) XRD pattern of the residue found in the chemical bath after film deposition at pH=10.2 (b) XRD pattern of 200nm thick CdS film deposited at pH of 10.2.

A hexagonal phase was seen after $CdCl_2$ heat treatment of as-deposited CdS regardless of the pH, initial phase or substrate type. The XRD pattern of $CdCl_2$ treated CdS was investigated using a whole pattern fitting approach, and a good match between experimental and simulated patterns was obtained for $CdCl_2$ treated CdS with lattice constants a= 0.414nm and c=0.672nm for the hexagonal phase. A decrease in peak width and an increase in peak height were seen after $CdCl_2$ treatments. These observations confirm the occurrence of grain growth and an increase in crystallinity during the $CdCl_2$ treatment process. The calculated grain size along <001> of $CdCl_2$ treated CdS was found to be around 90 (±20) nm. This value is comparable with the film thickness (i.e.100 nm) and this could be a result of columnar growth of hexagonal CdS. The other explanation is the coalescence of sub grains during the recrystallization.

Optical analysis

As deposited and $CdCl_2$ treated CBD CdS was characterised using a UV-visible double beam spectrophotometer. For direct band transitions, the relation between the absorption coefficient (α) and photon energy ($h\nu$) is given by the equation $(\alpha h\nu)^2 = A(h\nu - E_g)$ Where A is a constant and E_g is the band gap energy. The relationship between transmittance (T), film thickness (d) and absorption coefficient is approximated by $\alpha = -\frac{\ln T}{d}$, neglecting reflection loses and interference effects. The band gaps of as-deposited, annealed and $CdCl_2$ treated CBD CdS were estimated using above formulas (see table I). A significant non-linear variation was seen with the pH. Average values for the band gap are given in table I and the Student's T-test showed that there was no significant difference between the band gap of air annealed and air annealed with $CdCl_2$ films, both having a significantly lower band gap than the as deposited films.

Table I. The pH dependence of band gap Energy of CBD CdS films deposited under ultrasonic agitation.

Bath pH	Band gap energy (eV)		
	As deposited	Annealed in air	
		Without $CdCl_2$	With $CdCl_2$
8.3	2.45	2.36	2.30
9	2.50	2.37	2.30
10.2	2.43	2.31	2.33
10.5	2.26	2.25	2.30
Average	2.41	2.32	2.31

Film quality analysis

SEM micrographs of CdS are given in figures 3a and 3b and show the effect of ultrasonic agitation, seen when the film thickness is greater than 75 nm. Therefore, this might affect the junction characteristics and the uniformity of a solar cell having a CdS window layer greater than 75nm thick. The clusters appearing in figure 3b could be due to the surface adsorption of colloidal CdS particles, and gave an RMS roughness of 1.2 nm and 7.6 nm for 100 nm thick films, grown with and without ultrasonication respectively. The formation of colloidal particles of CdS was also minimised by choosing low temperatures and low reaction rates.

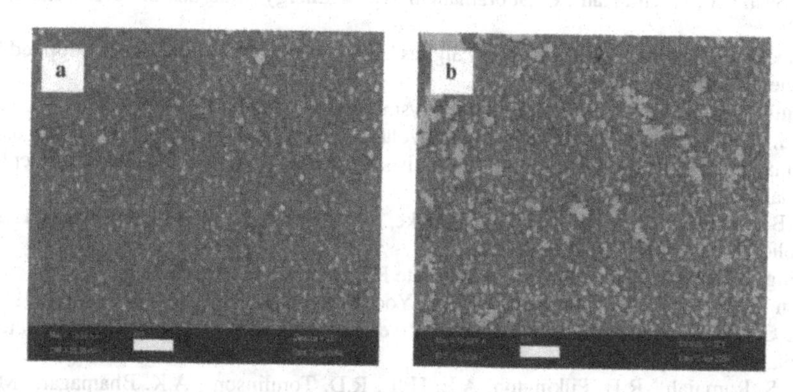

Figure 3. SEM micrographs of CBD CdS on conducting glass substrates (a) with ultrasonic agitation (b) without ultrasonic agitation (white bar = 2 μm)

CONCLUSION

In addition to the observations the effect of pH on CdS film properties, the preparative conditions for fabrication of high quality, pinhole-free thin films of CBD CdS with no apparent detrimental effect to other material properties were optimised. The other interesting feature of this method is that structure of CdS changes from hexagonal to the cubic phase by increasing bath pH while keeping all the other parameters fixed. The above documented experiments were repeated many times in order to assess their reproducibility, and remarkable consistency was achieved with the structure and film morphology. We have also confirmed here the formation of polytype CdS when 10.5 > pH >8.5, although some authors claim the same XRD pattern for cubic material. Any change in the stoichiometry of CdS against the pH should be investigated in detail to understand the variation in the band gap energy versus bath pH. However, the decrease in the band gap after annealing may be due to the dissociation of volatile sulphur from the CdS and creating sulphur vacancies.

ACKNOWLEDGMENTS

This work was funded by the EPSRC under grant ER/R50585/01

REFERENCES

1. M.A.Contreras, B.Egaas, K.Ramanathan, J.Hiltner, A.Swartzlander, F.Hasoon, R.Noufi, Prog. Photovoltaics:Res.Appl. 7 (1999) 331.
2. A.Catalano, Solar Energy Materilas and Solar Cells 41/42(1996)205.
3. R.B.Hall and J.D.Meakin , Thin Solid Films ,63,1979,p.203.
4. G.Sasikala , P. Thilakan , C. Subramanian, Solar Energy Materials & Solar Cells 62 (2000) 275.
5. Giedrius Laukaitis, Seppo Lindroos, Sigitas Tamulevicius, Markku Leskela, Applied Surface Science 185,(2001)134.
6. Paul O'Brien, Tahir Saeed , Journal of Crystal growth 158 (1996) 497.
7. F.S. Hasoon, M.M. Al-Jassim, A.Swartzelander, P. Sheidon, A.A.J. Al-Douri, and A.A. Alnajjar, 26th IEEE photovoltaic Specialists Conference, September 29- Octorber3, 1997, Anaheim, California.
8. N.B. Chaure, S. Bordas, A.P. Samantilleke, S.N. Chaure, J. Haigh, I.M. Dharmadasa, Thin Solid Films 437 (2003) 10.
9. Ting L. Chu and Shirley S.Chu, Solid-State Electronics 38 (1995) 533.
10. Jun Young Choi, Kang-Jin Kim, Ji-Beom Yoo, Donghwan Kim, Solar Energy 64 (1998) 41.
11. A. Cortes, H. Gomez, R.E. Marotti, G. Riveros, E.A. Dalchiele, Solar Energy Materials & Solar Cells 82 (2004) 21.
12. K. S. Ramaiah , R.D. Pilkington, A.E. Hill , R.D. Tomlinson , A.K. Bhatnagar, Materials Chemistry and Physics 68 (2001) 22.
13. T.L. Chu, S.S. Chu, N. Schultz, C. Wang, C.Q. Wu, J. Electrochem. Soc. 139 (1992) 2443.
14. H.R. Moutinho , D. Albin, Y. Yan , R.G. Dhere, X. Li, C. Perkins, C.S. Jiang, B. To, M.M. Al-Jassim, Thin Solid Films 436 (2003) 175.
15. P.N. Gibson, M.E. Ozsan, D. Lincot , P. Cowache, D. Summa, Thin Solid Films 361-362 (2000) 34.

Mater. Res. Soc. Symp. Proc. Vol. 836 © 2005 Materials Research Society L5.37

Absorber Films of Antimony Chalcogenides via Chemical Deposition for Photovoltaic Application

M. T. S. Nair, Y. Rodríguez-Lazcano, Y. Peña, S. Messina, J. Campos, and P. K. Nair,
Centro de Investigación en Energía, Universidad Nacional Autónoma de México,
Temixco, Morelos – 62580, México.

ABSTRACT

Antimony sulfide thin films (300 nm) have been deposited on glass substrates at 1-10°C from chemical bath. When heated these become crystalline and photoconductive with optical band gap (direct) of 1.7 eV. Thin films formed from chemical baths containing $SbCl_3$ and sodium selenosulfate are of mixed phase Sb_2O_3/Sb_2Se_3, which when heated in the presence of Se-vapor converts to single phase Sb_2Se_3 film with optical band gap of 1.1 eV. Such films possess dark conductivity of 10^{-8} $ohm^{-1}cm^{-1}$ and show photosensitivity of two orders. Reaction of Sb_2S_3-CuS in nitrogen at 400°C produces crystalline, photoconductive p-type $CuSbS_2$ with optical band gap (direct) of 1.5 eV. By controlling the deposition and heating condition, (i)Sb_2S_3-(p)$CuSbS_2$ layer is formed, which is utilized in a photovoltaic structure, (n)CdS:In-(i)Sb_2S_3-(p)$CuSbS_2$, with a V_{oc} of 345 mV and J_{sc} 0.18 mA/cm^2 under 1 kW m^{-2} tungsten halogen illumination. In the case of a structure, CdS:Cl-Sb_2S_3-$Cu_{2-x}Se$, V_{oc} of 350 mV and J_{sc} of 0.5 mA/cm^2 are observed.

INTRODUCTION

Antimony chalcogenides are known for their potential as photoconductors, optical storage media and thermoelectric cooling materials. Due to optical band gaps of 1.7-1.8 eV and 1.1 eV, respectively, the sulfides and selenides of antimony with V_2VI_3 composition are suitable for application as absorber materials in photovoltaic devices. The feasibility of chemically deposited thin films of Sb_2Se_3 and n-Sb_2S_3 as active components in photoelectrochemical [1] as well as Schottky barrier and in heterojunction type solar cells has been demonstrated [2]. Thin films of these materials obtained by chemical bath deposition are reported as amorphous in the as-prepared form [3]. Subsequent work on chemically deposited thin films of antimony chalcogenides by our group [4,5] has shown that after annealing in nitrogen at 300 °C, the x-ray diffraction (XRD) patterns of the films of antimony sulfide show well defined peaks similar to that of the mineral stibnite, Sb_2S_3, while those of antimony selenide show in addition to the ones corresponding to Sb_2Se_3 peaks also due to Sb_2O_3 as in the mineral senarmontite. The films are photosensitive and exhibit a high resistivity in the dark.

We have also shown that a p-type $CuSbS_2$ can be made from the reaction of thin films of intrinsic Sb_2S_3 with p-type CuS at 400 °C [6]. These films show a direct band gap of 1.52 eV and electrical conductivity around 0.03 $(\Omega\ cm)^{-1}$. The material has prospects for use as an absorber, similar to $CuInSe_2$, where the presence of In(III) in the lattice is believed to inhibit diffusion of Cu(I) ions through the junction to the n-side in a heterojunction solar cell.

The present paper deals with results of our continuing investigations on obtaining thin films of antimony chalcogenides of binary and ternary compositions by optimization of chemical bath deposition conditions to improve the thickness of thin films and post deposition treatments to modify their composition, structure and optical and electrical properties for use as absorber materials in photovoltaic structures.

EXPERIMENTAL DETAILS

Preparation of thin films: The deposition of thin films of antimony chalcogenides on clean microscope glass slides (75 mm x 25 mm x 1mm) from chemical baths reported earlier [4,5] from our laboratory, methods to obtain films of higher thickness, and post-deposition treatments to modify the structure and composition of the films are described below.

Sb_2S_3 thin films: The deposition bath was prepared by dissolving 0.65 g $SbCl_3$ in 2.5 ml of acetone followed by the addition of previously cooled 25 ml of 1 M sodium thiosulfate and the rest cold water to bring the volume up to 100 ml. The depositions were done at 10 °C as well as at 1°C by keeping the beakers containing the solutions with vertically placed glass substrates in a refrigerator and a temperature controlled bath, respectively for different durations of deposition. At the end of the deposition, the coated substrates were washed well with distilled water and dried. In the case of the films deposited at 1°C, care must be taken while washing to use adequately cooled water in order to avoid peeling of the films. Figure 1 shows the variation of thickness, as measured using an Alfa Step thickness measurement unit, with duration of deposition at 1°C. Since the films start dissolving in the bath after 8-10 h of deposition, multiple depositions in a fresh bath were done to get higher thin film thickness. Heating in a nitrogen atmosphere at 300°C of these films leads to the formation of crystalline photoconductive Sb_2S_3 thin films.

Antimony selenide thin films: In a 100 ml beaker 25 ml of 0.1 M potassium antimony tartrate was taken. To this was added with stirring 2 ml of approx. 3.7 M triethanolamine followed by 20 ml of 30 % ammonia (aq), 10 ml of 0.4 M sodium selenosulfate and the rest de-ionized water to take the volume to 100 ml. Sodium selenosulfate was prepared in the laboratory by refluxing 4 g of selenium powder and 12.5 g of sodium sulfite in 100 ml deionized water for 4 h, which resulted in a solution that is approximately 0.4 M in Na_2SeSO_3. Clean glass substrates were introduced vertically in the bath, supported against the wall of the beaker and the temperature maintained at 27 °C. Both sides of the substrates were coated with specularly reflecting brown films, characteristic color of antimony selenide. The thickness of the films was: 0.1 μm (2 h) - 0.4 μm (4 h). These films are of mixed phase containing antimony selenide/oxide films, which when heated in the presence of Se at 300°C leads to the formation of photoconductive crystalline Sb_2Se_3 films.

$CuSbS_2$ films from Sb_2S_3-CuS films: CuS thin films were deposited from chemical bath containing solutions of copper chloride, thiosulfate and dimethylthiourea on Sb_2S_3 thin films, which when heated at 400°C in nitrogen result in crystalline p-type $CuSbS_2$ thin films described in ref. [6].

$Cu_{2-x}Se$ thin films: These films were deposited from chemical baths containing solutions of copper sulfate, ammonia (aq), and sodium selenosulfate, as described in ref [7], which form crystalline highly p-type $Cu_{2-x}Se$ thin films, with crystalline structure similar to berzelianite and stable in heating nitrogen at temperatures up to 300°C, above which partial loss of Se lead to the presence of Cu_2Se in the films [8].

Photovoltaic structures: CdS thin films of thickness 100 nm were deposited from chemical bath containing cadmium-citrate complex, ammonia (aq.) and thiourea, as reported in ref [7] on commercial transparent conductive oxide films of SnO_2:F of sheet resistance 8-12 Ω supplied by Pilkington-LOF. In order to make the films *n*-type, an indium thin films of < 20 nm in thickness was evaporated first on the TCO. The heating of CdS-In thin films at 250°C in air is known to produce *n*-type CdS:In thin films [7]. Alternatively, CdS thin films are immersed in a dilute solution (0.01 M) of $HgCl_2$ for 15 min and heated at 200°C for 20 min, which forms *n*-type CdS:Cl thin films [7]. Sb_2S_3 thin films (200 nm) were deposited on these SnO2:F-CdS:In or CdS:Cl films and annealed in air at 300°C for 1 h to make it into a crystalline and near-intrinsic photoconductive layer. An Sb_2S_3 thin film (300 nm) and CuS or $Cu_{2-x}Se$ thin films of about 100 nm thickness were subsequently deposited on it. These were annealed in nitrogen at 300-400°C to form the *p*-type absorber film. In a separate experiment we found that electrodes of silver paint provide nearly ohmic contact on this top layer.

RESULTS AND DISCUSSION

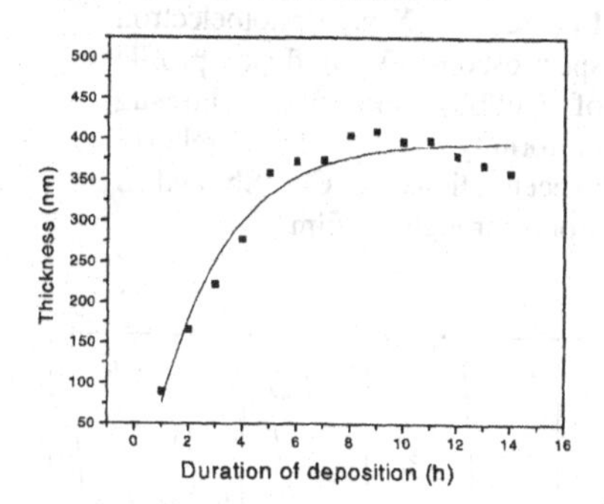

Figure 1. Variation of thickness of thin films of antimony sulfide with duration of deposition at 1 °C.

Figure 2. XRD patterns of thin films of antimony sulfide after annealing at 300 °C in 50 mTorr of nitrogen for 1 h and that of antimony selenide after annealing in presence of selenium at 300 °C in 50 mTorr for 30 min.

Structure, optical and electrical properties: Figure 2 shows the XRD patterns recorded using CuK_α radiation for antimony sulfide thin films (600 nm) heated at 300°C in nitrogen and on antimony selenide thin films formed through heating the chemically deposited film at the same temperature in nitrogen in presence of selenium vapor evolved from Se-powder placed along with. The results show that the films of antimony sulfide and antimony selenide obtained in the present study have similar crystal structures of orthorhombic lattices [9,10], the former being similar to that of the mineral stibnite, Sb_2S_3 and the latter to Sb_2Se_3. These films are photosensitive, as shown in Figure 3. In the case of $CuSbS_2$ formed by heating Sb_2S_3-CuS thin films at 350-400°C, crystalline structure similar to that of chalcostibite, (orthorhombic, PDF 44-1417) has been reported [6]. Figure 4 shows the atomic depth profile using argon-ion etch of the $CuSbS_2$ thin film for Cu, Sb and S, recorded in a VG Microtech Multilab ESCA 3000 using Mg-K_α radiation. A uniform composition is noted for these atoms along the depth of the film thickness. Figure 5 shows the optical transmittance spectra, corrected

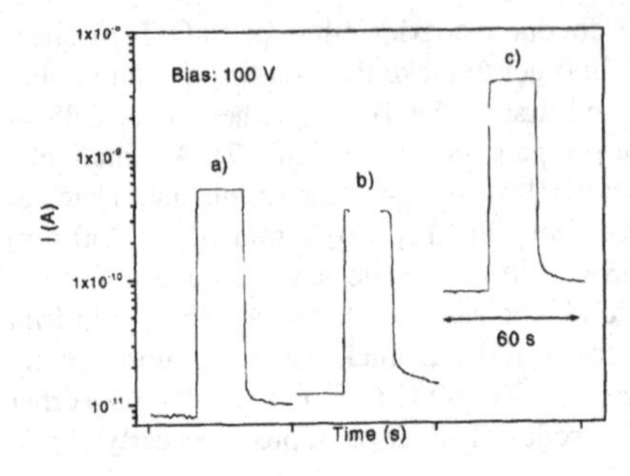

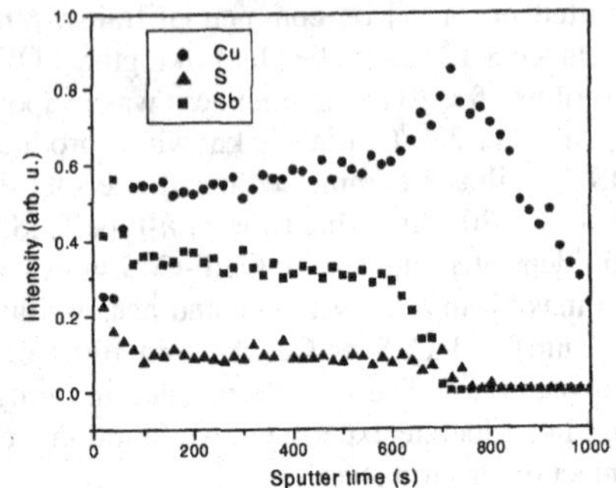

Figure 3. Photocurrent response curves of polycrystalline thin films of Sb_2S_3- a); 0.3 μm thickness, dep., 1 °C and b) 0.6 μm, 10 °C and Sb_2Se_3- c) 0.2 μm, dep. 30°C, resulting after the annealing described in Figure 2 .

Figure 4. X-ray photoelectron spectroscopy (XPS) depth profile of $CuSbS_2$ thin films showing uniformity in the relative concentrations of Cu, Sb and S atoms through the film.

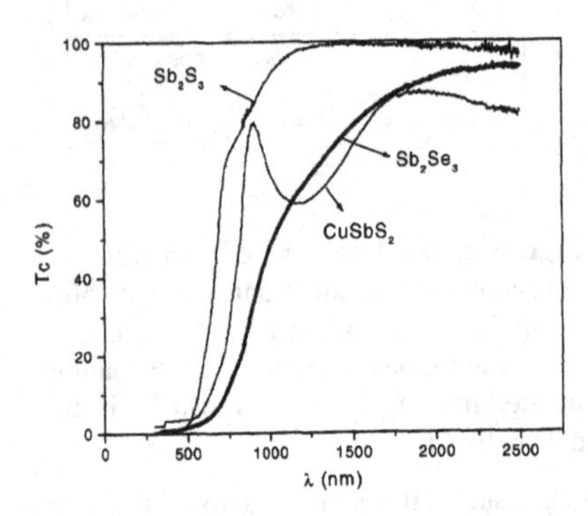

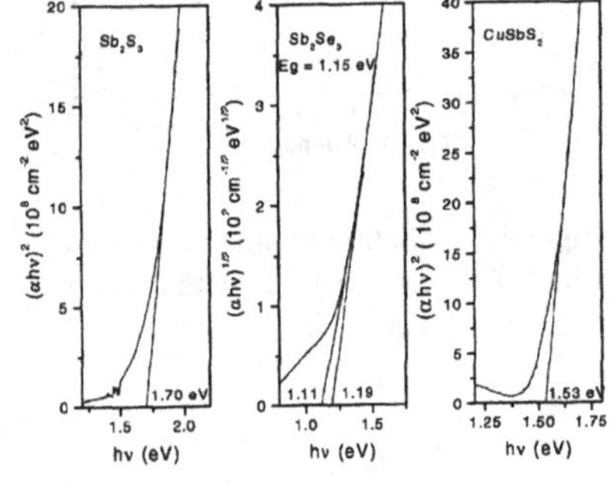

Figure 5. Optical transmittance (corrected for losses due to reflectance) spectra of polycrystalline thin films of Sb_2S_3, Sb_2Se_3 and $CuSbS_2$.

Figure 6. Plots of $(\alpha h\nu)^{1/n}$, where $1/n = 2$ or ½, against photon energy hν in the wavelength region near the absorption edge. The optical absorption coefficients α at different wavelengths are computed from the transmittance curves in Fig. 5.

for reflectance loss, of Sb_2S_3, Sb_2Se_3 and $CuSbS_2$ thin films formed in the above manner. An analysis of the optical absorption coefficients of these films, as shown in Figure 6, shows optical band gaps of 1.7 eV (direct), 1.15 eV (indirect) and 1.53 eV (direct), respectively, for these materials.

Characteristics of the photovoltaic structures: The J-V characteristics of the photovoltaic structures formed were measured in a computerized measurement system by recording the currents as a function of voltage. In the case of SnO_2:F-(n)CdS:In-(i)Sb_2S_3-(p)$CuSbS_2$-silver print, an active area of 3 mm^2 was involved. Figure 6 shows that a V_{oc} of 345 mV and J_{sc} 0.18 mA/cm^2 under 1 kW m^{-2} tungsten halogen illumination are obtainable in this structure. Optimization of individual film thickness and provision of a p^+ layer over the $CuSbS_2$ might lead to an improvement on these characteristics. In the case of SnO_2:F-(n)CdS:Cl-(i)Sb_2S_3-(p)$Cu_{2-x}Se$-silver print, an active area of 1 mm^2 was involved. Figure 7 shows V_{oc} of 350 mV and a short circuit current of 5 μA (J_{sc} 0.5 mA/cm^2) are observed. Here again, optimization of the film thicknesses is required to improve upon the characteristics.

The linear increase of the current with the reduction in the bias voltage is because the requirement of the parallel resistance (R_p) to be higher by at least two orders magnitude of the series resistance (R_s) is not fulfilled in either of the cells. The values of R_p in both cases are nearly 75 kΩ, and R_s is in the same range of values. In further research, this situation must be improved upon.

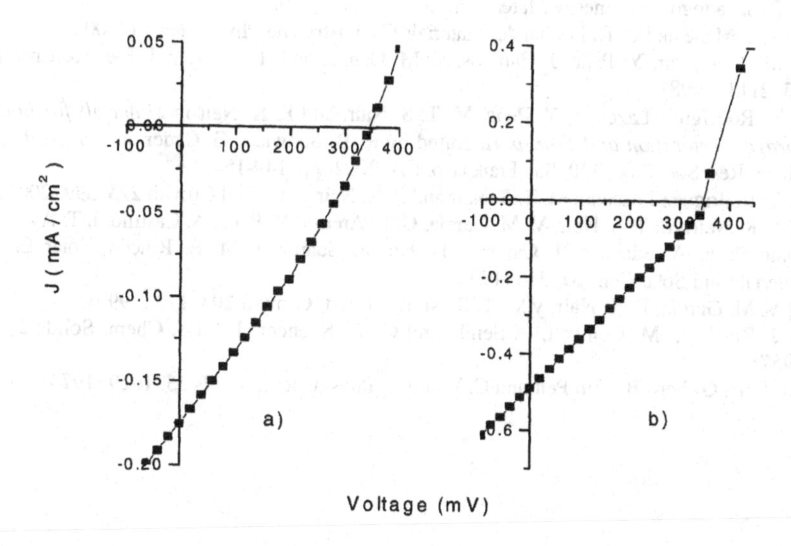

Figure 7. Electrical characteristics of photovoltaic structures incorporating the absorber films presented here: a) (n)CdS:In-(i)Sb_2S_3-(p)$CuSbS_2$ and b) (n)CdS:Cl-(i)Sb_2S_3-(p)$Cu_{2-x}Se$.

CONCLUSIONS

Polycrystalline thin films of Sb_2S_3 and Sb_2Se_3 have been obtained via chemical bath deposition followed by post deposition heating. The films are photosensitive and show band gaps of 1.7 eV corresponding to direct transitions in the case of Sb_2S_3 and 1.15 eV corresponding to indirect transitions in the case of Sb_2Se_3. In the formation of $CuSbS_2$ atomic depth profile using ESCA has shown uniform atomic composition through the film thickness. Photovoltaic effect in p-i-n structures using these absorber films has been demonstrated in: (n)CdS:In-(i)Sb_2S_3-(p)$CuSbS_2$, with a V_{oc} of 345 mV and J_{sc} 0.18 mA/cm^2 and (n)CdS:Cl-(i)Sb_2S_3-(p)$Cu_{2-x}Se$ with V_{oc} of 350 mV and J_{sc} of 0.5 mA/cm^2 are observed under 1 kW m^{-2} tungsten halogen illumination. Optimization of thickness and thermal processing of the films involved will be required to improve upon these characteristics.

ACKNOWLEDGMENTS

We acknowledge Lázaro Huerta (IIM-UNAM) for ESCA measurements, Maria Luisa Ramón García and Leticia Baños for XRD measurements, and CONACYT (Mexico), DGAPA-UNAM, DGEP-UNAM for graduate fellowships for Y. Rodriguez Lazcano, Y. Peña, and S. Messina and for other forms of financial support.

REFERENCES

[1] R. N. Bhattacharya and P. Pramanik, Solar Energy Materials **6**, 317 (1982)
[2] O. Savadogo, Sol. Energy Mater. Sol. Cells **52**, 361 (1998).
[3] R. S. Mane and C. D. Lokhande, Materials Chemistry and Physics **65**, 1 (2000) .
[4] M. T. S. Nair, Y. Peña, J. Campos, V. M. García, and P. K. Nair, J. Electrochem. Soc. **145**, 2113 (1998).
[5] Y. Rodríguez-Lazcano, Y. Peña, M. T. S. Nair, and P. K. Nair in *Materials for Energy Storage, Generation and Transport*, edited by R. B. Schwarz, G. Ceder, and S. A. Ringel, (Mater. Res. Soc. Proc. **730**, San Francisco, CA, 2002) pp. 149-154
[6] Y. Rodríguez-Lazcano, M. T. S. Nair and P. K. Nair, J. Crystal Growth **223**, 399 (2001).
[7] P. K. Nair, M. T. S. Nair, V. M. García, O. L. Arenas, Y. Peña, A. Castillo, I. T. Ayala, O. GomezDaza, A. Sánchez, J. Campos, H. Hu, R. Suárez, y M. E. Rincón, Solar Energy Materials and Solar Cells **52**, 313 (1998).
[8] V. M. García, P. K. Nair, y M. T. S. Nair, J. Cryst. Growth **203**, 113 (1999).
[9] J. Black, E. M. Conwell, L. Seigle and C. W. Spencer, J. Phys. Chem. Solids **2**, 240 (1957).
[10] L. R. Gilbert, B. Van Pelt and C. Wood, J. Phys. Chem. Solids **35**, 1629 (1974).

Mater. Res. Soc. Symp. Proc. Vol. 836 © 2005 Materials Research Society L5.38

$AgSbSe_2$ thin films for photovoltaic structures produced through reaction of chemically deposited selenium thin films with Ag and Sb_2S_3

K. Bindu, M. T. S. Nair, and P. K. Nair
Centro de Investigación en Energía, Universidad Nacional Autónoma de México, Temixco, Morelos -62580, México
bk@cie.unam.mx; pkn@cie.unam.mx

ABSTRACT

Selenium thin films (350 nm) deposited from a 0.01 M solution of Na_2SeSO_3 of pH 4.5 maintained at 10 °C for 13 h, have been used as a source of selenium vapour for reaction with vacuum deposited Ag thin film on chemically deposited Sb_2S_3+Ag layers. When a stack of Sb_2S_3+Ag is heated in contact with Se film, $AgSbSe_2$ is formed through solid state reaction of Sb_2S_3 and Ag_2Se. The latter is formed at 80°C through the reaction of Ag-film in Se-vapour. This thin film is photoconductive and p-type. The optical band gap is nearly 1 eV and dark conductivity, 10^{-3} $\Omega^{-1}cm^{-1}$. This thin film has been incorporated to form a photovoltaic structure, SnO_2:F-(n)CdS:In-(i)Sb_2S_3-(p)$AgSbSe_2$-silver print. V_{oc}> 400 mV and J_{sc}>12 mA/cm^2 have been observed in this under an illumination intensity of 1 kWm^{-2}.

INTRODUCTION

I-III-VI_2 compound thin films of chalcopyrite structure (I-Cu,Ag; III-In,Ga,Al; VI-Se,S) have been investigated during the past two decades motivated by the early success of $CuInSe_2$ in the role of p-type absorber films in high efficiency solar cells [1]. The optical and electrical properties of some of I-V-VI_2 compounds (I-Cu, Ag, V-As, Sb, Bi, VI-S, Se) are also promising [2]. The application of $CuSbS_2$ thin films (E_g, 1.5 eV) as p-type absorber layer in a photovoltaic structure has been demonstrated [3]. For $AgSbSe_2$ (NaCl structure), optical band gap of ~1 eV and p-type conductivity have been reported previously in thin films produced by thermal evaporation of powdered bulk material prepared by direct fusion of stoichiometric quantities of the elements [4].

In this paper, we report a novel method for producing $AgSbSe_2$ thin films by heating a Sb_2S_3-Ag stack, with the Ag-side in contact with a chemically deposited Se thin film [5, 6], serving as a planar source for Se-vapour. Thin films of Sb_2S_3 have been prepared by chemical deposition [7] and the Ag thin films were deposited on it by vacuum evaporation. A two-stage process, consisting of the formation of Ag_2Se films at about 80°C and its reaction with Sb_2S_3 film at 200-300°C lead to the formation of $AgSbSe_2$ films. X-ray diffraction studies illustrate the evolution of the crystalline structure. Optical and electrical properties of the films are presented. The application of the films as a photovoltaic absorber layers is illustrated in the configuration, (n)CdS:Cl-(i)Sb_2S_3-(p)$AgSbSe_2$.

EXPERIMENTAL DETAILS

Deposition of Sb_2S_3 thin films, optical and electrical characteristics: Thin films of Sb_2S_3 were deposited on glass substrates (Corning, 75 mm x 25 mm x 1 mm) by chemical bath deposition, as reported earlier [7]. For this, 0.65 g of $SbCl_3$ was dissolved in 5 ml of acetone in a 100 ml beaker. To this was added 25 ml (1 M) of $Na_2S_2O_3$ (kept at 10 °C), and 72.5 ml of chilled (10 °C) deionised water. The glass substrates were mounted vertically in the bath, using a slotted acrylic lid. The deposition was made at 10 °C for 4 h. Thickness of the films formed under this condition was about 150 nm, measured in Alpha Step 100 (Tencor Instruments, CA). In two successive depositions, films of 300 nm in thickness were prepared either on glass substrate or over an annealed Sb_2S_3 film. The optical transmittance of a 300 nm film shown in Figure 1 (a) testifies to the quality of these films: well defined interference peaks arise from the specularly reflecting thin film surface. The 'as-deposited' films are of low crystallinity. Heating in a nitrogen atmosphere at 300°C for 30 min make them polycrystalline. Such films are photosensitive, as shown in Figure 1 (b).

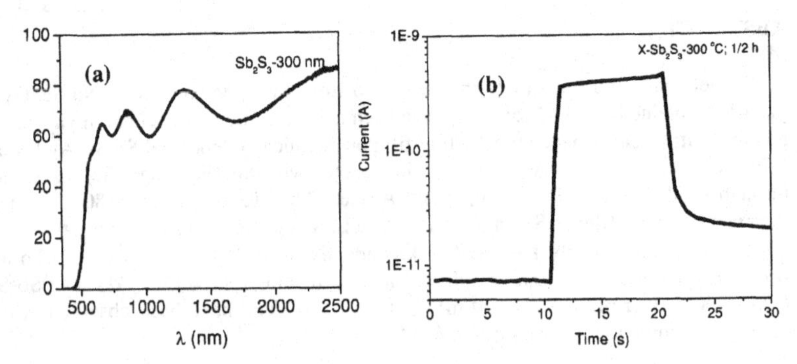

Figure 1. (a) Optical transmittance (T%) of as-prepared Sb_2S_3 thin film of 300 nm thickness and (b) photocurrent response recorded on the film after heating at 300 °C in a nitrogen atmosphere (300 mTorr) for 30 min; bias, 50 V applied across a pair of silver print electrodes (5 mm x 5 mm) and illumination, 500 W/m^2 tungsten halogen light (10-20 s).

Preparation of selenium thin films: Thin films of Se were deposited on microscope glass slides (Corning - 25 mm × 75 mm, 1 mm thickness) coated with a thin film of ZnS (150 nm thickness), following the chemical deposition method reported in [6, 8]. A 0.1 M solution of Na_2SeSO_3 with an excess of $(SO_3)^{2-}$ was prepared by refluxing 0.79 g of selenium powder (99.5%-Aldrich) with 100 ml of 0.3 M solution of Na_2SO_3 (Baker Analyzed Reagent) for 3 h at about 100°C. A volume of 10 ml of this solution was added to 70 ml of water (0.0125 M of Na_2SeSO_3 in the bath) in a 100 ml beaker. To this was added 2 ml of - 4.4 M acetic acid for keeping the bath at pH 4.5. The substrates were placed against the wall of the beaker containing the bath. After 10 h at 10 °C, a Se-thin film of 350 nm in thickness was deposited on the substrates. Figure 2 (a) shows the optical transmittance and specular reflectance of such a thin film. The thin film yield, which is the quantity of Se recovered from the bath as thin film may be estimated considering a deposition geometry shown in Figure 2 (b). The yield is about 95% [8]. These thin films would serve as a planar source of selenium for reaction with metal film.

Thermal Evaporation of Ag thin films and Reactions in Ag – Se and Sb_2S_3-Ag-Se

The reaction between Ag and Se and with Sb_2S_3 resulting in the formation of $AgSbSe_2$ involving a two-stage reaction may be written as:

$$2Ag + Se \rightarrow Ag_2Se$$

$$Sb_2S_3 + Ag_2Se \xrightarrow{Se} 2AgSbSe_2 + 3S\uparrow$$

For the overall reaction, the thickness required of the different layers may be estimated from the respective mass density and formula masses (Sb_2S_3: 4.56 g/cm^3, 339.72 g/mol; Ag: 10.49 g/cm^3, 107.86 g/mol; Se: 4.5 g/cm^3, 78.96 g/mol; and $AgSbSe_2$: 6.69 g/cm^3, 387.54 g/mol):

$$(300nm)Sb_2S_3 + (80nm)Ag \xrightarrow{Se} (228nm)AgSbSe_2 + S\uparrow$$

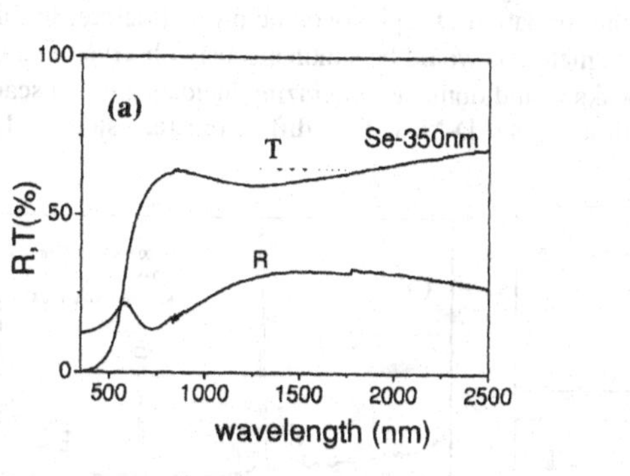

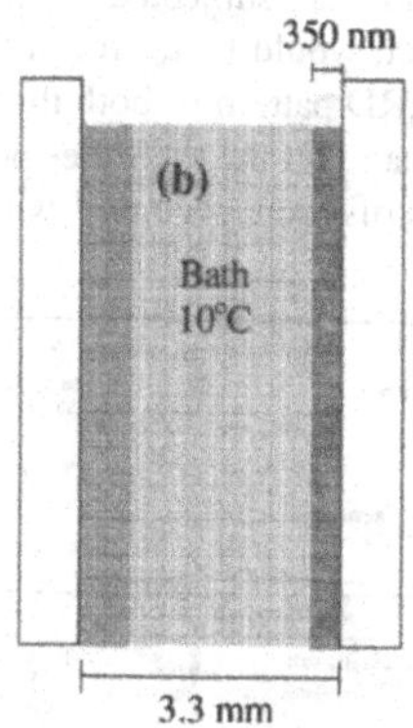

Figure 2. (a) The optical transmittance (T %) and reflectance (R %) spectra of Se thin film (350 nm); (b) Schematic diagram of the cross section of the deposition cell used for estimating the thin film yield of 95% at substrate spacing 3.3 mm [8]

It has been established that a Se-thin film of 350 nm would serve as planar source of 2×10^{-6} mol cm^{-2} for reacting with metal films [8]. Based on the estimate of thickness Ag-films of approximately 50 nm and 100 nm in thickness were deposited on Sb_2S_3 films by vacuum evaporation of Ag-wire (Aldrich Chem. Co, 99.99+%). Ag-films were held in contact with Se-films and heated at 80°C for 1 h in an evacuated oven at a pressure, 20 mTorr, which resulted in the formation Ag_2Se film. In order to form $AgSbSe_2$ films, the Sb_2S_3-Ag stack and Se-film were held together using a polyimide adhesive tape. This was first heated in vacuum (20 mTorr) at 80°C for 1 h and subsequently at 200 or 300°C for 1 h in a nitrogen atmosphere of 300 mTorr.

Solar cell structure: This was developed over 3 mm sheet glass with a coating of spray pyrolitically deposited SnO_2:F film of sheet resistance, 8-12 ohms, obtained from Pilkington-Libbey-Owens-Ford. Over this, a CdS thin film of 60-80 nm was deposited at 70°C in 150 min using a chemical bath involving cadmium citrate complex. This film was immersed in a 0.01 M $HgCl_2$ solution for 15 min and subsequently heated in air for 15 min at 250°C to produce n-type CdS:Cl films [9]. A thin film of Sb_2S_3, 150 nm in thickness, was deposited over the CdS:Cl film and heated at 300°C in nitrogen (300 mTorr) to make this film crystalline. A further deposition of Sb_2S_3 (300 nm) was made over the crystalline Sb_2S_3 film. An Ag-film of about 50 nm was deposited over this film using an appropriate mask for edge-isolation. A Se-film was held in contact with this multilayer structure and heated in a two-stage process mentioned above, which was expected to produce a structure: (n)CdS:Cl-(i)Sb_2S_3-(p)$AgSbSe_2$ with photovoltaic effect.

DISCUSSION

Figure 3 (a) illustrates the evolution of $AgSbSe_2$ thin films in a two stage reaction in Se-vapour, first forming Ag_2Se at 80°C, then a mixed phase Ag_2Se and $AgSbSe_2$ at 200°C and finally resulting in crystalline single phase $AgSbSe_2$, matching PDF 12-0379. If the Sb_2S_3 film is heated to 300°C to make it polycrystalline prior to thermal evaporation of Ag-film and then subjected to the two-stage heating as above, Sb_2S_3-$AgSbSe_2$ would result, which would be useful in the formation of a photovoltaic p-i-n structure. In this case, the XRD pattern of both these materials would be simultaneously observed in a θ-2θ XRD scan, but the $AgSbSe_2$ peaks would dominate in grazing incidence (0.5°) scan. These patterns were recorded with a Rigaku D-Max 2000 diffractometer using Cu-K_{α} radiation.

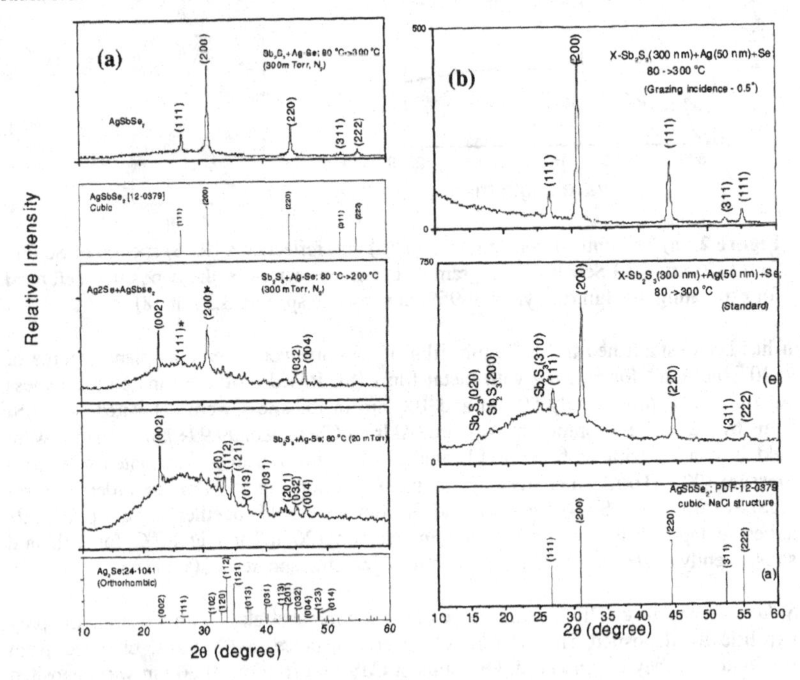

Figure 3. XRD patterns recorded for: (a) Sb_2S_3-Ag layers after close space reaction at 80, 200 and 300°C in Se-vapour evolved from a Se-thin film; (b) Sb_2S_3-Ag layer subjected to the same heat treatment, but with the Sb_2S_3 film heated at 300°C prior to Ag-film deposition.

In Figure 4, the optical characteristics of $AgSbSe_2$ thin films are illustrated. An optical band gap of 0.95 eV is observed for this material, which agrees with reported values of ~ 1 eV [4]. Photocurrent response measurement given in Figure 5 (a) shows that the electrical conductivity of this material in the dark is $4x10^{-3}$ $(\Omega\ cm)^{-1}$ and double this value under illumination of 500 Wm^{-2} (tungsten-halogen). These are in the range of values suggested for photovoltaic absorber materials. Silver print contacts make satisfactory non-rectifying contacts with $AgSbSe_2$, as illustrated in Figure 5 (b).

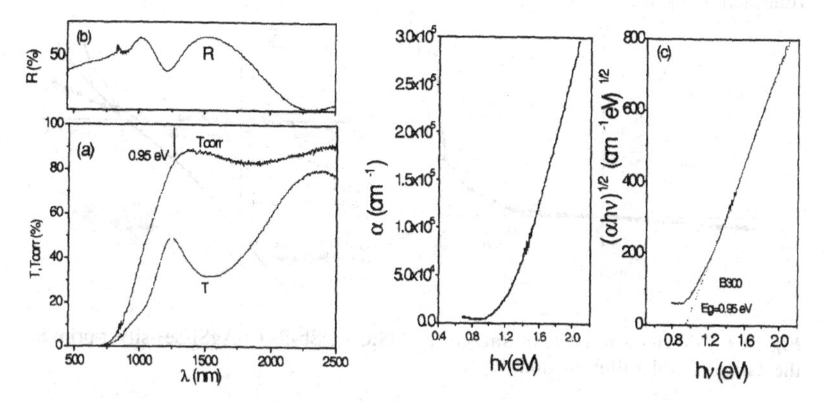

Figure 4. The optical transmittance, Tcorr%, in (a) was corrected for reflectance losses (b), to evaluate the optical absorption coefficient (α), which suggests an indirect gap of 0.95 eV for $AgSbSe_2$, in (c).

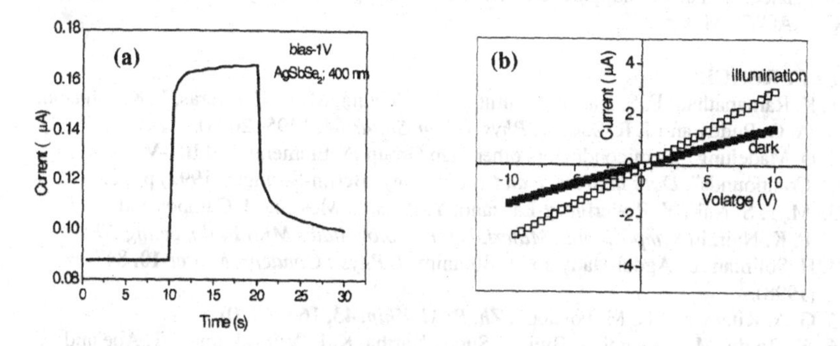

Figure 5. (a) Photo current response in $AgSbSe_2$; (b) I-V characteristics of Ag-$AgSbSe_2$-Ag, showing ohmic behaviour of the contacts.

Figure 6 shows the schematic diagram of the photovoltaic structure, (n)CdS:Cl-(i)Sb_2S_3-(p)$AgSbSe_2$, developed to illustrate the use of $AgSbSe_2$ thin films as well as the current density (J) versus voltage (V) characteristics in the dark and under two distinct intensities of illumination of tungsten-halogen light. Values of V_{oc}> 400 mV and J_{sc}>12 mA/cm^2 observed in this structure, with active area of 1 mm^2, suggest that in future investigation improved characteristics may be obtained through optimization.

CONCLUSIONS

Thin films of $AgSbSe_2$ and a heterostructure of Sb_2S_3-$AgSbSe_2$ are formed at 300°C through the reaction of Sb_2S_3-Ag and crystalline Sb_2S_3-Ag in Se vapor evolved from chemically deposited Se thin films. $AgSbSe_2$ films (cubic) are p-type with optical band gap of nearly 1 eV and electrical conductivity of ~ 10^{-3} $(\Omega cm)^{-1}$. A photovoltaic structure of SnO_2:F-(n)CdS:Cl-(i)Sb_2S_3-(p)$AgSbSe_2$-silver print was fabricated, which gives J_{sc}>12 $mAcm^{-2}$ and V_{oc}>400 mV under an illumination intensity of 1 kWm^{-2} (tungsten-halogen).

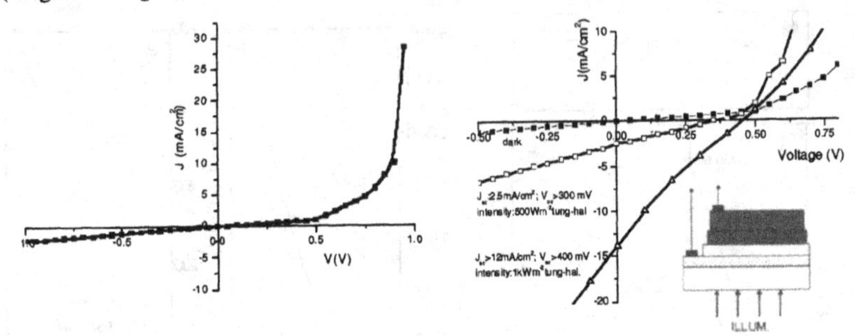

Figure 6. J-V characteristics of SnO_2:F-(n)CdS:Cl-(i)Sb_2S_3-(p)$AgSbSe_2$-silver print in the dark and under illumination.

ACKNOWLEDGMENTS

The authors acknowledge Letcia Baños and María Luisa Ramón for recording the XRD, José Campos for the electrical characterization and Oscar Gomez Daza for technical assistance. Financial support for the work was provided by DGAPA-UNAM and CONACYT-Mexico.

REFERENCES

1. K. Ramanathan, F. S. Hason, S. Smith, D. L. Young, M. A. Contreras, P. K. Johnson, A. O. Pudov and J. R. Sites, *J. Phys. Chem. Solids* **64,** 1495 (2003).
2. O. Madelung "Semiconductors other than Group IV Elements and III –V Compounds", *Data in Science and Technology* (Berlin-Springer, 1992) p. 72.
3. M. T. S. Nair, Y. Roderíguez-Lazcano, Y. Peña, S. Messina, J. Campos and P. K. Nair, in *Symposium L, Materials for Photovoltaics MRS Fall meeting 2004.*
4. H. Soliman, D. Agdel-Hady and E. Ibrahim, *J. Phys.: Condens. Matter* **10,** 847 (1998).
5. G. A. Kitaev and G. M. Fovaouv, *Zh. Prikl. Khim* **43,** 169 (1970).
6. K. Bindu, M. Lakshmi, S. Bini, C. Sudha Kartha, K. P. Vijayakumar, T. Abe and Y. Kashiwaba, *Semicond. Sci. Technol.* **17**, 270 (2002).
7. M.T. S. Nair, Y. Peña, J. Campos, V. M. Garcia and P. K. Nair, *J. Electrochem. Soc.* **145,** 2113 (1998).
8. K. Bindu and P. K. Nair, *Semicond. Sci. Technol.* **19,** 1348 (2004).
9. M. T. S Nair, P. K. Nair, R. A. Zingaro and E. A. Meyers, *J. Appl. Phys.* **75,** 1557 (1994).

Mater. Res. Soc. Symp. Proc. Vol. 836 © 2005 Materials Research Society L5.39

Structural studies of chloride-treated RF sputtered $Cd_{1-x}Mn_xTe$ films

S. L. Wang*, S. H. Lee, A. Gupta, and A. D. Compaan
Department of Physics & Astronomy, University of Toledo,
Toledo, OH, 43606, USA

ABSTRACT

$Cd_{1-x}Mn_xTe$ alloy films with band gaps of 1.6 ~ 1.8 eV have been deposited by RF magnetron sputtering for solar-cell applications. The films have been treated by chloride vapors to improve the photovoltaic performance. These as-deposited and chloride-treated CdMnTe films have been investigated by Raman spectroscopy, x-ray diffraction (XRD) and scanning electron microscopy (SEM). Raman results indicate that Te and/or TeO_2 exists in the annealed samples depending on anneal conditions.

INTRODUCTION

A practical conversion efficiency of 25% has been predicted for a two-junction polycrystalline thin-film tandem cell with an energy band gap of 1.14 eV for bottom cell and 1.72 eV for top cell [1]. CdTe-based ternary alloy materials, such as CdZnTe and CdMnTe are attractive due to their flexibility of controlling band gap with composition and have been considered as attractive wide-band-gap top cell candidates for such tandem cells [2, 3]. CdMnTe has been studied for several decades as a well-known dilute magnetic semiconductor [4-6]. However, there are few reports of this material on photovoltaic applications [3, 7].

We have previously reported [3] limited chloride annealing efforts and fabrication of CdS/CdMnTe cells. The cells were of poor efficiency maybe due to a shift in band gap towards lower energy, and nonuniformities and damage on the surface. Since chloride treatment is very important and essential in the fabrication of high efficiency CdTe solar cells [8, 9], in this paper we investigate the chloride treatment of RF sputtered CdMnTe films over a range of temperatures with various ambients. Raman, XRD and SEM are employed to characterize the properties of as-deposited and treated films.

EXPERIMENTAL DETAILS

CdMnTe films were magnetron sputtered on glass from a target composed of 13% MnTe and 87% CdTe. The composition of the as-deposited CdMnTe films was determined by the optical absorption edge from the transmission spectra. The CdMnTe film thickness was typically 1μm. To obtain p-type CdMnTe films, a post deposition treatment was carried out on a small piece of sample (2 cm × 3 cm) in a one inch diameter quartz tube. There were two approaches to the post deposition treatment, one was to directly carry out vapor chloride treatment on the films, and another was a two-step process in which a high annealing was carried out in a temperature range of 400 to 500 °C and then followed by the vapor chloride treatment. The high annealing process was under different ambients, including 2% H_2 / Ar, nitrogen and dry air. A chloride source plate of either $CdCl_2$ or a 10% $CdCl_2$ + 90% $MnCl_2$ ($Cd(Mn)Cl_2$) mixture was used in the vapor chloride treatments. The source was made by evaporating a methanol solution of the chlorides on a warm glass plate. The sample was placed on the source plate with the film side face down with a 1 mm space between the sample and the source plate. The chloride treatment temperature was in the range of 360 to 400 °C in dry air. Raman scattering measurements were carried out at room temperature with 30 mW of 514.5nm Ar ion laser excitation. The spectra were recorded with a step size of 4 cm^{-1} and a collection time of 5 seconds. All of films used for experiments had a band gap of 1.63 eV at room temperature.

RESULTS AND DISCUSSION

To obtain p-type CdMnTe films we adopted two approaches to the post deposition treatment, the hot probe measurement results indicated that the CdMnTe films presented p-type behavior by either direct vapor chloride treatment or annealing at high temperature followed by vapor chloride treatment. However, it is also recognized that both approaches somehow affect the morphology and optical properties

XRD characterization of CdMnTe films

Figure 1 shows the XRD data of as-deposited and two-step annealed samples. For the two-step process, the samples were annealed at 520 °C for 10 minutes in 2% H_2 + 98% Ar gas, and then were treated with chloride vapors at 380 °C in air with either a 2 minute (short) or a 30 minute (long) anneal. The as-deposited CdMnTe films on glass show very strong (111)

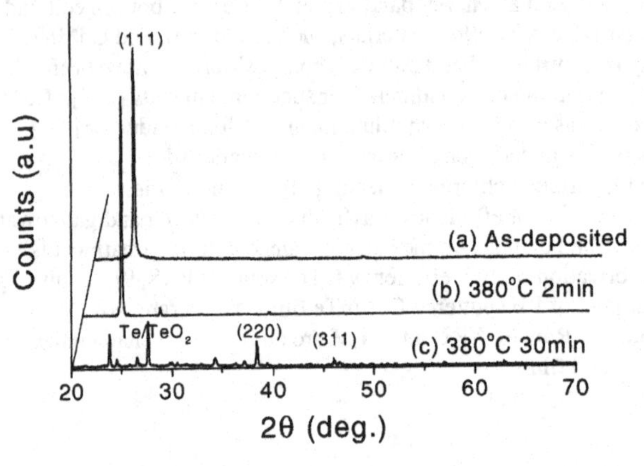

Figure 1. X-Ray diffraction pattern of as-deposited and two-step annealed CdMnTe samples.

preferred orientation with no other significant peaks; even after thermal annealing they still present this preferred orientation. Other orientation peaks appear in addition to the (111) orientation after thermal annealing. The short-time chloride-treated sample shows the appearance of other small peaks, and a higher intensity of the (111) peak than that of the as-deposited sample, indicating recrystallization and grain re-growth. A peak around 27.6° is attributed to Te (101) since it is close to the value of 27.525° given in the Te JCPDS file (file 36-

1452). For the long-time chloride-treated sample, the intensity of the (220) peak is almost the same as the (111) peak, and the (311) peak also significantly enhanced, the ratios of intensity of (220)/(111) and (311)/(111) are 0.77 and 0.26, respectively, the random orientation of film implies a strong recrystallization. Besides the peak at 27.6°, we also observe more peaks in the long-time Cl-treated sample, these peaks probably relate to Te or Cd oxides [10]. Similar cases were observed on some directly vapor chloride treated samples.

Raman scattering characterization of CdMnTe films

Figure 2 displays the Raman spectra of as-deposited samples excited with the 514.5 nm Ar ion laser. There were no any treatments on as-deposited samples. The TO-phonon and LO-phonon modes of CdTe sample are observed at 146 cm^{-1} and 170 cm^{-1}, respectively. For CdMnTe samples, the peaks at 141 cm^{-1} and 164 cm^{-1} correspond to CdTe-like TO and LO-phonons.

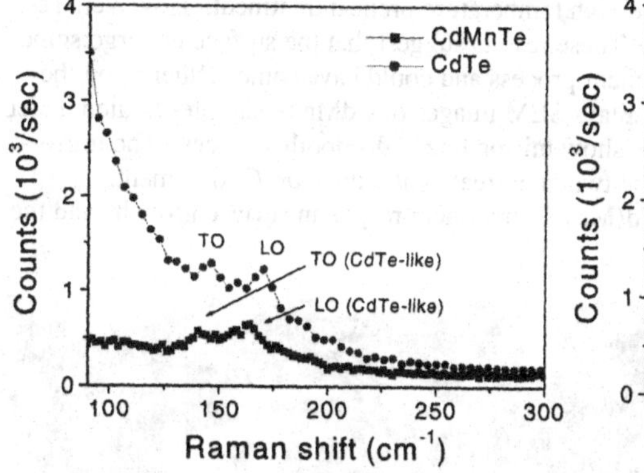

Figure 2. Raman spectra of as-deposited CdTe and CdMnTe samples.

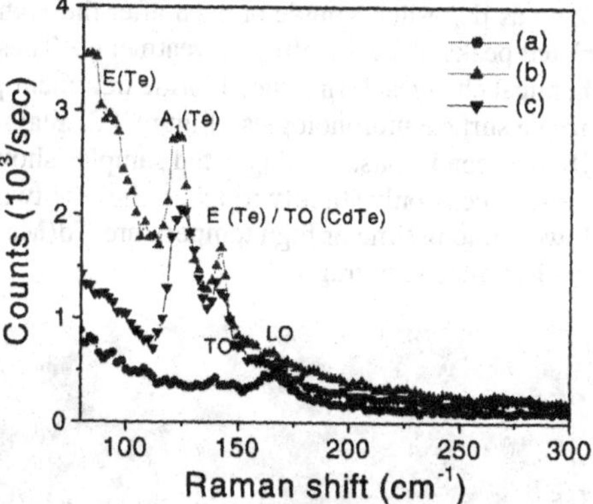

Figure 3. Raman spectra of (a) annealed at 520 °C 10min in 2% H_2; (b) 380°C 30 min vapor chloride treatment after the same process of (a); (c) 380°C 2 min vapor chloride treatment after the same process of (a).

Figure 3 shows the Raman spectra of samples at different chloride treatments. After 10 minutes annealing at 520 °C in 2%H_2 + 98% Ar gas, the Raman spectrum is not much different from that of an as-deposited sample. If the sample is subsequently annealed in $CdCl_2$ + $MnCl_2$ ambient, there are additional peaks appear at 124 cm^{-1} and 142 cm^{-1}. These peaks are related to Te precipitates in CdTe crystals or surface layers of Te. The peak at 124 cm^{-1} is due to the pho non with A_1 symmetry, and the peaks at 92 cm^{-1} and 142 cm^{-1} correspond to the phonons with E symmetry [11, 12]. We observed similar Raman spectra on the sample which was first annealed at 400 °C then using NH_4Cl instead of $CdCl_2$ + $MnCl_2$ to anneal the sample at 370 °C. The spectrum of the sample annealed at 400 °C for 10 minutes in H_2/Ar gas was similar to that of the as-deposited sample. Then the sample annealed at 370 °C for 15 minutes in NH_4Cl showed the Te-related phonon peaks.

We found Te-related peaks (124 cm^{-1}, 142 cm^{-1}) in the Raman spectra of Cl treated samples; this Raman shift corresponded to elemental Te precipitates. However, there were no Te-related peaks observed in the Raman spectra of as-deposited or high temperature annealed samples, this implied that the CdMnTe surface layer suffered from some reactions during the chloride treatment, there could be some oxides on the surface due to oxygen existed in chloride treatment. If TeO_2 exists on the top layer, there will be a peak at 648 cm^{-1} [13], indeed, we observed this peak on long-time (30 minutes) treated sample, but did not observe it on short-time (2 minutes) treated sample. This result is consistent with that we observed more peaks relating to oxide in the XRD patterns of Figure 1 (c).

Surface morphology characterization of CdMnTe films

In the XRD patterns of Figure 1, we observed a peak at 27.6° in the chloride treated samples and related to Te peak. From the Raman spectra, we noted that there were no Te- related peaks for an as-deposited sample or even after the high temperature preheat treatment. However, Te-related peaks observed after Cl treatments. These results suggest that the surface undergo some chemical changes during the chloride treatment process and could have some effluence on the sample surface morphologies. Figure 4 displays SEM images of CdMnTe samples treated under different conditions. As-deposited samples show mirror-like and smooth surfaces. The mirror-like surface is only slightly rougher after the two-step treatment with short Cl treatment. However, long time or high temperature $Cd(Mn)Cl_2$ treatment results in surface growths and the film becomes very fragile.

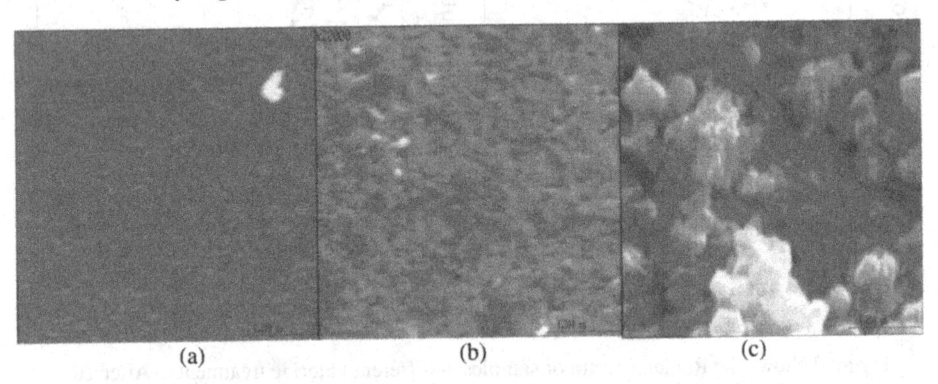

(a) (b) (c)

Figure 4. SEM images of CdMnTe at different treatment conditions (a) as-deposited sample, (b) preheated (520 °C, 10min) + 2 min 380 °C $Cd(Mn)Cl_2$ treatment, (c) 30 min 380 °C $Cd(Mn)Cl_2$ treatment.

Figure 5 shows energy dispersive spectroscopy (EDS) characterizations of as-deposited and chloride treated samples. EDS measurement was operated at 20 KV acceleration voltage, oxygen could not be observed due to instrument limitation. Most of the prominent peaks indicate the films are mainly composed of Cd, Te and Mn, and their intensity does not drastically vary from sample to sample. However, the high Cl peak observed in the 30 minute $Cd(Mn)Cl_2$ treated sample indicates a high Cl content in this sample. Due to the appearance of Cl in the EDS data, one may argue the presence of some Cl-related compounds, such as $TeCl_2O$ or $TeCl_4$. Niles et al [14] investigated the chemical reaction of Cl treatment of CdTe films (400 °C, 30 mins) by x-ray photoelectron spectroscopy, they proposed that CdO, TeO_2, $TeCl_2O$ are building blocks for the surface Cl residue. In our Raman spectra, we do not observe a $TeCl_2O$ related peak (~131 cm^{-1}) [15] or a $CdCl_2$-related peak (132 cm^{-1}) [16]. Additional studies of x-ray mapping on the 30

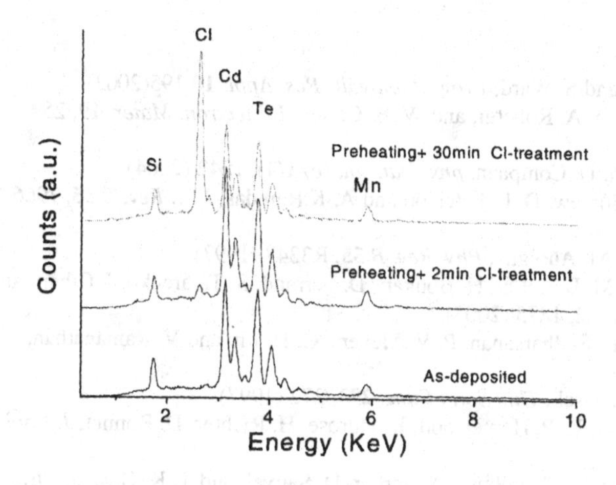

Figure 5. EDS of as-deposited and Cl-treated samples. (20KV acceleration voltage)

minute treated sample indicated some spots composed of Cd and Cl, these spots could be $CdCl_2$ sublimed on the surface. Although Raman intensities are very sensitive to resonant enhancement effects, however, Te-related peaks (124 cm^{-1}, 142 cm^{-1}) are observed in Raman spectra, that mean the top layer is mostly composed of Te or TeO_2, and its development depends on the annealing process.

CONCLUSIONS

We used a sintered, compound semiconductor target to sputter CdMnTe films with band gaps of ~1.6 eV on glass and studied the structural changes after post deposition treatment. As-deposited samples showed good quality in terms of XRD, SEM and Raman scattering analyses. The samples on which chloride treatments were employed indicated some other features. We observed Te-related peaks in the Raman spectra, probably due to Te precipitates or possibly TeO_2 components. EDS indicated that $CdCl_2$ sublimed on the surface. Comparing with Cl-treated CdTe films, CdMnTe films show some Te-rich characterizations after Cl treatment, and this demonstrates that the Cl treatment has a complex chemistry in this alloy system. Although CdMnTe/CdS cells have not yet produced good PV performance, this analysis of vapor Cl treatments is useful for understanding and optimizing the process, and hopefully will accelerate the use of CdMnTe/CdS solar cells in tandem cell applications.

ACKNOWLEDGMENTS

The authors would like to thank Pannee Burckel in the Instrumentation Center of the University of Toledo for SEM measurements. This project was supported by the National Renewable Energy Laboratory -High Performance PV Project (Martha Symko-Davies).

REFERENCES

1. T. J. Coutts, K. A. Emery and S. Ward, *Prog. Photovolt: Res. Appl.* **10**,195(2002).
2. S.A. Ringel, R. Sudharsanan, A. Rohatgi, and W. B. Carter, *J. Electron. Mater.* **19**, 259 (1990).
3. S. H. Lee, A. Gupta, and A. D. Compaan, *phy. stat. sol. (c)* **1**(4), 1042 (2004).
4. A. K. Arora, D. U. Bartholomew, D. L. Peterson and A. K.Ramdas, *Phy. Rev. B* **35**, 7966 (1987).
5. T. Dietl, A. Haury, and Y. M. Aubigne, *Phy. Rev. B* **55**, R3347 (1997).
6. W. Maslana, P. Kossacki, M. Bertolini, H. Boukari, D. Ferrand, S. Tatarenko, J. Cibert, and J. A. Gaj, *Appl. Phys. Lett.* **82**, 1875 (2003).
7. A. Rohatgi, S. A. Ringel, R. Sudharsanan, P. V. Meyers, C. H. Liu and V. Ramanathan, *Solar Cells*, **27**,219 (1989).
8. J. Touskova, D. Kindl, J. Tousek, *Thin Solid Films* **293**, 272 (1997).
9. M. A. Lourenco, Y. K. Yew, K. P. Homewood, K. Durose, H. Richter, D. Bonnet, *J. Appl. Phys.* **82,** 1423 (1997).
10. U. De, M. K. Chattopadhya, S. Chaudhury, A. Sarkar, D. Sanyal, and T. K. Dey, *J. Phys. And Chem. Of Solids* **61**, 1955 (2000)
11. S. H. Shin, J. Bajaj, L. A. Moudy and D. T. Cheung, *Appl. Phys. Lett.* **43**, 68 (1983).
12. P. M. Amirtharaj, and F. H. Pollak, *Appl. Phys. Lett.* **45**, 789 (1984).
13. A. S. Pine and G. Dresselhaus, *Phys. Rev. B*, **5**, 4087 (1972).
14. D. W. Niles, D. Waters, D. Rose, *App. Surf. Sci.*, **136**, 221 (1998).
15. G. H. Westphal, F. Rosenberger, P. R. Cunningham and L. L. Ames, *J.Chem.Phys.* **72**, 5192 (1980).
16. Y. I. Vesnin, *Russian Chemical Bulletin, International Edition*, **51**, pp.37 (2002).

Mater. Res. Soc. Symp. Proc. Vol. 836 © 2005 Materials Research Society L5.40

The Band Offsets of Isomeric Boron Carbide Overlayers

A.N. Caruso[1], P. Lunca-Popa[1,3], Y.B. Losovyj[2], A.S. Gunn[3] and J.I. Brand[3]
[1]Department of Physics & Astronomy and the Center for Materials Research and Analysis (CMRA) Behlen Laboratory of Physics University of Nebraska, Lincoln, NE 68588-0111
[2]Center for Advanced Microstructures and Devices (CAMD), Louisiana State University, Baton Rouge, LA 70806
[3]Department of Engineering University of Nebraska-Lincoln Lincoln, NE 68588-0511

ABSTRACT

Semiconducting boron carbide overlayers, formed from the decomposition of orthocarborane and metacarborane have been studied by angle resolved photoemission. The incurrence of surface photovoltage and the photovoltaic process, from the photoemission experiment, reveal band offsets in the orthocarborane multilayer configurations that are invereted relative to single layer configurations. Defect induced gap states which trap charge at the heterostructure interface is used as one explanation of these results. The role of defects is also used to help illuminate why opposite semiconducting type materials are formed from the decomposition of isomer carborane molecules.

INTRODUCTION

Recent experimental results [1,2] demonstrated that boron carbide fabricated from the precursor molecules *closo* – 1,2 dicarbadodecaborane (orthocarborane) and *closo* – 1,7 dicarbadodecaborane (metacarborane) form p-type and n-type semiconducting thin films. Based on these results an all boron carbide p-n junction diode [3] was constructed. These "heteroisomeric" diodes also demonstrated both photovoltaic [3,4] and neutronvoltaic [3] responses. The fundamental question arising from this discovery is why thin films composed of isoelectronic molecules, with the same chemical makeup and differing only by carbon position (Figure 1), would form opposite semiconducting types? To begin providing answers to this observed phenomenon, a systematic photoemission study of surface photovoltage and photovoltaic accumulation was completed with boron carbide overlayers. This experiment is a study of the electronic structure created at the metal/semiconductor and semiconductor/semiconductor interface.

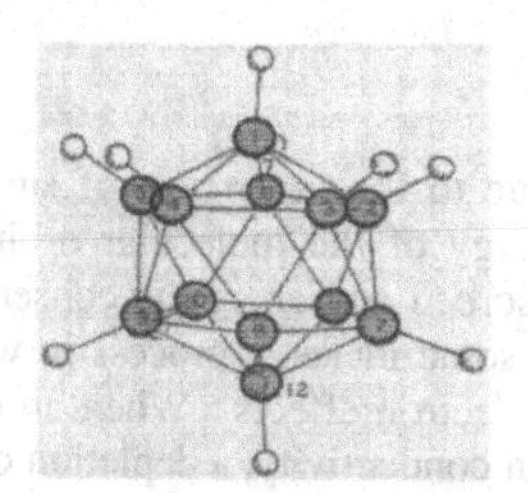

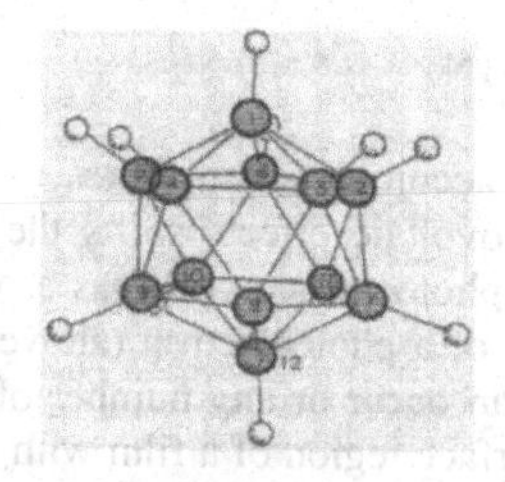

Figure 1. (LEFT) *closo* – 1,2 dicarbadodecaborane or metacarborane and (RIGHT) *closo* – 1,7 dicarbadodecaborane or orthocarborane. Both molecules are isoelectronic with the constiutients $C_2B_{10}H_{12}$.

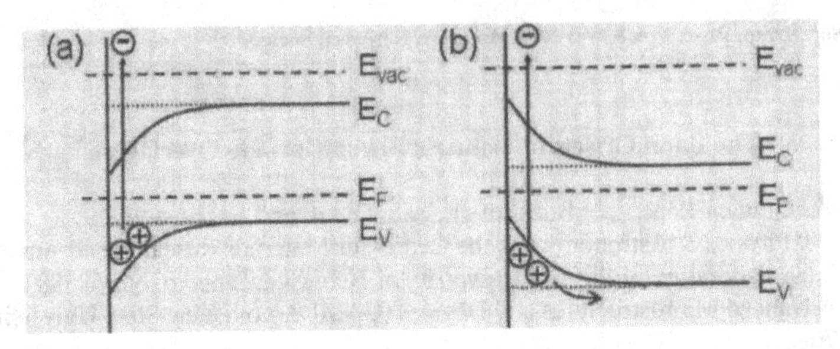

Figure 2. Photoemission based rectification of p-type (a) and n-type (b) semiconductors through the accumulation of surface photovoltage. The dotted lines represent the final band position upon rectification. The initial band bending is caused by the space charge layer at the vacuum interface.

EXPERIMENTAL DETAILS

The electronic structure studies were undertaken by angle resolved photoemission. Photoemission extracts a quantity which is proportional to the electron density of states (photoemission intensity) as a function of binding energy; we treat the Fermi level as zero binding energy. The photoemission experiments were completed at the Center for Advanced Microstructures and Devices synchrotron radiation facility in Baton Rouge, Louisiana.

Deposition of orthocarborane and metacarborane was completed through vapor adsorptions. The vapor was generated by sublimation from commercially purchased powder (Aldrich), leaked into the UHV characterizing chamber and adsorbed on a cooled (105K) polycrystalline silver substrate. Film thicknesses were approximately 30 molecular layers. Semiconducting boron carbide from the precursor molecules, indicated above, is formed by removing the exopolyhedral hydrogen from the icosahedral cage. The hydrogen removal is typically accomplish by PECVD (plasma enhanced chemical vapor deposition), however electron bombardment or zero order diffracted light (white light) from the synchrotron may be used [5]. For the studies described herein, the synchrotron light provided the foundation for both decomposition and characterization.

The Fermi level was calibrated by tantalum foil in intimate contact with the sample surface and the measurements described herein represent photoelectron collection normal to the substrate surface ($k_{\|}=0$) such that only the center of the Brillouin Zone was being probed. The spectra presented below were completed using p-polarized light (70º off normal) with a photon energy of 45 eV.

DISCUSSION

The accumulation of holes or electrons due to the incurrence of surface photovoltage and/or photovoltaic effects shifts the binding energy of the molecular orbitals (band offset). During the photoemission process a photon is absorbed and energy is conserved elastically by the ejection of a photoelectron (above E_{vac}) or by some inelastic process in which an interband transition can occur or any number of secondary electron effects. When an electron is ejected from the surface region of a film with poor electron conductivity, a depletion of electrons occurs so that there is a net concentration of holes. To satisfy neutrality, the holes migrate by diffusion (rather than electron replacement by earth ground). For a surface concentration of holes in a material that has a p-type bulk, but n-type space charge surface, a rectification occurs, so that the surface band moves toward the Fermi level (Figure 2a). For a material with an n-type bulk, but p-type space charge surface, the surface band moves away from the Fermi level (Figure 2b). There also exists a high probability [6] of photoelectron yield from and to surface or defect states as well as secondary electrons (inelastic). A more extensive description of these phenomena and the types of band bending they may produce can be found elsewhere [7-9]. We will use these band bending phenomena to describe the charging effects incurred during photoemission to overlayers of orthocarborane and metacarborane in their pristine and decomposed forms.

The spectra below (Figure 3) represent the molecular and decomposed forms of metacarborane deposited on silver, with molecular and decomposed forms of orthocarborane as overlayers. For the first decomposition of metacarborane, the common molecular orbital binding energies moved away from the Fermi energy as would be expected for a material with an n-type bulk. Without moving the sample position, a second layer of equal thickness (30 monolayers) was deposited. The binding energy of this fresh layer and its decomposed form, exist at a higher binding energy than the original film of the same carborane precursor. We attribute this shift to the locality of synchrotron induced decomposition and the inability for the photovoltage to reach equilibrium. Further evidence to support this claim is found by moving the sample position to a region which has not been irradiated, finding that the binding energy of the pristine and decomposed film matches the original deposition.

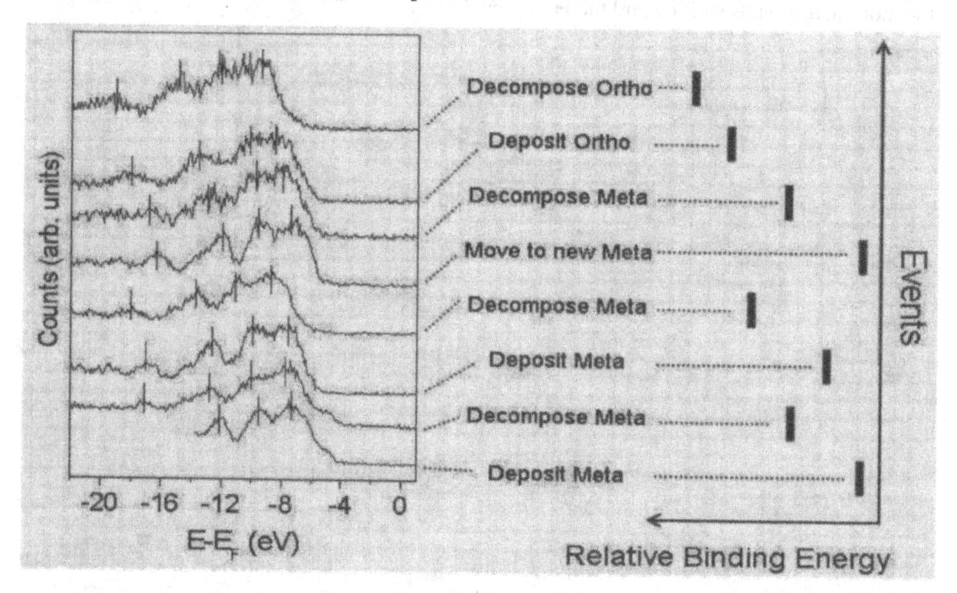

Figure 3. Photoemission spectra (LEFT) of metacarborane, boron carbide formed from the decomposition of metacarborane, with overlayers of orthocarborane and decomposed orthocarborane. An exaggeration of the common molecular orbital shifts of the carboranes due to the accumulation of photovoltage is shown right.

An overlayer of orthocarborane is then adsorbed on the decomposed metacarborane resulting in binding energies positions concurrently farther away from E_F. Although this result seems odd, it may be explained by the excess donor states which have been made available to this fairly intrinsic molecular state and follows the result obtained for depositing metacarborane on decomposed metacarborane. However, the decomposition of orthocarborane on decomposed metacarborane leads to a state where the common molecular orbitals exist at even higher binding energies. This observation is inverted with respect to the decomposition of a thin layer of orthocarborane on a metal in which the molecular orbital binding energies decrease or move toward the Fermi level [2,3] as expected of a p-type material.

A case of comparison is made, to help understand the above phenomena, for metacarborane and decomposed metacarborane overlayers on orthocarborane and decomposed orthocarborane (Figure 4). During the initial deposition and decomposition, the orthocarborane molecular orbital binding energies shift toward lower binding energies as would be expected for a bulk p-type material using the zeroth order model of Figure 2. This same behavior (Figure 4 bottom two spectra) has been observed for orthocarborane decomposition by itself in previous studies [2,5]. But, a further deposition and decomposition of orthocarborane reveals a shift of the common molecular orbitals toward higher binding energy, counter to the simple model and counter to the original orthocarborane deposition and decomposition. We again find an "inverted" charging effect with orthocarborane just as in the first scenario. Then the sample position is moved to a location undisturbed by radiation wherein the binding energy of the pristine molecule and its decomposed form repeat the original binding energy positions. That is, the molecular orbitals shift toward the Fermi level.

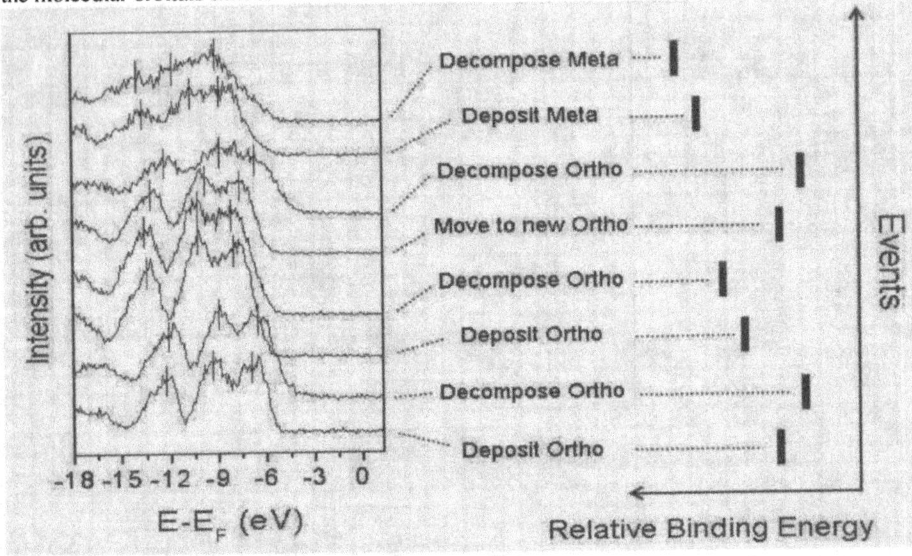

Figure 4. Photoemission spectra (LEFT) of orthocarborane, boron carbide formed from the decomposition of orthocarborane, with overlayers of metacarborane and decomposed metacarborane. An exaggeration of the common molecular orbital shifts of the carboranes due to the accumulation of photovoltage is shown right.

Hence, we feel confident that the charging is localized to the region of synchrotron irradiation and the sample surface and bulk have very poor conductivity, due to the re-establishment of binding energy position for both new orthocarborane and metacarborane overlayers (moved positions). This moved position for orthocarborane is key because it reflects the fact that an interface was not constructed and therefore no interface induced states were formed to cause a shift toward higher binding energy. The deposition and decomposition of metacarborane on decomposed orthocarborane reveal the result, analogous to the first overlayer scenario, that the molecule exists at higher binding energy than its underlayer and the decomposed molecular orbitals exist at even higher binding energy.

Thus, we may summarize the above findings as follows: the molecular orbital binding energies of one carborane molecular layer only and their decomposition result in surface photovoltage accumulation which can be described within the simple model provided by Figure 2, consistent with prior work [2,3]. What is novel is that with either the orthocarborane derived boron carbide on the metacarborane derived boron carbide, or vis versa, shifts the common molecular orbitals toward larger binding energies. We attribute these larger binding energies to accumulation of donors which most probably exist at the interface between layers where secondary electrons from photoemission are accumulated.

Furthermore, both metacarborane adsorbed on decomposed metacarborane and orthocarborane adsorbed on decomposed orthocarborane demonstrate a shift of the common molecular orbitals toward higher binding energy. When either of the carborane molecules is decomposed, the space charge layer at the vacuum interface is more greatly pronounced from the accumulation of donor states. Upon adsorption of the pristine molecule, that space charge layer is compensated where any prior surface states now turn into interface states. These interface states are opposite in character (free carrier density) to the surface states for a standard p-type material (decomposed orthocarborane) but analogous to the surface states formed from an n-type material (decomposed metacarborane). Thus, for the orthocarborane on metacarborane an accumulation layer is formed at the interface, whereas for the metacarborane on orthocarborane and inversion layer is formed [7]. In either case, donor states prevail. For orthocarborane on decomposed orthocarborane it is not clear which phenomenological layer is formed.

CONCLUSIONS

Photovoltage studies of orthocarborane and metacarborane overlayers, in their pristine and decomposed forms have been studied by ultraviolet photoemission spectroscopy. As sole layers, the metacarborane (n-type) and orthocarborane (p-type) films respond to photovoltage accumulation as would be expected; the metacarborane density of states shift away from the Fermi energy whereas orthocarborane density of states shift toward the Fermi energy. However, upon forming any overlayer structures, whether by orthocarborane on itself or on metacarborane, the spectra show a shift of the carborane molecular orbitals toward higher binding energies (away from the Fermi energy). We interpret these results through the accumulation of donor states, formed as a function of creating the carborane/carborane interface, regardless of the bulk semiconducting type found for sole carborane layers. With regard to the problem of isomer dependent semiconducting type, we find only more questions. It will be of interest to find if paracarborane (*closo* - 1,12 dicarbadodecaborane) forms donor states at any carborane based interface.

ACKNOWLEDGMENTS

This research was supported by the National Science Foundation through grants #ECS-0300018, and #EPS-0091909, the Nebraska Research Initiative and the U.S. Department of Energy National Nuclear Security Administration Office of Nonproliferation Research and Engineering (NA-22) through Pacific Northwest National Laboratory. The authors wish to thank S. Balaz for help with some of the measurements.

REFERENCES

1. A.N. Caruso, L. Bernard, B. Xu and P.A. Dowben, *J. Phys. Chem. B* **107** (2003) 9620-9623
2. A.N. Caruso, Snjezana Balaz, B. Xu and P.A. Dowben, A.S. McMullen and J.I. Brand, Y.B. Losovyj, D.N. McIlroy, *Appl. Phys. Lett.* **84** (2004) 1302
3. A.N. Caruso, Ravi B. Billa, Snjezana Balaz, J.I. Brand and P.A. Dowben, *J. Phys. Condens. Matt.* **16** (2004) L139-L146
4. Ravi B. Billa, A. N. Caruso, and J. I. Brand, *Mat. Res. Soc. Symp. Proc.* **799** (2003) Z3.10.1
5. Dongjin Byun, Seong-don Hwang, P.A. Dowben, F. Keith Perkins, F. Filips and N.J. Ianno, *Appl. Phys. Lett.* **64** (1994) 1968
6. J.G. Endriz, *Phys. Rev. B* **7** (1973) 3464
7. Lüth, Hans, Surfaces and Interfaces of Solid Materials 3rd Ed. (Springer, Berlin, 1995)
8. H.C. Gatos and J. Lagowski, *J. Vac. Sci. Tech.* **10** (1973) 130
9. L.J. Brillson, *Surface Science Reports* **2** (1982) 123

Mater. Res. Soc. Symp. Proc. Vol. 836 © 2005 Materials Research Society L5.44

Crystal Growth of Photovoltaic Polycrystalline $Si_{1-x}Ge_x$ by Die-Casting Growth

H. Hirahara, T. Iida, Y. Sugiyama, T. Baba, Y. Takanashi, and S. Sakuragi[1]
Department of Materials Science and Technology, Tokyo University of Science, 2641 Yamazaki, Noda-shi, Chiba 278-8510, Japan
[1]Union Material Inc., 1640 Oshido-jyoudai, Tone-Machi, Kitasouma, Ibaraki 300-1602, Japan

ABSTRACT

Coin-shaped multicrystalline $Si_{1-x}Ge_x$ crystals were grown using a Brigdman method combined with die-casting growth. $Si_{1-x}Ge_x$ alloy is known as a candidate material for producing Auger generation, which creates more than one electron/hole pair per absorbed photon. Since $Si_{1-x}Ge_x$ alloy shows a complete series of solid solutions, precipitating crystals with a certain composition of silicon or germanium by conventional selective growth methods is burdensome. Using die-casting combined with Bridgman growth brought about $Si_{1-x}Ge_x$ precipitation in a form completely different from that predicted by the Si-Ge phase diagram. By combining this growth with subsequent heat treatment of the precipitated $Si_{1-x}Ge_x$ sample, $Si_{1-x}Ge_x$ ($x= 0.5 \pm 3$ %) could be obtained. Indirect band-gap energy was estimated by measuring room-temperature optical absorption coefficient of the grown samples.

INTRODUCTION

With the spread of solar-cell power generation, the demand for solar cells with higher conversion efficiency is increasing. Currently, prospective candidates for highly efficient (≥20%) solar-cell materials are seen to be gallium arsenide (GaAs) and chalcopyrite-type compounds such as copper indium diselenide ($CuInSe_2$). However, these materials are not environmentally friendly photovoltaic semiconductors. That is, first, arsenic and selenium exhibit toxicity when they are emitted; second, indium is a scarce element. It is necessary that the materials which can be used for energy-conversion applications should be environmentally friendly semiconductors, namely, they possess the advantages of abundance of their constituent elements in the earth's crust and non-toxicity of their processing by-products.

Alloys of silicon and germanium ($Si_{1-x}Ge_x$) are important materials not only for microelectronic devices but also for solid-state power generators such as solar cells and thermoelectric devices. Chemical stability, mechanical strength at elevated temperatures, and a close match of the n-/p-type alloys in terms of their thermal and electrical characteristics enable better device operation with no noticeable variation in efficiency. Accordingly, $Si_{1-x}Ge_x$ has been studied as a possible high-conversion-efficiency photovoltaic material. Such alloys can attain a conversion efficiency of 43% when the germanium content is about 60%. This is because of the carrier-multiplication mechanism that comes from the Auger generation [1]. However, because the Si-Ge system shows a complete series of solid solutions with a phase relationship, it is not easy to precipitate crystals possessing a certain composition of silicon or germanium selectively by using conventional Bridgman or Czochralski methods [2-7]. In the case of the Zone-melting method, the precipitating crystals at a certain composition may be formed with a slight compositional change, but the grown ingot includes many defects that may degrade the photovoltaic properties [8]. We have applied a die-casting growth technique combined with an advanced version of the Brigdman method for growing coin-shaped $Si_{1-x}Ge_x$ polycrystalline crystals with a diameter of 25 mm and a thickness of 2 mm [9-10]. We report here on the preparation of these crystals by the die-casting growth process. Effects of post-annealing heat treatment of the grown samples, in terms of changes in the compositional fraction and optical properties, are also examined.

AUGER GENERATION

Auger generation, a carrier-multiplication mechanism in semiconductors, can create two electron/hole pairs derived from the incidence of a single photon, hν. The best conditions to realize the Auger generation are a semiconductor with a fundamental band gap $E_g < h\nu/2$ and a specific band structure. An incident photon with an energy $h\nu > E_g$ causes an optical transition that in turn produces a primary electron/hole pair, and the excess energy of the primary excitation can be transferred to create a secondary electron/hole pair. For creating of the secondary pair, not only must $h\nu \geq 2Eg$ hold, but the excess energy has to exceed E_g. Additionally, to excite the secondary pair by the excess energy of the primary excitation, the photo-excitation of the primary pair within the band structure E(k) of the semiconductor has to take place at energy E and wave vectors k, which are either close to the valence-band maximum or to the conduction-band minimum. It can thus be said that indirect-bandgap semiconductors such as $Si_{1-x}Ge_x$ are suitable to meet this requirement. Minimum onset energy at $h\nu = 2Eg$ for Auger generation in $Si_{1-x}Ge_x$ alloys is explained in Fig. 1. The most favorable composition, x, which is derived from the crossover point between the $2E_g$-line with optical transitions at the direct gaps and the E_0-line, is expected to be obtained when the germanium content is about x=0.62 to 0.70.

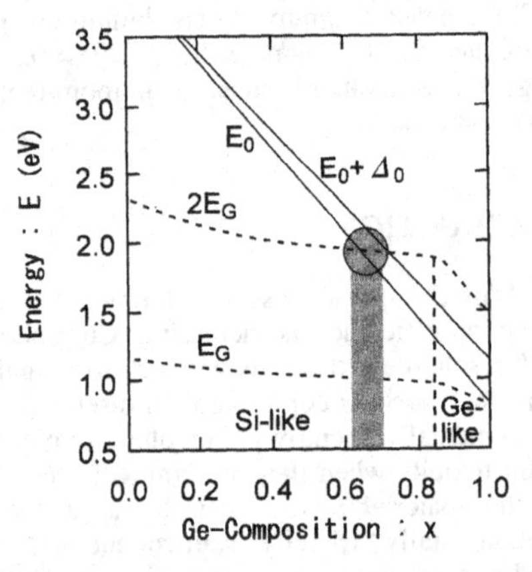

Figure 1. Onset energy for Auger generation in $Si_{1-x}Ge_x$ alloys. The crossover area between $2E_G$ and E_0 or $E_0+\Delta_0$ provides a promising band structure at germanium concentration of x = 0.6 to 0.7.

EXPERIMENTAL DETAILS

Crystals of $Si_{1-x}Ge_x$ were grown using the vertical Bridgman method combined with die-casting growth. A crystallization was performed from a non-stoichiometric melt during incongruent phase transformation [11] using a die-cast growth technique, which is described elsewhere in detail [9-10]. Mixtures of silicon (99.99999%) and germanium (99.995%) with Si:Ge ratios (at.%) of 20:80, 30:70, and 40:60 were put into a crucible, which was then put into a quartz ampule. The charged ampule was evacuated and then pressurized to 0.08 MPa with argon/hydrogen forming gas. The crucible used for the growth was made of isotropic graphite covered with dense chemical-vapor-deposited (CVD) SiC with a thickness of 40 μm, and the SiC surface of the crucible was coated with a SiN_x. For vertical Bridgman crystal growth, the ampoule was heated by an electric furnace to 1460 K (which is above the melting point of the constituent alloy) and then cooled at 4.8 K/h. The axial temperature range can be approximated by a linear gradient of 65 K/cm near the melt-solid interface for a control set point of 1390 K. Post-annealing heat treatment of the grown samples were carried out in a argon/hydrogen pressurized ampoule at 1223 K for 10-20 h. The coin-shaped grown crystal was polished to form wafers 600-700 μm thick. The Si/Ge compositions were determined by electron-probe microanalysis (EPMA). Room temperature optical absorption was performed.

RESULTS AND DISCUSSION

Die-cast $Si_{1-x}Ge_x$ crystal growth

Multicrystalline $Si_{1-x}Ge_x$ crystals were grown in a modified Bridgman electric furnace using SiN_x and CVD-SiC coated graphite crucible. The grown crystals obtained by die-casting were 20 mm in diameter and 3 mm thick, and were shaped like a coin, as shown in Fig. 2. Although the reactivity of molten $Si_{1-x}Ge_x$ is rather high when silicon content is beyond 30 % (i.e. x < 0.7), indicating sticking of molten $Si_{1-x}Ge_x$ with the inner wall of the crucible, using the die-casting growth method provides a sticking-free growth condition even if the concentration of the silicon is high. The use of the die-casting method allows various shapes of crystals to be grown by adapting to the inner shape of the crucibles. the die-casting method is based on the purification of materials used for the inner wall of the crucibles, and surrounding

Figure 2. Graphite crucible used for the die-cast growth and the grown crystal of $Si_{1-x}Ge_x$ (x~ is about 0.5).

atmosphere for Bridgman crystal growth, resulting in the a non-wetting and non-reactive relation between the molten material and the container surface. Since the Si-Ge system has a complete series of solid-solution phase diagrams, forming a uniform composition of $Si_{1-x}Ge_x$, for example, where x=0.5, is generally difficult by the conventional Brigdman technique. It is suggested from the Si-Ge phase diagram that the precipitating $Si_{1-x}Ge_x$ (x~0.6) crystals, as candidate for producing Auger generation, should be initiated from the melt, which is composed of silicon-germanium ratios (at.%) of 15 to 85. However, the initial melt composed of Si-Ge ratios of 30:70 and 20:80 was used for the present die-casting crystallization. This is mainly because, in the present experiment, multicrystalline $Si_{1-x}Ge_x$ (x~0.6) samples were assumed to be processed by the combination of the die-cast growth and subsequent heat treatment to reduce a slight compositional fraction of the grown crystal.

Post annealing

Figure 3 shows the results of EPMA measurements of the as-grown and post-annealed (1223 K) samples for the concentration mapping of the constituent element of germanium as a function of the annealing duration. In an central area of the as-grown sample, a spread of germanium content ranging 30 to 60% was observed, while in the rim of the sample there are portions with less than 30% germanium, as shown in Fig. 3(a). Subsequent annealing at 1223 K made distribution of the constituent elements uniform with increasing annealing time, as shown by the decrements in the shade of the gradation shown in Figs. 3(b) and (c). After annealing, the distribution of the germanium concentration in the central area of the samples was 50 ± 10% and 51 ± 5% for 10-hr and 20-hr annealing, respectively. Figure 4 represents the concentration variation in the annealed sample at 1223 K for 20 hr. As seen in Fig. 3, the deviation of the obtained germanium content was within ± 5%. As for the maximum annealing temperature for a germanium-rich ($x \geq 0.5$) condition, annealing at higher temperature gives rise to evaporation of germanium in the annealing ampoule, which

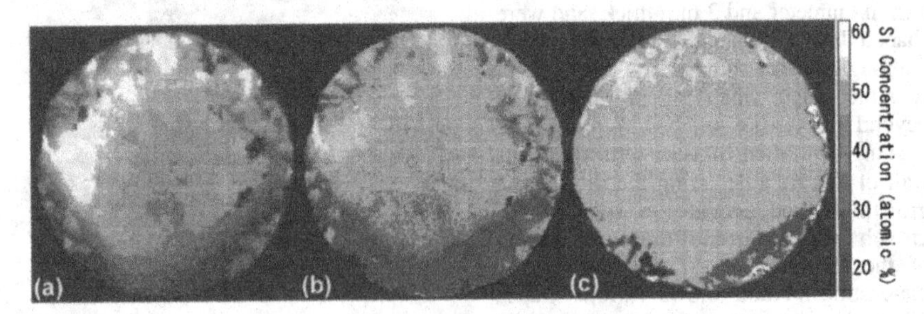

Figure 3. Results of EPMA concentration mapping of germanium for the as-grown and post-annealed samples: (a) as-grown, (b) annealed at 1223 K for 10 h, and (c) 20 h.

causes a noticeable degradation of the sample surface. Thus annealing at about 1223 K seems to be the maximum, so longer annealing may obtain more uniform sample at this temperature. At the moment, however, the grown $Si_{1-x}Ge_x$ (x of 0.5) seems to be insufficient to start Auger generation. Systematic and more precisely controlled growth feedstocks of silicon and germanium and post processing may form $Si_{1-x}Ge_x$ crystal at the desirable x in the range of 0.6 to 0.7.

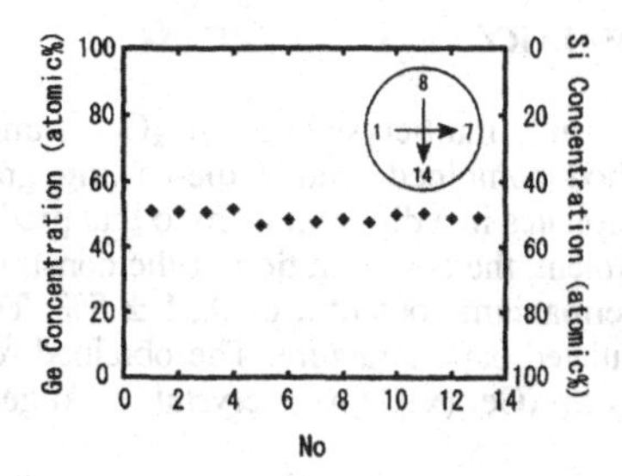

Figure 4. Variation in the germanium concentration determined by EPMA in the post-annealed sample at 1223 K for 20 h.

Optical properties

Figure 5 shows the room-temperature optical absorption coefficient of the annealed and as-grown samples. The measured portion was the central part of the sample with a germanium content of x = 0.51 ± 5%. The estimated indirect band-gap energies were 1.03 eV and 0.97 eV for the as grown and post-annealed samples, respectively. Below the fundamental optical-absorption edge, the absorption tail associated with the Urbach tail was not created, meaning that the grown specimen did not contain high concentrations of crystalline imperfections or defects such as impurities, dislocations, bond defects, or grain boundaries. Indirect band-gap versus composition of the grown $Si_{1-x}Ge_x$ samples is given in Fig. 6. The solid line corresponds to the analytical band structure calculated according to the molecular coherent potential approximation from [12]. It is clear from the figure that the experimentally observed energy gaps as a function of germanium content are in excellent agreement with the calculated values.

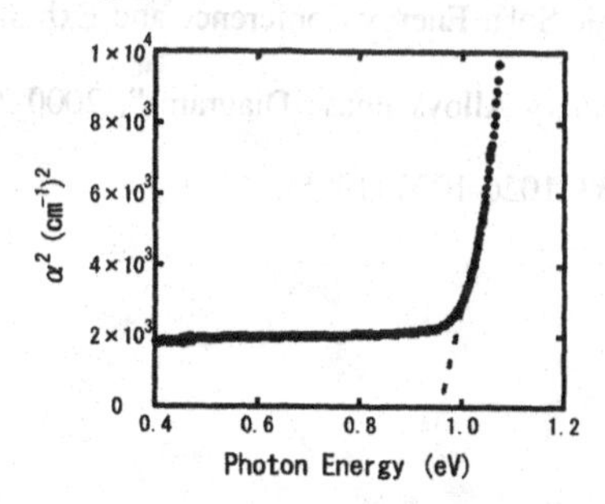

Figure 5. Room temperature optical absorption coefficients (α) vs. photon energy (hν) for the post-annealed sample at 1223 K for 20 h.

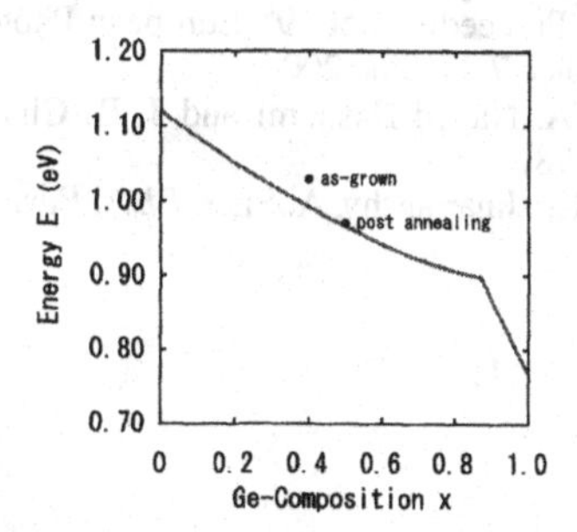

Figure 6. Band gap determined by room-temperature absorption measurements versus composition for grown $Si_{1-x}Ge_x$ alloys. The solid line corresponds to the analytical expression from [12].

SUMMARY

Bulk multicrystalline $Si_{1-x}Ge_x$ samples were grown using the vertical Bridgman method combined with a die-casting growth process. In this growth method, $Si_{1-x}Ge_x$ precipitates in a different form to that predicted by the Si-Ge phase diagram. After post heat treatment, the concentration of the constituent silicon and germanium become homogenous at germanium content x of 0.51 ± 5%. The estimated energy gap was consistent with the calculated band structure. The obtained results imply that this combination technique can grow $Si_{1-x}Ge_x$ (x~0.6-0.7) crystal for Auger generation. [my best guess]

REFERENCES

1 S. Kolodinski, J. H. Werner, H. –J. Queisser, *Appl. Phys.*, **A61**, 535 (1995).
2 C. Marin, A. G. Ostrogorsky, *J. Crystal Growth*, **211**, 378 (2000).
3 N. V. Abrosimov, S. N. Rossolenko, W. Thieme, A. Gerhardt, W. Schroder, *J. Crystal Growth*, **174**, 182-186 (1997).
4 P. Dold, A. Barz, S. Recha, K. Pressel, M. Franz, K. W. Benz, *J. Crystal Growth*, **192**, 125-135 (1998).
5 Yonenaga, *J. Crystal Growth*, **198/199**, 404-408 (1999).
6 K. Nakazima, T. Kusunoki, Y. Azuma, N. Usami, K. Fujiwara, T. Ujihara, G. Sazaki, T. shishido, *J. Crystal Growth*, **240**, 373-381 (2002).
7 K. Nakazima, S. Kodama, S. Miyashita, G. Sazaki and S. Hiyamizu, *J. Crystal Growth*, **205**, 270-276 (1999).
8 D. Bliss, B. Demczyk, A. Anselmo, J. Bailey, *J. Crystal Growth*, **174**, 187-193 (1997).
9 S. Sakuragi, T. Shimasaki, G. Sakuragi, H. Nanba, "Poly-silicon sheets for solar cells prepared by die-casting" To be published in the Proceedings of 19th European Photovoltaic Solar Energy Conference and Exhibition, France 7-11 June 2004
10 S. Sakuragi, "Liquinert – A new concept for shaped crystal growth" To be published in the Proceedings of 19th European Photovoltaic Solar Energy Conference and Exhibition, France 7-11 June 2004
11 A. A. Nayed-Hashemi and J. B. Clark, "Binary Alloys Phase Diagrams", 2000-2001 (1998)
12 S. Krishnamurthy, A. Sher, *Phys. Review B* **33**, 1026-1035 (1986).

Mater. Res. Soc. Symp. Proc. Vol. 836 © 2005 Materials Research Society L5.45

Characterization of polycrystalline Si silicon sheet grown by die casting combined with the Bridgman technique

K. Saito, T. Iida, D. Akimoto, A. Nose, Y. Takanashi, S. Sakuragi[1], H. Nanba[1], G. Sakuragi[1], and T. Shimazaki[1]
Department of Materials Science and Technology, Tokyo University of Science, 2641 Yamazaki, Noda-shi, Chiba 278-8510, Japan
[1]Union Material Inc., 1640 Oshido-Jyoudai, Tone-Machi, Kitasouma, Ibaraki 300-1602, Japan

ABSTRACT

Die-casting growth was used for manufacturing the multicrystalline silicon sheet with a size of 100 x 120 x 0.5 mm. During the growth, incorporation of contaminants such as iron, cobalt, nickel and chromium was well suppressed. The average etch-pit density values ranged from $1x10^4$ cm^{-2} to $4x10^6$ cm^{-2} for growth rates of 5 to 60 mm/h, respectively. Measurement of minority-carrier lifetime bye microwave-photoconductivity-decay (μ-PCD) method was 0.5 μs for as-grown specimens, suggesting that defects and residual strain exist in the grown sheet. Moreover, post heat treatment at 1473 K reduced the etch-pit density and improved carrier lifetime up to 2.2 μs.

INTRODUCTION

During the past decade, photovoltaic power-generation systems have become widespread as consumer products. However, the reduction of the fabrication cost of their solar-cell module has not followed the plan as laid out. This seems mainly because of the low utilization coefficient of the crystal wafers grown for the photovoltaic cells. Namely, in the case of the cast method, which is the current dominant fabrication technique for the wafer production, more than 50% of the grown ingot is discarded during the sawing process. To avoid the slicing of the grown crystal, methods for forming direct sheet from molten silicon have been tried. Accordingly, edge-defined film-fed growth (EFG) and string ribbon methods, promising direct sheet formation, have been developed and have successfully made solar cells achieving conversion efficiency of 16 to 18% [1].

As one more promising growth technique for preparing multicrystalline silicon wafers, die-casting from molten silicon has been developed. This is based on "shaped crystal technology" that can grow crystals of various shapes in a crucible. This technique produces low-cost silicon due to the high utilization of the silicon feedstock and the absence of ingot sawing. Since molten silicon is one of the most reactive elements, it reacts with almost all elements and compounds resulting in stable ceramics, alloys, and many kinds of compounds. An important aspect of the die-cast growth method is to obtain a non-wetting (non-reactive) chemical condition for the molten silicon during growth at elevated temperature. We report here the results of a basic examination performed on multicrystalline silicon sheet by die-cast growth method.

EXPERIMENTAL

Crystal growth

Multicrystalline silicon wafers were grown in a modified Bridgman electric furnace using the preparation sequences of the die-casting growth process as shown in Fig. 1. In brief, semiconductor-grade raw silicon (99.99999 %) in the crucible are heated up to 1723 K and placed in a separator, which is 110 mm wide by 160 mm high, in argon ambient, shown in Fig. 1(a). To obtain a p-type substrate, boron is incorporated at a concentration of $1x10^{15}$ cm^{-3}. Molten silicon is separated when the piled separators are pushed down, forcing molten silicon into a thin gap 0.5 mm in width (Fig. 1(b)). This die-casting process of the molten silicon forms silicon sheets (Fig. 1(c)). The molten silicon in the separated gap is crystallized by the modified vertical Bridgman method at growth rate ranging from 5 to 60 mm/h (Fig. 1(d)).

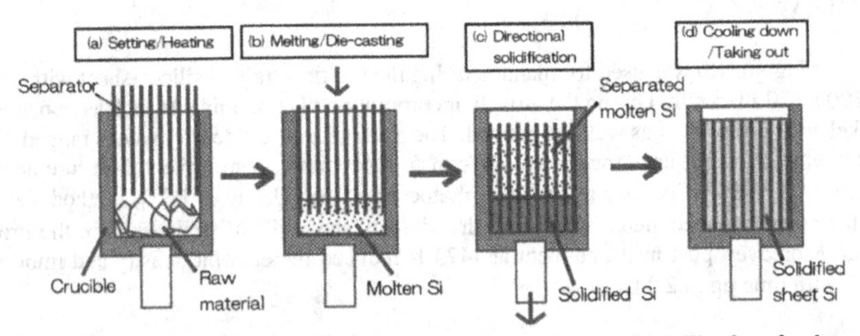

Figure 1. Schematic illustration of the die-casting preparation method. The four fundamental processes to form silicon sheet are shown.

The solidified silicon is then cooled down to room temperature in about 6 h. The use of this die-casting method can realize a totally different silicon-sheet fabrication process from the traditional ingot sawing, thereby curtailing sawing waste. This method seems to be more realistic solution for low-cost silicon-sheet manufacturing and mass production [2-3].

Characterization

The wafers were polished by chemical mechanical polishing (CMP), after which a piece was cut out of the wafer. Some of the wafers were exposed to Secco etching for 75 sec to reveal crystal defects such as grain boundaries and dislocations. Unintentional incorporation of, especially, transition-metal impurities during growth was analyzed by glow-discharge mass spectroscopy (GDMS). The minority carrier lifetime was measured by microwave photoconductivity decay (μ-PCD) using SEMILAB WT-2000. Some samples were annealed subsequently at temperature ranging from 1373 to 1573 K in argon/hydrogen forming gas [4].

RESULTS AND DISCUSSION

Grown silicon sheet

Figure 2 shows the grown silicon sheet, which is 100 mm long by 120 mm wide and 0.5 mm thick. It is clear that the configuration of grains reveals the crystallization by directional solidification in the growth direction of the vertical Bridgman method. While releasing the separators, no sticking was observed between the silicon sheet and the separator surface, so it was easy to get sheets from the growth crucible. The grain sizes of the sheets were larger (i.e., from several square millimeters to a few square centimeters) than those produced in the typical casting process. The external appearance of the grown sheet is similar to EFG ribbon in terms of columnar grains along the growth direction.

Figure 2. External appearance of the silicon sheet grown by the die-casting method. The sheet is 0.5 mm thick with an area of 100 x 120 mm.

Residual impurities and etch-pit density

Impurities included in the grown crystals analyzed by GDMS are summarized in Table 1. Transition metal impurities such as iron (Fe), cobalt (Co), nickel (Ni) and chromium (Cr) are well known to degrade solar cell performance. Although the impurity levels for these anxiety impurities of feedstock silicon is less than the 10^{12} cm^{-3} level, the grown sheets were contaminated by Fe, Co and Cr at approximately $1x10^{14}$ cm^{-3}, $1x10^{13}$ cm^{-3}, and $3x10^{12}$ cm^{-3}, respectively. Ni was not detected, so its concentration must be less than the order of 10^{13} cm^{-3}. The typical contamination levels of EFG[5] wafer over Fe, Co, Ni, and Cr are $6x10^{14}$ cm^{-3}, $<1x10^{14}$ cm^{-3} , $2x10^{12}$, and $2x10^{12}$ cm^{-3}, respectively. Thus, as compared with the typical EFG wafer, the concentrations of Fe was lower, and Co and Cr were slightly more in the grown sheet. In general, the allowed impurity levels for solar-cell fabrication concerning

Table 1. Concentration of the critical impurities incorporated in the die-casting multicrystalline silicon sheet analyzed by GDMS measurement.

Element	BaySix (cm^{-3})	EFG (cm^{-3})	This work (cm^{-3})
Fe	4.0×10^{14}	6.0×10^{14}	$1\sim4\times10^{14}$
Ni	$<3\times10^{14}$	$<1\times10^{14}$	---------
Co	2.1×10^{13}	1.7×10^{12}	$<1\times10^{13}$
Cr	1.0×10^{13}	1.7×10^{12}	$<3\times10^{13}$

above-mentioned elements are less than 10^{14} cm^{-3}; therefore, it can be said that no serious degradation in carrier performance such as lifetime is caused by the impurities included in the sheets[6-8].

In the case of multicrystalline silicon grown by conventional casting, there are many kinds of grown-in defects that shorten the carrier lifetime and reduce the performance of solar cells. Undoubtedly, the present die-cast technique shows a clear trend toward introducing grown-in defects, because the sheet growth is carried out in a limited spatial sphere of solidification in the molten silicon. The evaluated etch-pit density for the as-grown sheets by the die-cast method with different growth rates of 5 mm/h and 60 mm/h is shown in Figs. 3 and 4. The average etch-pit density values are $1x10^4$ cm^{-2} and $4x10^6$ cm^{-2} for these growth rates, respectively. It is noted that the etch-pit density values are very consistent with the growth rate, indicating that the observed etch pit density for the slower growth rate is low enough for the solar-cell application. Figure 5 shows the observed etch-pit density for the post-annealed sample grown at 60 mm/h. At this growth rate, the sample contains high etch-pit density across the whole sample and the respective grains showed divergent characteristics of etch-pit density. However, it is clearly seen that the subsequent heat treatment reduced the etch-pit density, indicating an annealing out of the defects included in the as-grown condition. The effect of annealing on the defect annihilation was more

Figure 3. Micrograph of the as-grown die-casting multicrystalline silicon sheet after Secco etching. The growth rate was 5 mm/h.

Figure 4. Micrograph of the as-grown die-casting multicrystalline silicon sheet after Secco etching. The growth rate was 60 mm/h.

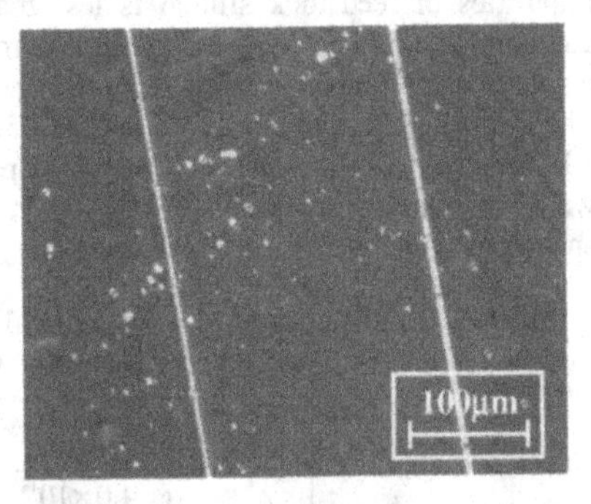

Figure 5. Micrograph of the post-annealed multicrystalline silicon sheet after Secco etching at 1473 K for 20 h.. The growth rate was 60 mm/h.

noticeable for larger grains, while the larger grains exhibited better crystalline quality than the smaller ones. In a part which indicates fewer etch pit density, the effect of the post-annealing was large, resulting in decreasing the etch-pit density down to $2x10^5$ cm^{-2}, which was one order of magnitude smaller than that observed for the as-grown sample. However, it was observed that etch-pit density values and spatial distributions were apparently affected by the state of the directional solidification, which is basically controlled by a linear temperature gradient along the growth direction in the furnace and the growth rate. It seems that more precise control of the directional solidification during die-cast growth in the present growth configuration and slower growth rate may help reduce the defect density [9].

Minority-carrier lifetime

The minority-carrier lifetime, as an important characteristics to conjecture solar-cell performance, of the as-grown and annealed samples (20 x 20 mm^2 in area) are shown in Fig. 6. For the carrier-lifetime measurement, iodic surface passivation was performed for all samples. For the as-grown condition, typical carrier lifetime was 0.5 μs. Tentative post annealing at 1373 K improved carrier lifetime up to 2.2 μs. It appeared that the lower carrier lifetime of the wafer formed by the die-cast method as compared with the substrate fabricated by other growth techniques can be attributed to the growth configuration of the present process; that is, a limited spatial sphere of solidification of the molten silicon during growth can contain defects and leave a residual strain in an the as-grown sample. These may give rise to the deterioration of carrier lifetime, however, it is implied that the optimization of the post-annealing conditions and subsequent gettering process achieve the manufacturing of die-casted silicon sheet [10-11].

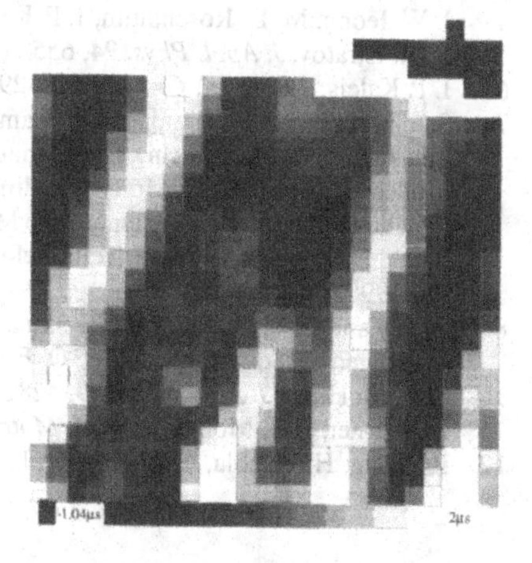

Figure 6. Mapping of the measured carrier lifetime for the post-annealed multicrystalline silicon sheet at 1373 K for 10 h. The surface was passivated with iodine.

SUMMARY

Multicrystalline silicon wafers were grown by die-casting growth and the modified Bridgman method. The unintentional impurity levels of the grown specimens sufficiently fulfilled the specification for solar-cell fabrication. Etch-pit density is dependent on the growth rate and indicates that a slower growth rate can provide enough quality for solar-cell application. Residual defects and strain of crystals in the as-grown condition influenced the minority-carrier lifetime. Further systematic post processes such as heat treatment and gettering with phosphorus [12] seem to be fruitful.

REFERENCES

1 P. Kalejs, *Solid State Phenomena* **95-96**, 174 (2004).
2 S. Sakuragi, T. Shimasaki, G. Sakuragi, H. Nanba, Proc. of 19th European Photovoltaic Solar Energy Conf, Paris (2004) (to be published).
3 T. Saito, *IEEE*, 576-580, (1981).
4 J. W. Jeong, M. D. Rosenblum, J. P. Kalejs, A. Rohatgi, *J. Appl. Phys* **87**, 7551 (2000).
5 A. A. Istratov, *J. Appl. Phys.* **94**, 6552 (2003).
6 J. P. Kalejs, *J. Crystal. Growth*, **128**, 298 (1993).
7 S.A. McHugo, A.C. Thompson, G. Lamble, A. MacDowell, R. Celestre, H. Padmore, M. Imaizumi, M. Yamaguchi, I. Perichaud, S. Martinuzzi, M. Werner, M. Rinio, H.J. Moller, B. Sopori, H. Hieslmair, C. Flink, A. Istratov, and E.R. Weber in Applications of Synchrotron Radiation Techniques to Materials Science IV, edited By: S. M. Mini, S. R. Stock, D. L. Perry, and L. J. Terminelo, (Mater. Res. Soc. Symp. Proc. **524**, Pittsburgh, PA, 1998) pp. 297-302.
8 Bailey and E. R. Weber, *Physica Status SolidiA*, **137** pp. 515 (1993).
9 O. Kruger, W. Seifert, M. Kittler, O. F. Vyvenko, *Physica Status Solidi*, **222**, 367 (2000).
10 S. A. McHugo, *J. Appl. Phys*, **89**, 4282 (2001).
11 M. Werner, H. J. Moller, E. Wolf, *Mater. Res. Soc. Symp. Proc.* **469**, 89 (1997).
12 P. Zhang, H. Vainola, A. A. Istratov, E. R. Weber, *Appl. Phys. Lett.*, **83**, 4324 (2003).

Mater. Res. Soc. Symp. Proc. Vol. 836 © 2005 Materials Research Society L5.47

Current Transport Study of Schottky and P-N Junction Solar Cells Using Metal-Induced Growth Poly-Si Thin Films

Chunhai Ji, Joon-Dong Kim and Wayne A. Anderson
University at Buffalo, The State University of New York, Dept of Electrical Engineering, Buffalo, NY

ABSTRACT

Poly-Si thin films deposited at low temperature by using the metal-induced growth (MIG) method have the advantage of less metal impurity contaminations and relative large grains with preferred crystal orientation of (220). In recent research, the Schottky solar diode made of MIG poly-Si shows J_{sc} of 12 mA/cm^2 and V_{oc} of 0.214V. In this paper, current transport mechanisms were studied by current-voltage-temperature (I-V-T) testing from 100 K to 400K. For the samples deposited by a one-step sputtering process, the large value of ideality factor (n) and abnormal increase of barrier height with the temperature implies that the current transport mechanism does not follow the pure thermionic-emission theory, which was proven to be thermionic-field emission due to the highly doped Si film. By using a two-step sputtering process, the ideality factor and Au-Schottky barrier height at room temperature were about 1.5 and 0.7 eV, which was improved from one-step sputtering. Hydrogenation by electron cyclotron resonance (ECR) plasma can further improve the Schottky diode ideality factor and barrier height. Although a low-level Phosphors-doped Si target was used for poly-Si thin film deposition, a thermionic-field emission mechanism was still found by plotting the activation energy (E_0) versus the testing temperature range. Capacitance-voltage (C-V) analysis revealed an unexpected carrier density of 10^{17} cm^{-3} level, which is 1000 times higher than the doping density level in the Si film. "Oxygen thermal donor" effect was assumed due to high oxygen level (10^{20} cm^{-3}) detected by SIMS and processing at ~ 600 °C. Increasing of the total carrier density due to the oxygen donor may cause the transport mechanism change from pure thermionic emission to thermionic-field emission. Reducing oxygen in the Si film by filtering the sputtering gas to 50 ppb oxygen level was proven to be effective. C-V results gave ~10^{16} cm^{-3} level of carrier density after using oxygen filtering. P-N junction solar cells were made by B-ion implantation into n-type Si film and dopant activation at 700 °C. I-V-T study showed similar curves for P-N junction as for Schottky junction devices. This implies that the current transport was dominated by the Si films instead of the junctions for both Schottky and P-N junction devices.

INTRODUCTION

In the metal induced growth (MIG) technique, the poly-Si thin film was deposited by using Co and Ni seed layers, 25 ~ 50 nm thick, at elevated temperatures of 550 °C ~ 625 °C. The deposited crystalline Si thin film has the grain size of 1 μm [1,2]. Previous studies focused on the grain structure and device fabrication [3,4]. Recent research on Schottky junction solar cell devices fabricated from MIG poly-Si films showed that the J_{sc} reached 12 mA/cm^2 and V_{oc} reached 0.214 V [5]. Current-voltage-temperature (I-V-T) analysis measures the I-V characteristics of the junction diode with temperature change. The measurement is important to identify the junction transport mechanisms and defect properties in the film and at the junction. In this research, the I-V-T study was applied to

Schottky junction and P-N junction devices at different processing stages which include: single-stage sputtering, two-step sputtering, ECR hydrogen passivation (hydrogenation) and P-N junctions. Current transport mechanisms were analyzed based on the current-voltage changing with the temperature from 100 K to 400 K. Poly-Si and device quality were discussed by combining the current transport analysis with the capacitance-voltage (C-V) analysis.

Experimental

The MIG poly-Si thin films were prepared under different conditions as follows:

1. Single-step sputtering: The Si film was deposited by using a 0.02 Ω-cm n-type Si target. The deposition conditions were as follows: sputtering power of 50 watts, sputtering pressure of 1.8 mTorr, using 50 nm Co as a seed layer, substrate temperature 700 °C annealing in forming gas.
2. Two-step sputtering: In the two-step sputtering process, the poly-Si film was deposited from a 5~15 Ω-cm n-type Si target at 0.5 mTorr at 625 °C. In the first-stage of sputtering, the sputtering power was controlled at 50 watts and deposition rate was between 60 Å/min ~70 Å/min. In the second stage, the sputter power and rate were increased to 200 watts and 300 Å/min, respectively.
3. Poly-Si with ECR passivation: The poly-Si thin film deposited by two-step sputtering was hydrogenated by using an ECR hydrogen plasma reactor. The ECR hydrogenation process parameters were as follows: 600 Watts, 4 mTorr H_2, 350 °C and 30 min. In ECR hydrogen passivation, the ionized hydrogen passivates the surface Si dangling bonds, the bulk Si defects such as dangling bonds on grain boundaries [6,7,8,9] and impurities such as oxygen [10].
4. Poly-Si with oxygen control: To decrease the oxygen level in the poly-Si film, an organic oxygen filter was added to the sputtering gas pipeline in order to filter the oxygen in the sputtering gas. According the specification of the oxygen filter, the oxygen level after filtering is below 50 ppb.
5. Device fabrication: The Schottky junction devices were made in a metal-insulator-semiconductor (MIS) structure by first growing a 20 Å-thick SiO_x (insulator) in a muffle furnace at 500ºC. The P-N junctions were fabricated by ion-implanting boron into the n-type Si film.

Analysis and Discussion

1. I-V-T analysis on a Schottky junction sample from single-step sputtering

Figure 1 shows the I-V-T results for a single-step sputtering deposited Si film. For region I as shown in the figure, the current density tends to be saturated with the voltage change and is independent of temperature change. This indicates the effect of series resistance or space-charge-limited-current (SCLC). The linear portion in region II was found to be thermionic-field-emission (TFE) dominated according to the equation below:

$$E_{00}/kT = \frac{qh}{4\pi kT}\sqrt{\frac{N_A}{\varepsilon_s m_k^*}} = 18.5\times10^{-12}/kT\sqrt{\frac{N_A}{m_r\varepsilon_r}}eV = 0.19, \quad (1)$$

where N_A is the doping concentration ($2\text{x}10^{17}$ cm^{-3}) in the semiconductor, ε_s is the dielectric constant, ε_r is the relative dielectric constant of the Si, m^*_k is the effective hole mass, and the rest of the symbols have their usual meanings. E_{00}/kT is a measure of the ratio of thermionic-emission (TE) current to field-emission (FE) current. FE becomes dominant if $E_{00}>>kT$, however TFE dominates when $E_{00}\sim kT$ and TE for $E_{00}<<kT$. In Region IV, the I-V relation can be described with

$$J = kV^{m} \tag{3}$$

where m is found to be 1 by calculating the slope of the linear portion in log J ~ log V. So, the current transport in region IV and region III, which are symmetrical about the y-axles, is controlled by shunt resistance (R_{sh}) since both of them are under low applied voltages. According to the thermionic-emission theory, the reverse current density of an ideal Schottky diode should saturate at the value $J_0=AR^*T^2e^{-q\phi B/kT}$. Region V shows the departure from this ideal behavior. Field dependence of the barrier height, tunneling and generation in the depletion region are the possible causes.

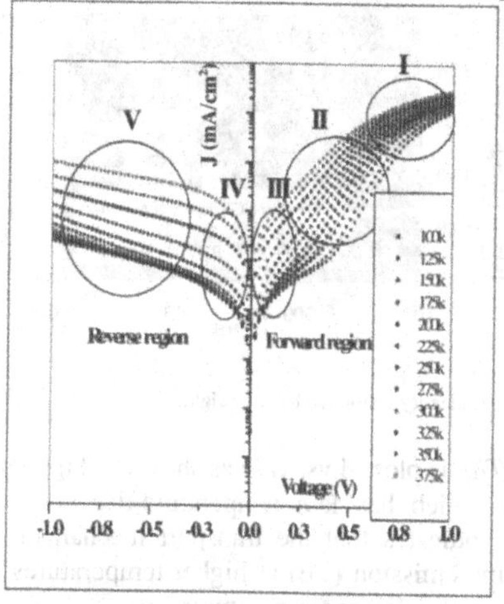

Figure 1. I-V-T plots of the forward current region for a Schottky junction sample (single step sputtering).

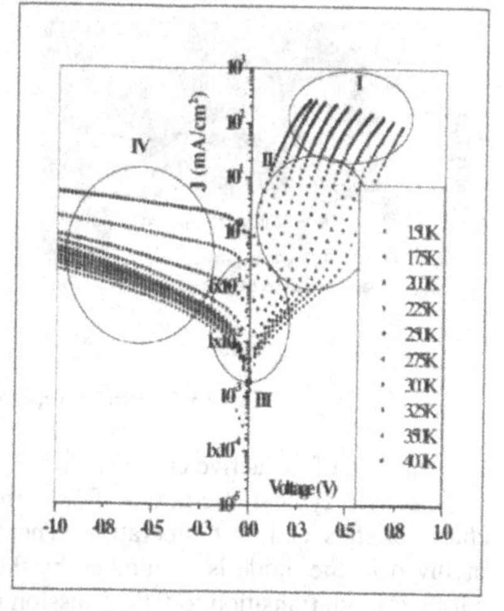

Figure 2. Dark J-V characteristics with temperature change for an Au/n-Si Schottky junction (two-step sputtering).

2. I-V-T analysis on a Schottky junction (two-step sputtering)

As reported previously, two-step sputtering produces much larger grain structures than single-step sputtering. Figure 2 shows the I-V-T curves in forward and reverse current regions for a device prepared by the two-step sputtering process. Similar to the curves in Figure 1, region I indicates series resistance or SCLC controlled current transport. In region III, the J-V curves are symmetric about the y-axis and follow the $J = kV$ relationship. Thus, the current transport was controlled by shunt resistance in this region. In region IV, the reverse currents increase with the reverse voltage instead of saturating at J_0, which implies current transport controlled by thermionic-field emission. Compared with the sample from the single-step sputtering process, the value of ideality factor was improved which shows a number closer to the ideal value of 1. The barrier height of the Schottky junction was also improved from 0.69 eV to 0.71 eV. The improvements can be attributed to the grain size increase and surface condition improvement by using the two-step sputtering process. However, the increase of the ideality factor and abnormal decrease of Φ_B with the temperature change implies that the current transport in the exponential region (II) deviates from pure thermionic-emission and other processes control the current transport, which will be discussed in the next section.

3. I-V-T analysis on a Schottky junction with ECR hydrogenation (two-step sputtering)

I-V-T curves for the sample prepared as titled show similarity to the sample without ECR passivation (section 2). However, with the hydrogenation, the Au Schottky device has a higher ϕ_B over the whole temperature range from 150 K to 350 K than the sample without hydrogen passivation. Also, with hydrogenation, the ideality factors at different temperature are more close to 1 which is more significant at the low temperature, as shown in Figure 3. This can be attributed to film passivation by ECR hydrogenation.

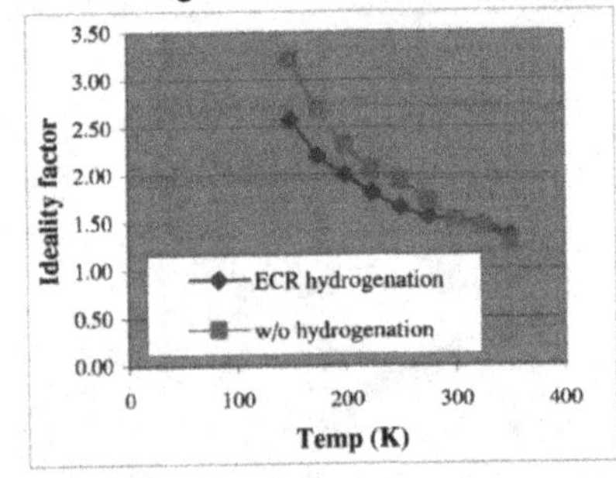

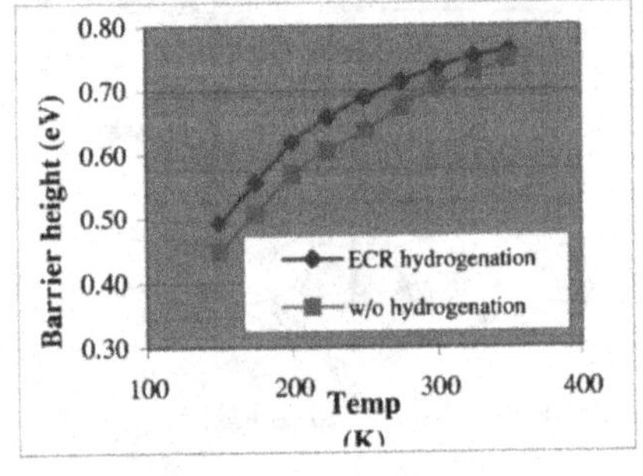

Figure 3. Temperature dependence of the ideality factor and barrier height.

Analysis of the active energy result, $E_0 = nkT/q$, is plotted vs. kT/q as shown in Figure 4. It shows typical thermionic-field emission which has less temperature-dependent characteristics at low temperature. The figure indicates that the transport mechanism involved in the diode is dominated by thermionic emission (TE) at higher temperatures with a smooth transition to field emission (tunneling) at lower temperatures.

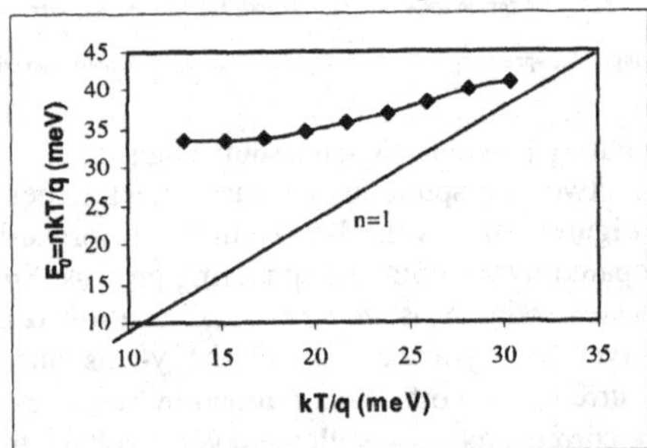

Figure 4 Experimental values of nkT/q as a function of temperature for the diode with ECR H_2 passivation.

C-V analysis revealed that the total carrier density for the sample discussed in section 2, which is without ECR hydrogenation, was 1.64×10^{17} cm^{-3} at room temperature. After ECR hydrogenation treatment, the charge state density measured by C-V was 1.25×10^{17} cm^{-3} at room temperature. These values are much higher than the doping density of the Si target. As reported [11,12], oxygen thermal donor effect should be considered based on SIMS result of 10^{20} cm^{-3} oxygen level in the Si film. With such high level n-type doping density, tunneling will strongly affect current transport. This might be the reason that thermionic-filed emission dominates the transport.

4. I-V-T analysis on the Schottky junction made from a Si film with oxygen control

Oxygen control on the sputtering gas was realized by adding an oxygen filter to the Ar gas line. The C-V result showed a three times lower level of the total charge density in the Si film. I-V-T results are shown in Figure 5. The reverse direction (region III) shows a big difference from previous diodes without oxygen control. In this regime, the current is fairly independent of the voltage, especially at high temperatures (300K~400K). Instead, the currents saturate at the value close to the saturation current, which indicates less tunneling to be involved. This should be attributed to the decreasing of the total charge density.

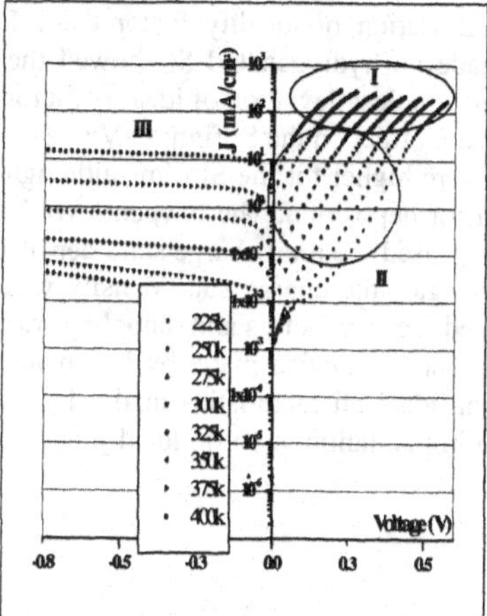

Figure 5. Dark J-V characteristics with temperature change for a sample with oxygen control.

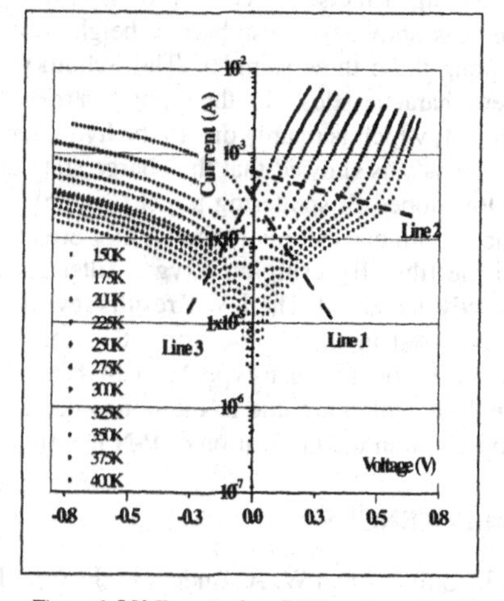

Figure 6. I-V-T results for a P-N junction solar cell sample prepared by ion implantation.

5. I-V-T analysis of a P-N junction solar cell

After ion-implantation, the sample was annealed in forming gas environment for 1 hr to activate the implanted dopant and cure the lattice damage during the ion-implantation. The sample was also hydrogenated by ECR plasma. From the I-V-T curves in Figure 6, the region over line 2 (high voltage region) shows the typical diffusion characteristics. The straight line portion between line 1 and line 2 has a large ideality factor value which implies a generation-recombination mechanism. The portion below line 1 (low voltage region) was symmetric with the corresponding part of the reverse voltage regime. From the LogJ-LogV plot, the J-V relation was found to follow: $J = kV^m$, with $m \approx 1$. So, the current transport in regime between line 1 and line 2 may be controlled by shunt resistance (R_{sh}) since both of them are under low applied voltages. Compared with the Schottky junction device, the P-N junction sample has higher reverse saturation current density and larger value of ideality factor which deviates more from the ideal value of 1. This shows the non-ideality of the P-N junction made by ion implantation. This might be due to the lattice damage by the ion implantation and non-ideal junction formation due to boron dopant diffusion along the poly-Si grain boundaries with the high temperature (700°C) annealing.

CONCLUSION

Current transport mechanisms were studied by using I-V-T results for the MIG poly-Si thin films with different processing methods. For the Schottky diodes made on single-step deposited Si (small grains), two-sep deposited Si (large grains) and hydrogenated two step deposited Si, I-V-T analysis shows that thermionic field emission dominated the current transport in the forward current region, while in the reverse current region tunneling affects the current transport. Single step sputtering deposited Si Schottky devices showed lowest barrier height and largest deviation of ideality factor from 1 among these three samples. The Schottky diode made on hydrogenated Si showed the best characteristics with the highest barrier height and smallest deviation of ideality factor from 1, which is mainly due to the hydrogen passivation effect on the Si film. C-V and C-V-T analysis showed that the charge state densities were higher for the Si films although a low doped Si sputtering target (source) were used for deposition. The oxygen level in the Si film caused the thermal donor effect which increased the total charge state density in the film. By using an oxygen filter for sputtering gas, the charge state density was greatly decreased. The I-V-T results revealed improved ideality factors and tunneling was suppressed in the reverse current region. The P-N junction device made by boron ion implantation in the n-type MIG Si film showed non-ideal characteristics in the I-V-T analysis which are due to the lattice damage by ion implantation and non-ideal junction formation in the fabrication of P-N junction devices.

REFERENCES

1 E. Guliants and W. A. Anderson, J. Appl. Phys., **87**, 3532

2 E. A. Guliants and Wayne A. Anderson, J. Appl. Phys. **89**, 4648

3 Chunhai Ji and Wayne A. Anderson, IEEE Electron Devices, v 50, No. 9, September 2003.

4 Chunhai Ji and Wayne A. Anderson, Solar Energy Materials and Solar Cells, In Press, Corrected Proof, Available online 16 June 2004

5 Chunhai Ji and W. A. Anderson, Mater. Res. Soc. Symp. Proc. 808, A4.21 San Francisco, CA, 2004

6 R.A. Ditizion, S.J. Fonash and B.C. Hseih, J. Vac. Sci. Tech. A. **10**, 1, 59, 1992.

7 N.H. Nickel, N.M. Johnson and W.B. Jackson, Appl. Phys. Lett., **62**, 25, 3266, 1993.

8 E.S. Cielazyk, K.H.R. Kirmse, R.A. Stewart and A.E. Wendt, Appl. Phys. Lett., **67**, 21, 3099, 1995

9 R.A. Ditizion, S.J. Fonash, B.C. Hseih, and D.W. Greve, Appl. Phys. Lett., **56**, 12, 1140, 1990

10 Y. Nasuno, M. Kondo, A. Matsuda, Appl. Phys. Lett. **78**, 2330, 2001.

11 M. Komoda, K. Kamesaki, A. Masuda, H. Matsumura, Thin Solid Films 395, 198, 2001.

12 T. Kamei, T. Wada, A. Matsuda, Mater. Res. Soc. Symp. Proc. 664, A10.1.1. 2001

III-V Semiconductors

Mater. Res. Soc. Symp. Proc. Vol. 836 © 2005 Materials Research Society

III-V Multi-Junction Materials and Solar Cells on Engineered SiGe/Si Substrates

Steven A. Ringel,[1] Carrie L. Andre,[1] Matthew Lueck,[1] David Isaacson,[2] Arthur J. Pitera,[2]
Eugene A. Fitzgerald[2] and David M. Wilt[3]
[1]Department of Electrical and Computer Engineering
The Ohio State University
2015 Neil Avenue, Columbus, OH 43210 USA
[2]Department of Materials Science and Engineering
Massachusetts Institute of Technology
77 Massachusetts Avenue, Cambridge, MA 02139 USA
[3]Photovoltaic and Space Environments Branch
NASA Glenn Research Center
21000 Brook Park Road, Cleveland, Ohio 44135 USA

ABSTRACT

The monolithic integration of high efficiency III-V compound solar cell materials and devices with lower-cost, robust and scaleable Si substrates has been a driving force in photovoltaics (PV) basic research for decades. Recent advances in controlling mismatch-induced defects that result from structural and chemical differences between III-V solar cell materials and Si using a combination of SiGe interlayers and monolayer-scale control of III-V/IV interfaces, have led to a series of fundamental advances at the material and device levels, which establish that the great potential of III-V/Si PV is within reach. These include demonstrations of GaAs epitaxial layers on Si that are anti-phase domain-free with verified dislocation densities at or below $1x10^6$ cm^{-2} and negligible interface diffusion, minority carrier lifetimes for GaAs on Si in excess of 10 ns, single junction GaAs-based solar cells on Si with open circuit voltages (V_{oc}) in excess of 980 mV, efficiencies beyond 18%, and area-independent PV characteristics up to at least 4 cm^2. These advances are attributed in large part to the use of a novel "engineered Si substrate" based on compositionally-graded SiGe buffers such that a high-quality, low defect density, relaxed, "virtual" Ge substrate could be developed that can support lattice-matched III-V epitaxy and thus merge III-V technology based on the GaAs (or Ge) lattice constant with Si wafers. This paper focuses on recent results that extend this work to the first demonstration of high performance III-V dual junction solar cells on SiGe/Si. Open circuit voltages in excess of 2 V at one-sun have been obtained for the conventionally "lattice-matched" $In_{0.49}Ga_{0.51}P$/GaAs dual junction cells on inactive, engineered SiGe/Si; to our knowledge is the first demonstration of > 2V solar power generation on a Si wafer. Comparisons with identical cells on GaAs substrates reveal that the V_{oc} on engineered Si retains more than 94% of its homoepitaxial value, and that at present both DJ/GaAs and DJ/SiGe/Si cells are similarly limited by current mismatch in these early cells, and not fundamental defect factors associated with the engineered Si substrates.

INTRODUCTION

Recent work using relaxed buffer layers consisting of step-graded SiGe has shown enormous potential for achieving large area, low-cost direct integration of III-V compound semiconductor photovoltaics based on the GaAs lattice constant [1], and III-V optoelectronics based on the AlGaInP-GaAs material system onto large area Si wafers [2]. This has been established over a period of more than 10 years, first with the demonstration of very low threading dislocation density (TDD) in fully relaxed Ge layers grown on SiGe buffers on Si [3], followed by a series of fundamental material advances that established a growth methodology to achieve ideal, low-mismatched GaAs growth on these Ge-capped SiGe/Si wafers with material quality consistent with that necessary to support high performance, minority carrier III-V devices on Si for the first time [4],[5]. Figure 1 shows the range of lattice constants bridged by the use of SiGe relaxed buffers moving from the Si to Ge lattice constants, indicating not only the ability to move technologies based on Ge or GaAs substrate to a Si substrate, but also the general utility in achieving any lattice constant between Si and Ge on an initial Si substrate. This paper briefly reviews the general approach of achieving a high quality GaAs layer on a Si wafer using this SiGe lattice engineering approach specific to the application of III-V/SiGe/Si heterostructures for single junction photovoltaics, followed by discussions of recent results that have established the first $In_{0.49}Ga_{0.51}P$/GaAs dual junction solar cell grown and tested on SiGe/Si substrates. It is shown that the high performance recently demonstrated for single junction GaAs cells on SiGe/Si [6] is replicated for these dual junction cells, with the first report of solar-generated power on Si with greater than a 2 volt output [7].

EXPERIMENT

Engineered SiGe/Si substrates and GaAs overlayers

All SiGe/Si substrates were formed by using n-type, relaxed, step-graded SiGe layers with an increasing Ge mole fraction from 0% Ge, at growth initiation on Si, to 100% Ge at completion. All growths were performed on (100) oriented Si substrates with a 6° off-cut toward the nearest {111} by ultra-high vacuum chemical vapor deposition. A chemical mechanical polish was employed at $Si_{0.5}Ge_{0.5}$ followed by growth of the remainder of the $Si_{1-x}Ge_x$ layers up to 100% Ge, in order to achieve a lower residual TDD [3]. Two separate sets of SiGe/Si substrates were used in this study with the same, nominal buffer design. The first set of SiGe/Si substrates were grown on 0.03 Ω-cm 4" Si wafers with a doping concentration in the epitaxial SiGe layers of $\sim 1 \times 10^{18}$ cm^{-3} and resulted in a TDD of $\sim 0.9 \pm 0.2 \times 10^{6}$ cm^{-2}. These substrates were used for single junction GaAs cells. The second set of SiGe/Si substrates were grown on 0.01 Ω-cm 6" Si wafers; the SiGe layers were again doped $\sim 1 \times 10^{18}$ cm^{-3}, but the final TDD was $1.8 \pm 0.2 \times 10^{6}$ cm^{-2}. These substrates were used for the dual junction cells. TDD values in the final Ge layers were measured using etch pit density (EPD) studies prior to III-V growth. The factor of ~2 variation in final TDD is typical of our SiGe growth process at this stage of development, and the influence of such variations on the minority carrier lifetimes of subsequently grown GaAs have been reported [5], [8].

The ability to achieve an ideal, low mismatched GaAs/Ge interface on the SiGe/Si substrates is a critical component for success in this process. The nucleation of any new dislocations at this interface must be negligible compared to the residual TDD of the SiGe/Si substrate. In addition, elimination of antiphase domain (APD) disorder at the interface scale due to the polar/nonpolar GaAs/Ge interface without use of thick GaAs buffers is also a necessity, as is the suppression of interdiffusion and autodoping. Each of these goals must be accomplished simultaneously to generate successful minority carrier devices such as solar cells on SiGe/Si.

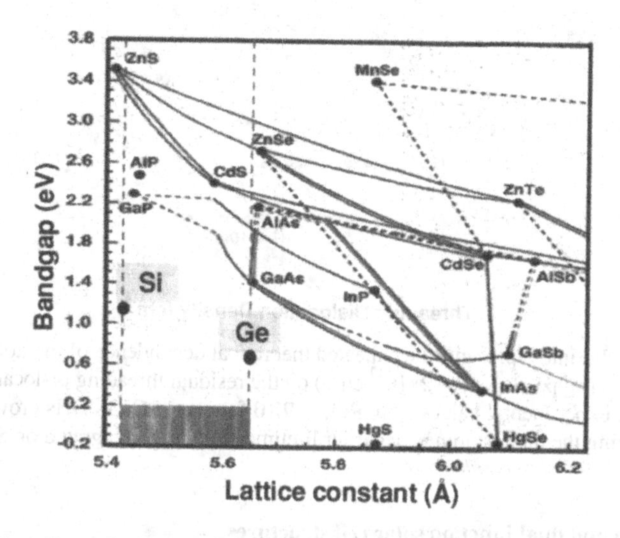

Figure 1. Semiconductor lattice constant versus energy gap. The solid lines represent direct bandgaps, where as the dotted lines represent indirect transitions. An inset of a cross sectional TEM image of a SiGe step graded buffer spanning the Si – Ge lattice constant range is shown.

To achieve this, a GaAs nucleation layer by solid source molecular beam epitaxy (SSMBE) on the Ge termination layer of the SiGe/Si substrates was grown prior to the growth of the solar cell structures. The SSMBE GaAs initiation conditions include the deposition of an epitaxial Ge layer (~30 nm), a substrate anneal at 640°C, 10 periods of migration-enhanced epitaxy of GaAs at 350°C beginning with an As pre-layer, and concludes with a 0.1 μm layer of GaAs grown at a rate of 0.1μm/hr and a substrate temperature of 500°C. (Further details concerning the SSMBE GaAs initiation can be found in [4, 5].) This initiation procedure completely eliminates APD formation and minimizes interdiffusion, which allows the use of thin GaAs buffer layers of less than 200 nm prior to cell growth and produces TDD values in thick GaAs layers grown on these substrates that match the TDD values in the Ge cap of the SiGe/Si substrate prior to III-V growth [5].

A very useful measure of the electronic material quality is provided by time resolved photoluminescence (TRPL) studies of the minority carrier lifetime, which can be used to predict ultimate solar cell performance. Figure 2 shows a summary of measured TRPL lifetimes plotted against measured TDD values for GaAs/Si using SiGe, and also GaAs/Si formed using other, previously reported approaches such as III-V graded buffers, strained layer superlattices and thermally-cycled annealing [9], [10]. As seen, minority carrier lifetimes achieved for GaAs/Si using SiGe are extremely high, in excess of 10 ns at a TDD of $\sim 1 \times 10^6$ cm^{-2}. Not only does this confirm the direct impact of the low TDD process enabled by SiGe grading on achieving the highest minority carrier quality GaAs on Si, but it also confirms that the GaAs/Ge interface nucleation process introduces negligible amounts of additional defects that would otherwise have reduced material quality.

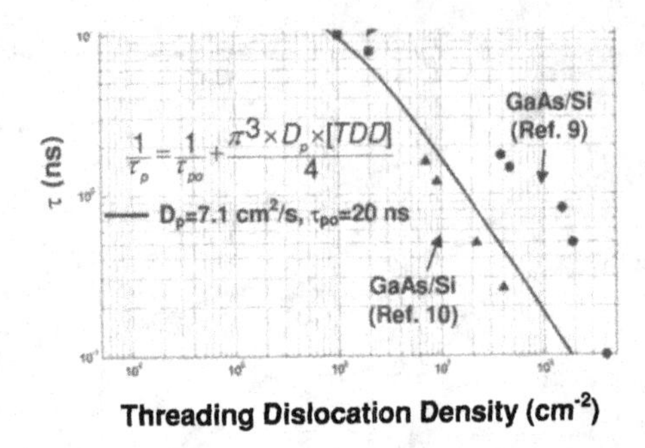

Figure 2. The solid line represents the expected theoretical dependence of the hole minority carrier lifetime in n-type GaAs (n~ 2×10^{17} cm^{-3}) on the residual threading dislocation density. Also plotted are experimental values from Ref. 5, 9,10 for n-type GaAs DHs grown on Si and SiGe/Si, indicating the highest minority carrier lifetimes for GaAs integrated on Si using SiGe/Si substrates.

Single junction and dual junction solar cell structures

The single junction (SJ) p+/n GaAs solar cells presented in this paper were grown in a horizontal geometry low-pressure metal-organic chemical vapor deposition (MOCVD) reactor which accommodates a single 2" wafer, maintains a pressure of 190 torr, and uses RF heating with a thermocouple placed below the susceptor [11]. The GaAs and $In_{0.49}Ga_{0.51}P$ layers were grown at a substrate temperature of 620°C, a growth rate ~ 2 μm/hr, and a V/III ratio of 100. Silane and diethylzinc were used as n-type and p-type dopant sources, respectively, and the doping concentrations were calibrated by electrochemical capacitance voltage measurements. Figure 3a shows the standard GaAs heteroface solar cell structure that was used for all devices. Figure 3b shows a cross-sectional transmission electron microscopy (X-TEM) image of the cell structure grown on a SiGe/Si substrate; this image shows the III-V device layers after the removal of the contact layer, the Ge termination layer of the SiGe/Si substrate, and some of the SiGe step-graded buffer layers. No impact from the transfer from MBE growth to MOCVD growth is seen from TEM; moreover, higher magnification images confirm the suppression of APD disorder at the GaAs/Ge interface. (Further details concerning the influence of MOCVD over growth on these SSMBE initiation layers can be found in [8, 12])

p++ GaAs contact layer (1000Å)	~1x10^{19} cm^{-3}
p+ $In_{0.49}Ga_{0.51}P$ window (500 Å)	~1x10^{18} cm^{-3}
p+ GaAs emitter (5000 Å)	~2x10^{18} cm^{-3}
n GaAs base (2.0 μm)	~1x10^{17} cm^{-3}
n+ $In_{0.49}Ga_{0.51}P$ back surface field (1000 Å)	~1x10^{18} cm^{-3}
n+ GaAs buffer (1000 Å)	~1x10^{18} cm^{-3}
n+ GaAs buffer by MBE (1000 Å)	~2x10^{19} cm^{-3}
Ge layer (300 Å)	-
n SiGe substrate	-

a)

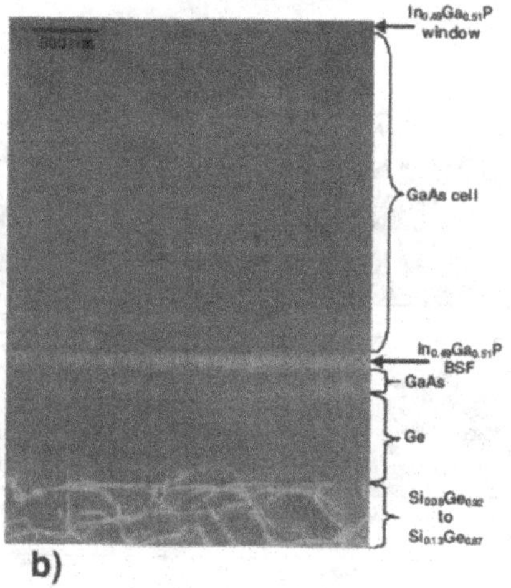

b)

Figure 3. a) The p+/n GaAs solar cell structure grown by MOCVD on SiGe /Si substrates. b) X-TEM image of a p+/n GaAs solar cell grown by MOCVD on a SiGe /Si substrate.

The $In_{0.49}Ga_{0.51}P$/GaAs p+/n dual junction (DJ) solar cell device layers were grown by solid source molecular beam epitaxy (SSMBE) on GaAs and SiGe/Si substrates in an identical manner. The target device structure and an X-TEM image of the entire grown device is shown in figure 4. The SSMBE growth rates used for the bottom cell were 1.0 μm/hr and 0.6 μm/hr, for the GaAs and $Al_{0.7}Ga_{0.3}As$ layers, respectively. N-type GaAs and $Al_{0.7}Ga_{0.3}As$ layers were grown at 610°C with an As_2:III beam equivalent pressure (BEP) ratio of 14:1 with reference to the Ga BEP, while the p-type GaAs and $Al_{0.7}Ga_{0.3}As$ layers were grown at 575°C with an As_2:III ratio of 24:1. After growth of the bottom cell, there was a growth stop to adjust the In and Ga source temperatures to achieve the proper $In_{0.49}Ga_{0.51}P$ composition. The GaAs tunnel junction layers were then grown at 550°C with an As_2:III ratio of 24:1 and a growth rate of 0.6 μm/hr. Subsequently, the $In_{0.49}Ga_{0.51}P$ and $In_{0.47}(Al_{0.7}Ga_{0.3})_{0.53}P$ layers were grown at a rate of ~ 1.15 μm/hr, a growth temperature of 490°C, and a P_2:III BEP ratio of 9:1 with reference to the In BEP. Finally, the p+ GaAs contact was grown in the same manner the GaAs tunnel junction. By using two Ga sources, there was no growth stop at the heterostructure interfaces for either the GaAs or $In_{0.49}Ga_{0.51}P$ sub-cells.

p++ GaAs contact layer (1000 Å)	~1x10^{19} cm^{-3}
p+ In$_{0.47}$(Al$_{0.7}$Ga$_{0.3}$)$_{0.53}$P window (300 Å)	~2 x10^{18} cm^{-3}
p+ In$_{0.49}$Ga$_{0.51}$P emitter (2000 Å)	~2 x10^{18} cm^{-3}
n In$_{0.49}$Ga$_{0.51}$P base (4000 Å)	~2 x10^{17} cm^{-3}
n+ In$_{0.49}$Ga$_{0.51}$P back surface field (300 Å)	~2 x10^{18} cm^{-3}
n+ In$_{0.47}$(Al$_{0.7}$Ga$_{0.3}$)$_{0.53}$P back surface field (300 Å)	~2 x10^{18} cm^{-3}
n++ GaAs TJ (250 Å)	~2 x10^{19} cm^{-3}
p++ GaAs TJ (300 Å)	~2 x10^{19} cm^{-3}
p+ Al$_{0.7}$Ga$_{0.3}$As window (400 Å)	~3 x10^{18} cm^{-3}
p+ GaAs emitter (5000 Å)	~2 x10^{18} cm^{-3}
n GaAs base (20,500 Å)	~2 x10^{17} cm^{-3}
n+ Al$_{0.3}$Ga$_{0.7}$As back surface field (1000 Å)	~2 x10^{18} cm^{-3}
n+ GaAs buffer (2000 Å)	~2 x10^{18} cm^{-3}
Ge (300Å)	uid
n+ SiGe substrate	~1 x10^{18} cm^{-3}

a)

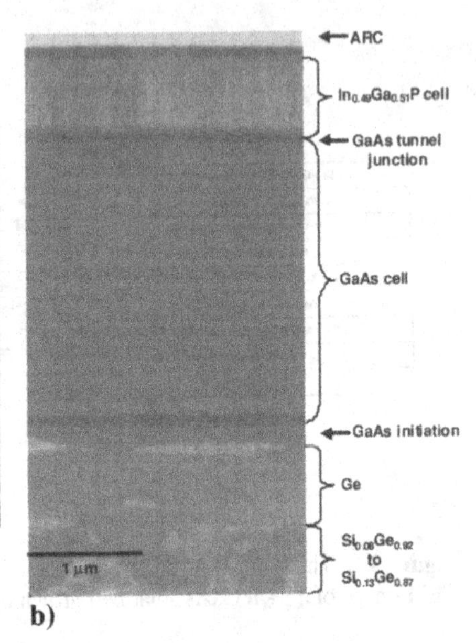

b)

Figure 4. a) p+/n In$_{0.49}$Ga$_{0.51}$P/GaAs dual junction solar cell structure grown on a SiGe/Si substrate. The DJ solar cell grown on GaAs has an identical structure with the exception of the Ge epitaxial layer. b) X-TEM micrograph of the dual junction solar cell grown on a SiGe/Si substrate.

The devices were processed using conventional photolithography and wet chemical etching. Wet chemical etching was used for device isolation; $NH_4OH:H_2O_2$:DI (2:1:50) was used for GaAs cell and HCl:DI (1:1) was used for the In$_{0.49}$Ga$_{0.51}$P cell. Ohmic contacts were made to the Ge coated back of the SiGe/Si substrate and the n-type GaAs substrate using Ni-Ge-Au (50 Å/ 328 Å/ 1000 Å) metallization with a 4 min anneal at 400°C and to the p-type GaAs contact layer using Cr-Au (100 Å/ 2 µm) metallization. The anti-reflection coating (ARC) consisted of MgF_2 / ZnS / MgF_2 and was deposited in a thermal evaporator. Due to uncontrolled variations in the thickness of ARC layers, the increase in short-circuit current density for solar cells with different ARC depositions varies between 30 – 37% and thus the changes in measured short-circuit densities for the cells presented in this paper do not necessarily indicate changes in material quality. Illuminated current density versus voltage (J-V) measurements under an AM0 spectrum were measured at the NASA Glenn Research Center and under an AM1.5-G spectrum at the National Renewable Energy Laboratory were performed in order to determine the short-circuit current density (J_{sc}), the open-circuit voltage (V_{oc}), the fill-factor (FF), and the efficiency (η) of the solar cells for space and terrestrial applications. The solar cells produced had device areas of 0.044 cm^2 with metal coverage of 10 percent.

RESULTS AND DISCUSSION

Single Junction GaAs/SiGe/Si

Figure 5 shows the illuminated J-V response for a p+/n single junction GaAs solar cell grown on a SiGe/Si substrate. AM1.5-G energy conversion efficiencies reached 18.1%. This is the highest independently verified one-sun efficiency reported to date for a single p^+/n or n^+/p junction GaAs solar cell grown on a Si-based substrate [13], [14]. The AM0 efficiency for the same cell shown in figure 5 was 15.5% with a V_{oc} of 980 mV, and is also the highest independently verified efficiency and V_{oc} reported to date under this spectrum [14]. The high performance obtained, in spite of the large lattice mismatch (4%) between the fully relaxed GaAs cell and the Si substrate, is attributed to the achievement of a high V_{oc}, 973 mV, which is the consequence of maintaining a low TDD value of ~ 1×10^6 cm^{-2} while simultaneously eliminating other mismatch related defects such as anti-phase domains and auto-doping, which could otherwise lower V_{oc} [4], [5]. These already high efficiencies are limited by a large, 10%, grid finger shadowing. A more accurate comparison with previous results is obtained by scaling the data in figure 5 for a metal coverage 4.5%, commensurate with the GaAs/Si cell from [13] and [14]. This yields efficiencies for GaAs/SiGe/Si of 18.8% and 16.5% at AM1.5-G and AM0, respectively, more than 1% absolute higher than other verified reports.

The high V_{oc} values obtained in this work are responsible for the high cell efficiency. Compared with [13] where a TDD of ~ 8×10^6 cm^{-2} produced an AM1.5-G V_{oc} of 891 mV, the reduction in the TDD in SiGe/Si substrates to ~1×10^6 cm^{-2} reduced the dislocation-mediated depletion region recombination in GaAs/SiGe/Si solar cells and thus increased the V_{oc} to 973 mV. The inability to reduce the TDD below ~ 5×10^6 cm^{-2} using other direct III-V/Si integration methods [15] had been the primary impediment to achieving higher efficiency GaAs-on-Si solar cells which stalled research in this area. Modeling of this recombination process has predicted theoretical V_{oc} values that track the experimental results as a function of TDD for various GaAs/Si and GaAs/SiGe/Si cell structures. These results directly track the lifetime versus TDD dependence shown in figure 2, and predict that a TDD of ~ 5×10^5 cm^{-2} is needed for heteroepitaxial GaAs p+/n cells to achieve V_{oc} values in excess of 1000 mV, which would closely match state of the art homoepitaxial GaAs cells.

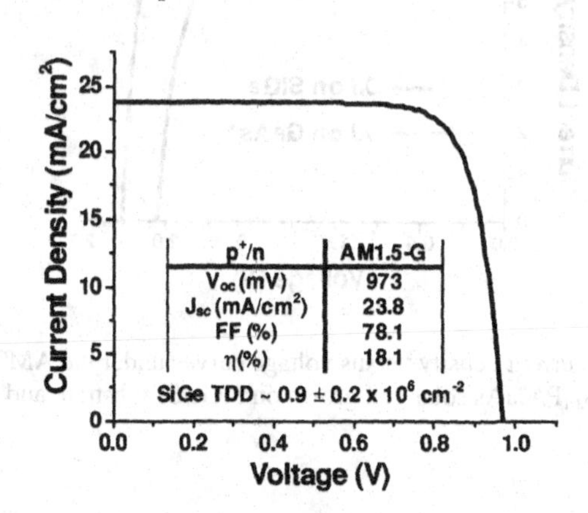

Figure 5. AM1.5-G performance for SJ GaAs cell grown on Si using inactive SiGe interlayers.

$In_{0.49}Ga_{0.51}P$ /GaAs Dual Junction cell on SiGe/Si

Figure 6 shows the solar cell performance measured under AM0 illumination for the first dual junction solar cell grown on a compositionally step graded SiGe/Si substrate. Results for an identical SSMBE-grown p+/n DJ cell on GaAs is also shown to make a direct assessment of the impact, if any, of substituting the novel SiGe/Si substrate. An open-circuit voltage of 2.023 V for the DJ solar cell grown on SiGe/Si is demonstrated. To our knowledge this is the first demonstration of solar-generated power that exceeds a 2 V on Si. The ability to achieve such high V_{oc} values has been one of the major hurdles for advancing a photovoltaics technology based on III-V/Si solar cells due to the strong dependence of V_{oc} on dislocation-related minority carrier recombination in highly lattice mismatched systems as discussed above [15]. For comparison, the only other reported $In_{0.49}Ga_{0.51}P$/GaAs DJ solar cell epitaxially grown on a Si-based substrate achieved a V_{oc} value of 1.70 V under AM0 illumination [16].

The high V_{oc} obtained here is attributed to maintaining the low TDD and high interfacial quality within the DJ cell on SiGe/Si, and is a direct extension from the SJ GaAs cells on Si using SiGe, which demonstrated the highest V_{oc} values for a SJ GaAs cell epitaxially integrated on Si. Similar material quality conclusions can be made from figure 6 concerning the J_{sc}. We find that the measured J_{sc} for the DJ solar cells grown on SiG/SiGe retains 99% of J_{sc} measured for the companion homoepitaxial cells. Hence, the impact of the threading dislocation density on the collected current is negligibly small. However, while the J_{sc} values for cells on both substrates are virtually identical, confirming substrate independence between GaAs and low TDD SiGe/Si, they *both* are well below the anticipated value of ~ 16 mA/cm^2 based on the $In_{0.49}Ga_{0.51}P$ top cell thickness [17]. Since this is clearly not a TDD effect, we investigated current collection within each sub-cell of the DJ cell to identify the source for current loss in the DJ solar cells. Figure 7a shows the measured external quantum efficiencies (EQE) of each sub-cell. Based on the EQE measurements, the integrated AM0 J_{sc} values were 7.3 and 14.3 mA/cm^2

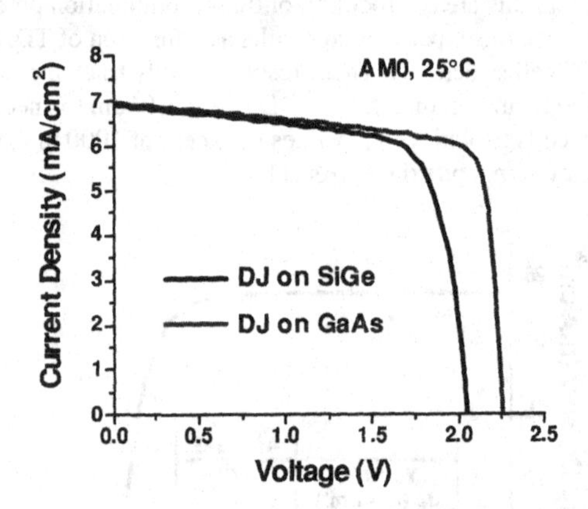

Figure 6. Illuminated current density versus voltage curves under the AM0 spectrum for a p+/n dual junction $In_{0.49}Ga_{0.51}P$/GaAs solar cell grown on SiGe/Si substrate and on a GaAs substrate.

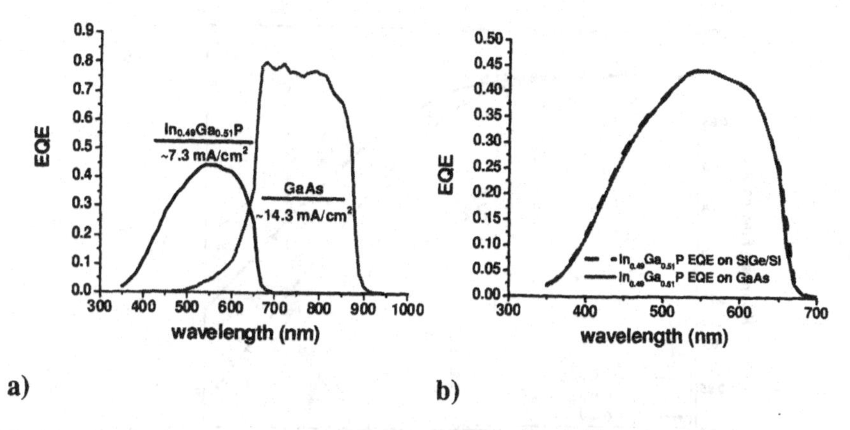

Figure 7. a) External quantum efficiency (EQE) measured for the DJ cell on GaAs. The integrated J_{sc} for an AM0 spectrum for each sub-cell is 7.3 mA/cm^2 and 14.3 mA/cm^2 for $In_{0.49}Ga_{0.51}P$ and GaAs, respectively. b) The EQE for the $In_{0.49}Ga_{0.51}P$ top cell of the DJ solar cells grown on GaAs and SiGe/Si shows no change in collection for either substrate.

for the $In_{0.49}Ga_{0.51}P$ and GaAs sub-cells, respectively. Since the calculated J_{sc} for $In_{0.49}Ga_{0.51}P$ top cell (7.3 mA/cm^2) is close to the value obtained for the DJ cell on *either* substrate, we conclude that the J_{sc} is limited by the current collection in the $In_{0.49}Ga_{0.51}P$ sub-cell of the series connected DJ device, and that this is *not* related to the substrate or TDD; i.e., the $In_{0.49}Ga_{0.51}P$ top cell design and/or baseline quality limited the total J_{sc} for these cells. This is confirmed in figure 7b by the comparison of the $In_{0.49}Ga_{0.51}P$ top cell responses for the DJ devices grown on GaAs and SiGe/Si.

DJ cell analysis

Since the above discussion shows that the DJ results on both substrates appear to be limited by the quality and immaturity of our current SSMBE process to generated $In_{0.49}Ga_{0.51}P$ sub-cells, we are currently working to optimize the $In_{0.49}Ga_{0.51}P$ material. Steps include thinning the emitter and an in-situ anneal. An improvement in J_{sc} by a factor of 1.6 over that shown in figure 6 has already been achieved for p+/n SJ $In_{0.49}Ga_{0.51}P$ devices on *both* GaAs and SiGe/Si [18], and thus the expected J_{sc} values for DJ/Si devices should be attainable once the modified $In_{0.49}Ga_{0.51}P$ cells are incorporated into the DJ structure (in progress). It is therefore useful to measure the impact of increased total current collection on V_{oc} by adjusting the incident photon flux, to obtain a J_{sc} value of 16 mA/cm^2, consistent with commercial DJ solar cells and our initial design assumption. This increases the V_{oc} from 2.023 V to 2.140 V, which, when compared with the homoepitaxial DJ cell measured under the same conditions, is 94% of the homoepitaxial value. Hence, DJ cells on SiGe/Si with an optimized $In_{0.49}Ga_{0.51}P$ top cell should realistically result in high efficiency devices.

Moreover, the $In_{0.49}Ga_{0.51}P$ top cell is expected to be more TD tolerant than the GaAs bottom cell stemming from the overall lower diffusion coefficients of minority carriers in

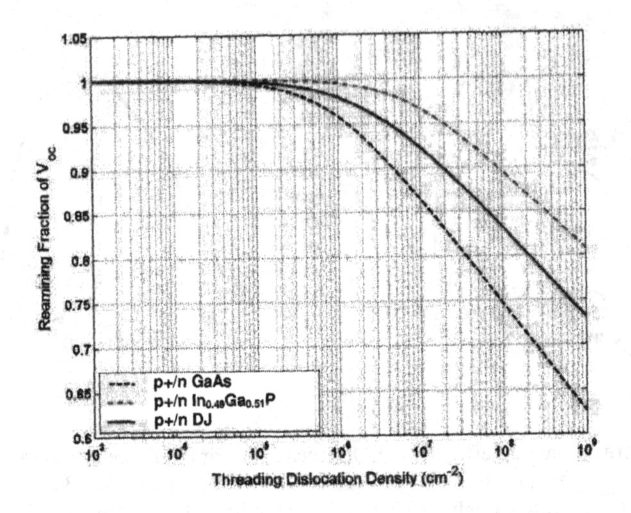

Figure 8. Remaining fraction of V_{oc} as function of TDD for the individual GaAs and $In_{0.49}Ga_{0.51}P$ sub-cells and the combined DJ solar cell for p+/n polarity devices assuming a short-circuit current density of 16 mA/cm^2 and a TDD dependent minority carrier lifetime models.

$In_{0.49}Ga_{0.51}P$ compared with GaAs.[18] Figure 8 compares the remaining fraction of V_{oc} as function of TDD for DJ $In_{0.49}Ga_{0.51}P$/GaAs solar cells to the remaining fraction of V_{oc} in the $In_{0.49}Ga_{0.51}P$ and GaAs sub-cells assuming a J_{sc} value of 16 mA/cm^2 for the p+/n device polarity. Here the remaining fractions are defined as the ratios of the V_{oc} values at a particular TDD to the V_{oc} values obtained in the absence of TDs. The minority carrier diffusion coefficient impacts the onset of degradation for both the minority carrier lifetimes and the V_{oc} and the rate of degradation in the TDD dominated regime for both parameters. By defining a 95% reduction as the "onset of degradation", degradation occurs at a TDD of 2x10^7 cm^{-2} for the p+/n $In_{0.49}Ga_{0.51}P$ top cell and TDD of 1x10^6 cm^{-2} TDD for the p+/n GaAs bottom cell. The combination of these sub-cells produces a p+/n DJ device with an "onset of degradation" occurring at a TDD of 4x10^6 cm^{-2}. In fact, current state of the art SiGe/Si substrates with a TDD of 1x10^6 cm^{-2} are already suitable for high performance p+/n DJ devices, with calculations showing that V_{oc} values will maintain 98% of their homoepitaxial V_{oc} value. These results imply an important conclusion; $In_{0.49}Ga_{0.51}P$/GaAs DJ cells should be more tolerant to residual TDD than SJ GaAs cells.

CONCLUSIONS

GaAs solar cells grown on SiGe/Si substrates with efficiencies of 18.1% under AM1.5-G illumination are demonstrated. High V_{oc} values are achieved by the reduction in depletion region recombination that results from the lower residual TDD compared with previous GaAs/Si integration methods. The low TDD achieved in the "virtual Ge" substrates was maintained in the GaAs solar cell via controlled III-V epitaxy with no introduction of other efficiency limiting defects. Thus, improvements in cell performance should continue to track decreases in TDD with future advancements in SiGe/Si substrate technology. Moreover, although not shown here, an increase in cell area by a factor of ~ 100 without fundamental degradation in J_{sc} and V_{oc} has also been achieved for constant substrate TDD values.[1] This data supports the eventual applicability of SiGe metamorphic buffers for industrially compatible III-V/Si photovoltaics.

Moreover, the first DJ III-V cells on Si using SiGe are also reported.[7] A low defect density is maintained in spite of the large lattice mismatch, resulting in solar cell V_{oc} values that exceed 2 V on Si for the first time under a standard solar spectrum. High efficiencies are realistically achievable once the now-optimized $In_{0.49}Ga_{0.51}P$ top cell is incorporated into the DJ structure so that the current mismatch, which is unrelated to substrate, is alleviated. Moreover, preliminary modeling, to be presented elsewhere, suggests that an optimized p+/n $In_{0.49}Ga_{0.51}P$/GaAs DJ cell on a SiGe/Si with a TDD of $\sim 1 \times 10^6$ cm^{-2} will achieve a V_{oc} that is greater than 98% of the homoepitaxial device value compared with 95% for SJ p+/n GaAs device structures with the same TDD [18]. This enhanced TD tolerance coupled with the possibility of higher efficiencies and greater radiation tolerance of multi-junction cells not only makes multi-junction solar cells on SiGe/Si of substantial technological potential, but also serves to demonstrate ideal devices for III-V/Si integration.

ACKNOWLEDGEMENTS

The authors acknowledge D. Scheiman at the NASA Glenn Research Center for AM0 characterization and K. Emery and T. Moriarty at the National Renewable Energy Laboratory for AM1.5 characterization. This work was supported in part by the National Aeronautics and Space Administration under Grant NCC3-974, National Renewable Energy Laboratory under Grant ACQ-1-30619-06 and XAT-4-33624-14-PVP40102, the Army Research Office under Grant (DAAD 19-01-0588), the National Science Foundation under FRG grant DMR-0313468, and by an Ohio Space Grant Consortium Fellowship (C.L.A).

REFERENCES

1. S. A. Ringel,. C. L. Andre, M. K. Hudait, D. M. Wilt, E. B. Clark, A. J. Pitera, M. L. Lee, E. A. Fitzgerald, M. Carroll, M. Erdtmann, J. A. Carlin, and B. M. Keyes, *Proc. 3rd World Conf. on Photov. Energy Conv.*,Vol. 1 , pp. 612- 615 (2003).
2. O. Kwon, J. Boeckl, M. L. Lee, A. J. Pitera, E. A. Fitzgerald, and S. A. Ringel in Progress in Compound Semiconductor III - Materials Electronic and Optoelectronic Applications, edited by D. Friedman, M.O. Manasreh, I. Buyanova, F.D. Auret, A. Munkholm, (*Mater. Res. Soc. Symp. Pro.*, **799**, Boston, MA, 2004) Z3.4.1.
3. M. T. Currie, S. B. Samavedam, T. A. Langdo, C. W. Leitz, and E. A. Fitzgerald, *Appl. Phys. Lett.*, **72**, 1718 (1998).
4. R. M. Sieg, S. A. Ringel, S. M. Ting, S. B. Samavedam, M. T. Currie, T. A. Langdo, and E. A. Fitzgerald, "Toward device-quality GaAs growth by molecular beam epitaxy on offcut Ge/$Si_{1-x}Ge_x$/Si substrates," *J. Vac. Sci. Technol.* B, **16**, no. 3, 1471 (1998)

5. J. A. Carlin, S. A. Ringel, E. A. Fitzgerald, M. Bulsara, and B. M. Keyes, "Impact of GaAs buffer thickness on electronic quality of GaAs grown on graded Ge/GeSi/Si substrates," *Appl. Phys. Lett.*, **76**, 1884 (2000).
6. C. L. Andre, J. A. Carlin, J. J. Boeckl, D. M. Wilt, M. A. Smith, A. J. Pitera, M. L. Lee, E. A. Fitzgerald, and S. A. Ringel, "Investigations of high performance GaAs solar cells grown on Ge/$Si_{1-x}Ge_x$/Si substrates", submitted to *IEEE Tran. in Electron Dev. (2004).*
7. C. L. Andre, A. J. Pitera, M. L. Lee, E. A. Fitzgerald, and S. A. Ringel, "Dual junction $In_{0.49}Ga_{0.51}P$/GaAs solar cells grown on metamorphic SiGe substrates", submitted to *IEEE Electron Device Lett. (2004).*
8. C. L. Andre, D. M. Wilt, A. J. Pitera, M. L. Lee, E. A. Fitzgerald, B. M. Keyes, and S. A. Ringel, *Appl. Phys. Lett.*, **84**, pp. 3447 (2004).
9. R. K. Ahrenkiel, M. M. Al-Jassim, B. M. Keyes, D. Dunlavy, K. M. Jones, S. M. Vernon, and T. M. Dixon, *J. Electrochem. Soc.*, **137**, 996 (1990).
10. M. Yamaguchi, C. Amano, and Y. Itoh, *J. Appl. Phys.*, **66**, 915 (1989).
11. D. M. Wilt, N. S. Fatemi, R. W. Hoffman, P. P. Jenkins, D. J. Brinker, D. Scheiman, R. Lowe, M. Fauer, and R. K. Jain, *Appl. Phys. Lett.*, **64**, 2415 (1994).
12. J. J. Boeckl, "Structural Characterization of GaAs grown on Ge and SiGe substrates" Ph.D. dissertation, Dept. Elect. and Comp. Eng., The Ohio State Univ., Columbus, OH, 2005.
13. S. M. Vernon, S. P. Tobin, V. E. Haven, Jr., C. Bajar, and T. M. Dixon, *Proc. 20th Photov. Spec. Conf.*, pp. 481-485 (1988).
14. S. P. Tobin, "Progress in Gallium Arsenide Solar Cell Research," *Proc. of the Photov. Science and Engineering Conf., Australia*, 1989, Paper 4.1.
15. M. Yamaguchi and C. Amano, *J. Appl. Phys.*, **58**, 3601 (1985)
16. C. Chu, *Proc. 28th IEEE Photov. Spec. Conf.*, pp. 1250-1252 (2000).
17. K. A. Bertness, S. R. Kurtz, D. J. Friedman, A. E. Kibbler, C. Kramer, and J. M. Olson,, *Appl. Phys. Lett.*, **65**, 989 (1994).
18. C. L. Andre, "III-V semiconductors on SiGe substrates for multi-junction photovoltaics" Ph.D. dissertation, Dept. Elect. and Comp. Eng., The Ohio State Univ., Columbus, OH, 2004.

Mater. Res. Soc. Symp. Proc. Vol. 836 © 2005 Materials Research Society

Metamorphic GaInP-GaInAs Layers for Photovoltaic Applications

A.W. Bett[1], C. Baur[1], F. Dimroth[1] and J. Schöne[1,2]
[1]Fraunhofer Institute for Solar Energy Systems ISE,
Heidenhofstrasse 2, 79110 Freiburg, Germany, andreas.bett@ise.fraunhofer.de
[2]Technische Fakultät der Christian-Albrechts-Universität zu Kiel, Kaiserstrasse 2, 24143 Kiel

ABSTRACT

$Ga_xIn_{1-x}As$ and $Ga_yIn_{1-y}P$ layers were grown lattice mismatched to GaAs and Ge by low-pressure metal organic vapor phase epitaxy (LP-MOPVE). These materials are very promising for further increasing the efficiency of monolithic triple-junction solar cells . Different buffer layer structures were realized. Transmission electron microscopy and x-ray diffraction analysis were used to characterize the quality of the crystal . Both linear and step-graded buffers in $Ga_xIn_{1-x}As$ were successfully used under an active solar cell structure. $Ga_yIn_{1-y}P$ as buffer material showed a worse performance. Excellent solar cell performance was achieved for lattice mismatched single-, dual- and triple-junction solar cells.

INTRODUCTION

Monolithic III-V multi-junction solar cells are widely used for power production in space satellites today. The main reasons are the high efficiency close to 30 %, recently achieved under the AM0 solar spectrum [1], and the excellent electron and proton radiation hardness, which is essential in the space environment. Another potential high-volume application for III-V multi-junction solar cells is terrestrial concentrators working at high illumination intensities. Recently excellent efficiencies above 37 % under the terrestrial AM1.5 solar spectrum were reported [2,3].

At present, the best III-V solar cells for both terrestrial and space applications are based on the lattice matched material combination of $Ga_{0.51}In_{0.49}P/Ga_{0.99}In_{0.1}As/Ge$. These monolithic triple-junction solar cell structures are grown by metal organic vapor phase epitaxy (MOVPE). High quality with respect to the electrical performance, i.e mobility, minority carrier lifetime and diffusion length, is obtained in this lattice matched configuration. However, theoretical calculations show that the chosen band-gap combination of the triple-junction cell is not ideal. Higher efficiencies can be achieved by reducing the band-gap of the middle cell. Theoretical calculations have been performed using the program EtaOpt [4]. EtaOpt simulates the radiative limit efficiency. Figure 1 shows the results for such modeling assuming AM0 and 500xAM1.5direct with low optical aerosol value of 0.085 [5] (referred as AOD-spectrum in the following) as illumination conditions. Ge, with a band-gap of 0.66 eV, was assumed as the bottom cell of a monolithic triple-junction solar cell. Figure 1 shows the calculated efficiencies versus the band-gap of the top and the middle cells.

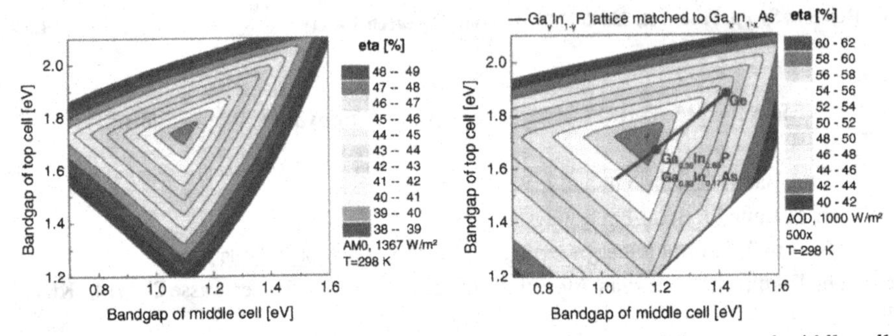

Figure 1: Calculated cell efficiency as a function of the band-gaps of the top and middle cells. Ge with a band gap of 0.66 eV is assumed as the bottom cell. The illumination conditions are: left: AM0 spectral condition: 25°C , 1367 W/m²; right: 500xAM1.5direct with the low optical aerosol depth value (=AOD), 25°C, 500 kW/m².

Increasing the In-content of the $Ga_xIn_{1-x}As$ or $Ga_yIn_{1-y}P$ layers lowers the bandgap energies (see also figure 1). Unfortunately, this also affects the lattice constant of the ternary compounds leading to lattice mismatched or metamorphic growth conditions. Only quaternary compounds would allow a reduction of the band-gap compared to GaAs while maintaining the lattice matching condition. A good example for a quaternary compound is GaInNAs which is considered an interesting semiconductor material for 4-junction solar cells. However, this material suffers from a very low diffusion length [6,7] and research is necessary to improve the quality. Thus, 4-junction solar cells are not envisaged in the near future making lattice mismatched multi-junction solar cells [2,8-11] one of the more promising methods of raising efficiency. One of the most critical issues of lattice mismatched growth is the formation of misfit and threading dislocations. This has to be carefully controlled and is correlated to the buffer growth. In this paper we discuss results obtained on different buffer layer structures grown on GaAs and Ge substrates. The epitaxially grown buffer layers are examined by transmission electron microscopy (TEM) and x-ray diffraction analysis and ultimately the performance of solar cells grown on top of them.

EXPERIMENTAL DETAILS

The epitaxial layers used in this work were grown in a production size AIX2600-G3 LP-MOVPE (low pressure metal organic vapor phase) reactor. TMGa, TMAl, TMIn in solution and DMZn from EPICHEM and the gas sources AsH_3, PH_3, SiH_4 with Millipure purifiers have been used as precursors. The reactor pressure was 100 mbar with the growth temperature typically in the range of 650°-700°C. (100)-oriented Ge or GaAs substrates with 6° off-orientation towards the [011] direction were used.

Different structures for the buffer layers were grown. Linear and step-graded buffers were grown in $Ga_{1-x}In_xAs$ material. Also, $Ga_yIn_{1-y}P$ was investigated as buffer material. For the growth on the n-doped Ge substrate only a step-graded $Ga_{1-x}In_xAs$ buffer was used. Table I summarizes the different buffer structures.

Table I: Structures of the buffers investigated in this work

buffer1	buffer2	buffer3	buffer 4
330 nm GaAs	330 nm GaAs	330 nm GaAs	nucleation layer
200 nm Ga(0.98)In(0.02)As	1400 nm linear graded	1500 nm linear graded	750 nm Ga(0.985)In(0.015)As
200 nm Ga(0.96)In(0.04)As	Ga1-xInxAs;	Ga(y)In(1-y)P;	200 nm Ga(0.95)In(0.05)As
200 nm Ga(0.94)In(0.06)As	start x: 0.02	start y: 0.51	200 nm Ga(0.93)In(0.07)As
200 nm Ga(0.92)In(0.08)As	end x: 0.17	end y: 0.35	200 nm Ga(0.90)In(0.10)As
200 nm Ga(0.90)In(0.10)As	200 nm Ga(0.83)In(0.17)As	50 nm Ga(0.83)In(0.17)As	200 nm Ga(0.88)In(0.12)As
200 nm Ga(0.88)In(0.12)As			230 nm Ga(0.86)In(0.14)As
200 nm Ga(0.86)In(0.14)As			240 nm Ga(0.845)In(0.015)As
200 nm Ga(0.83)In(0.17)As			240nm Ga(0.83)In(0.17)As
			240 nm Ga(0.81)In(0.19)As
			150 nm Ga(0.83)In(0.17)As

The samples were analyzed by cross-section and plan-view TEM in order to reveal the misfit dislocation structure. A Phillips XPertPro in triple axis configuration was used for performing reciprocal space mapping. This gives information about residual stress in the layers. The morphology of the surface was investigated by a Normarski interference contrast microscope and the surface roughness was determined by a Tencor alpha-step 200.

Solar cells were fabricated using a standard technology: Ti/Pd/Ag p-type and Ni/AuGe/Ni n-type contacts were evaporated on the solar cells. Both contacts were Au-electroplated to a thickness of about 2 μm. The GaAs cap-contact layer was removed selectively between the grid fingers using a citric acid/H_2O_2 mixture of 3:1. The area of the cell was finally defined through a mesa etching. An adapted double-layer antireflection coating consisting of either MgF_2/TiO_x or MgF_2/Ta_2O_3 was evaporated to reduce the reflectivity of the cell.

RESULTS AND DISCUSSION

Misfit or threading dislocations have to be expected during the heteroeptiaxial growth of $Ga_xIn_{1-x}As$ on GaAs or Ge if the critical thickness is exceeded, see for example [12]. However, there is no fully accepted theory concerning how the stress due to the difference in the lattice constant is released . In general it is believed that, after achieving the critical layer thickness, the stress is released by forming a misfit dislocation network at the interface between the substrate and the layer. The misfit dislocations are one of the origins for threading dislocations which propagate through the epitaxially grown layer. These threading dislocation are known to act as recombination centers for the minority carriers and reduce the performance of the solar cell. Therefore, the primary goal of this work is to reduce the threading dislocation density. One of the key parameters in controlling the threading dislocations is the structure of the buffer. In the literature linear as well as step-graded buffers have been investigated [13-16]. However, there is no clear indication which type of buffer is the best for solar cell applications. Moreover, not only the structure and growth conditions of the buffer but also the doping may influence the formation of the misfit dislocation network, as for example shown in [17].

Single-junction solar cells

In order to examine the influence of the buffer structure on solar cell performances we have fabricated nip single-junction $Ga_{0.83}In_{0.17}As$ on a 8 step-graded buffer (buffer 1 in table I) and

Figure 2: Typical surface morphology of $Ga_{0.83}In_{0.17}As$ grown on GaAs as observed with a Normarksi interference contrast microscope.

linear-graded buffer (buffer 2 in table I). The total buffer thickness was about 2 μm. Once the growth was completed we analyzed the x-ray rocking curve of the $Ga_{0.83}In_{0.17}As$/GaAs structure. The $Ga_{0.83}In_{0.17}As$ full width half maximum (FWHM) values for the (004)-reflection were slightly less for the step-graded buffer compared to the linear-graded buffer. This gives some evidence that the crystal quality of the $Ga_{0.83}In_{0.17}As$ grown on a step-graded buffer is superior to that on a linear graded buffer.

The surface morphology of all samples showed a typical cross hatch pattern, as shown figure 2. This cross hatch pattern can be directly attributed to the formation of the misfit dislocation network at the interface [18]. No difference was found for the growth on linear and step-graded buffers. Both structures showed the same surface roughness of about 5 nm. Single-junction nip-$Ga_{0.83}In_{0.17}As$ solar cells were fabricated and analyzed. Excellent efficiencies of 22.6 % (AM1.5g) were determined. No dependence of the cell performance on the buffer layer structure could be observed. Thus one has to conclude that, in spite of differences in the x-ray rocking analysis, the buffer structure has a negligible influence on solar cell performance..

Dual-junction solar cells

In comparison to single-junction solar cells the total thickness of the epitaxially grown layers in a dual-junction solar cell is larger. Therefore one may expect the buffer structure to have a stronger influence. In our investigations we examined nip-$Ga_{0.35}In_{0.65}P/Ga_{0.83}In_{0.17}As$ dual-junction cells on GaAs substrates. The epitaxially grown layers are lattice matched to each other but lattice mismatched to the underlying GaAs substrate. Again, a linear buffer scheme (buffer 2) and step-graded buffer scheme (buffer 1) were investigated, as described in table I. Besides the linear grading in $Ga_xIn_{1-x}As$ we also investigated $Ga_yIn_{1-y}P$ as a possible buffer material (buffer 3 in table I). Figure 3 shows TEM cross-section images of a sample using $Ga_yIn_{1-y}P$ as buffer material. Harmful threading dislocations are clearly seen in the active solar cell layers. The x-ray rocking analysis showed one peak for the $Ga_{0.35}In_{0.65}P/Ga_{0.83}In_{0.17}As$ layers with a FWHM-value of 104 arcsec for the (004)-reflection. A bad solar cell performance was determined. $Ga_yIn_{1-y}P$ buffer material appears to be unsuitable for solar cell applications. One reason might be that hardness of $Ga_yIn_{1-y}P$ decreases with increasing In-content [19]. This may support the propagation of dislocations into the growing layers.

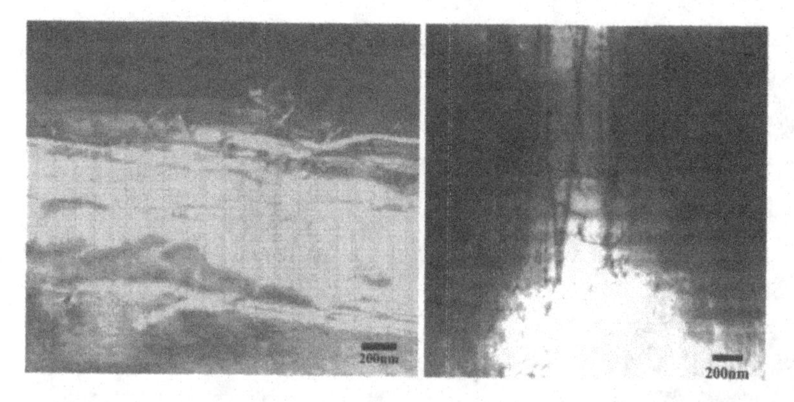

Figure 3: TEM cross sections images of the linear- graded $Ga_yIn_{1-y}P$ buffer (left) grown on GaAs substrate. In the right image threading dislocations are visible which have propagated through the active layers of the $In_{0.17}Ga_{0.83}As$ solar cell.

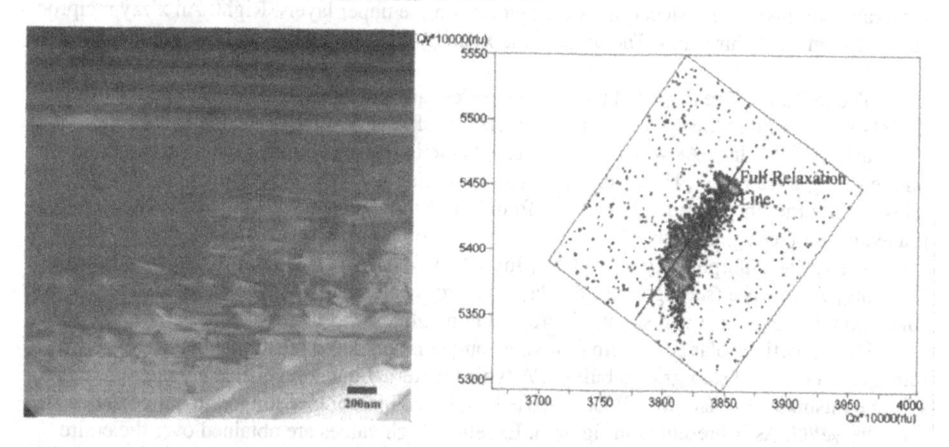

Figure 4: Left: TEM cross-section of a $Ga_xIn_{1-x}As$ step-graded layer buffer grown on a GaAs substrate. Right: An x-ray reciprocal space map in (224)-direction. The units of the x- and y-axis are reciprocal lattice units (rlu). The full-relaxation line is also shown.

Figure 4 shows for comparison the TEM cross-section of a step-graded $Ga_{1-x}In_xAs$ buffer. Here the dislocations are located at the interfaces of the different steps. A small number of threading dislocations was detected in the active solar cell layers.

The misfit dislocation network reduces the stress in the grown epitaxial layers. However, the stress is not totally released as can be seen in the x-ray reciprocal space map shown in figure 4. The FWHM-value of the $Ga_{0.35}In_{0.65}P/Ga_{0.83}In_{0.17}As$ (004)-reflection was 66 arcsec. Thus, a much better value compared to the similar $Ga_yIn_{1-y}P$ buffer was obtained.

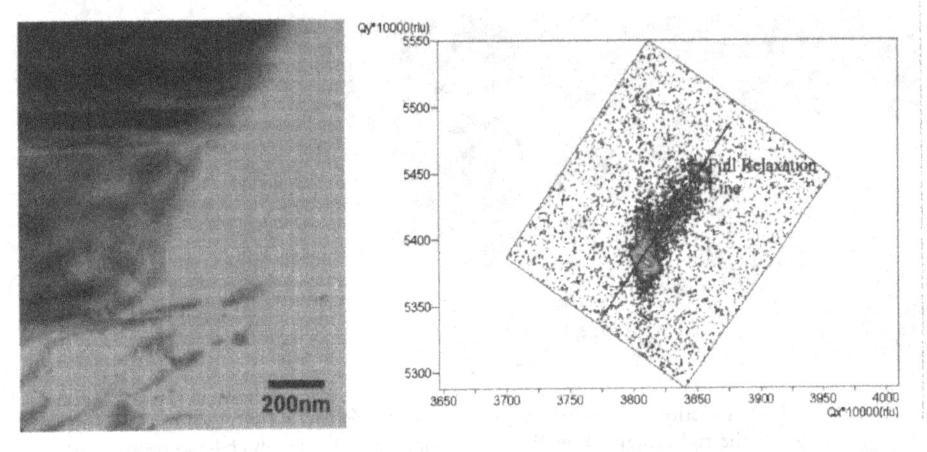

Figure 5: TEM cross section of a $Ga_xIn_{1-x}As$ linear-graded layer buffer layer grown on a GaAs substrate. No threading dislocations were detected in the upper layers. Right: An x-ray reciprocal space map in (224)-direction. The units of the x- and y-axis are reciprocal lattice units (rlu).

The results of linear-graded $Ga_xIn_{1-x}As$ buffers are shown in figure 5. Here, only a few defects were found in the active solar cell layers. The FWHM-value of the (004)-reflection of the $Ga_{0.35}In_{0.65}P$/ $Ga_{0.83}In_{0.17}As$ was 76 arcsec. This value is worse compared to the structure grown on the step-graded buffer. However, this result is similar as for the single-junction cells. The relaxation of the linear-graded buffer was found to be more advanced compared to the step-graded buffer (see figure 4 and 5).

The surface morphology of the $Ga_{0.35}In_{0.65}P/Ga_{0.83}In_{0.17}As$ is similar to that found for the $Ga_{0.83}In_{0.17}As$ surface (see figure 2), displaying a cross hatch pattern . The undulations at the cell surface are more distinct and slightly increased in height.

Dual-junction solar cells were fabricated and compared. No mentionable difference was found for linear- and step-graded buffers. A typical external quantum efficiency measurement of a lattice mismatched $Ga_{0.35}In_{0.65}P/Ga_{0.83}In_{0.17}As$ solar cell in comparison to the lattice matched $Ga_{0.51}In_{0.49}P/GaAs$ is presented in figure 6. Excellent high values are obtained over the entire wavelength response, even close to the band-edge. This indicates that the generated minority carriers can be collected effectively. These kind of devices (active solar cells lattice matched to each other and lattice mismatched to the substrate) are applied in concentrator systems. Therefore a 0.13 mm² concentrator solar cell has been fabricated and evaluated under high-illumination intensity. Figure 7 shows the dependence of the efficiency on the concentration ratio under AM1.5direct conditions. The maximum efficiency value was 31.2 % obtained at a concentration level of 300x. This is among the best reported values for any dual-junction cell [20]. However, even at concentration levels beyond 1000x an efficiency of 30 % was measured. Our $Ga_{0.35}In_{0.65}P/Ga_{0.83}In_{0.17}As$ dual-junction concentrator cells were also tested by another group under ultra-high sun concentrations of 3500-4000x. Efficiencies between 24 % and 28 % were determined [21]. These results are very promising. However, long-term reliability is another issue which has to be investigated. A highly dislocated buffer may alter under thermal and current cycles. A study performed recently at Fraunhofer ISE revealed the sudden death of some of the

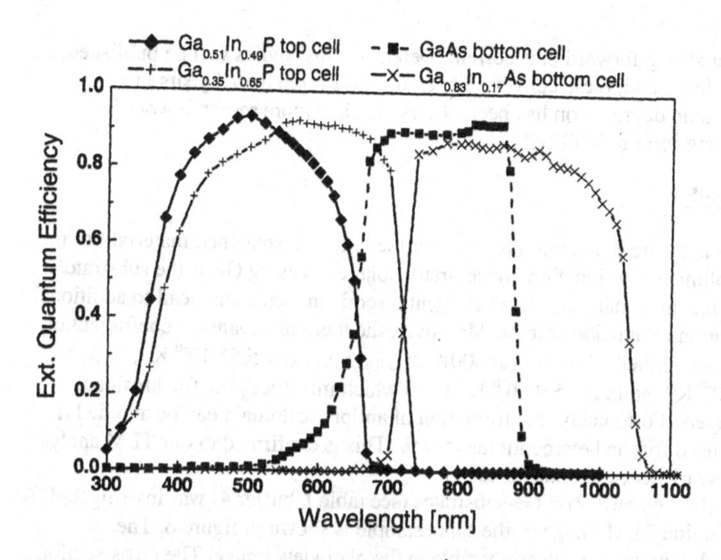

Figure 6: External quantum efficiency measurements of dual-junction cells. For comparison the lattice matched $Ga_{0.51}In_{0.49}P/GaAs$ and lattice mismatched $Ga_{0.35}In_{0.65}P/Ga_{0.83}In_{0.17}As$ structures are shown.

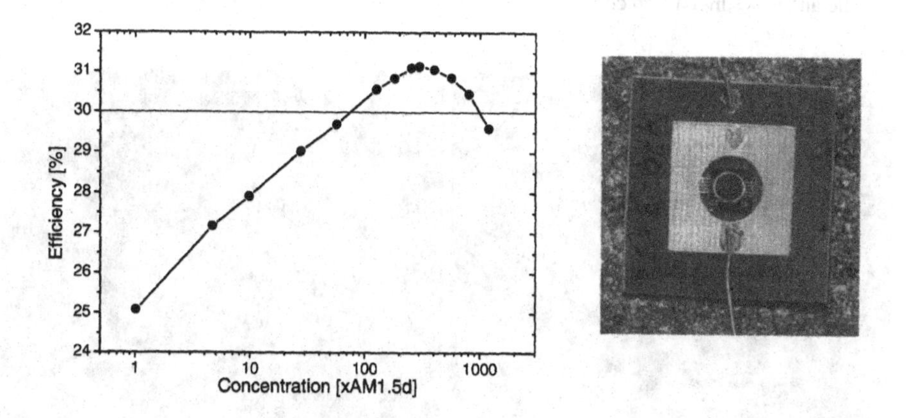

Figure 7: Efficiency versus AM1.5d concentration of a lattice mismatched dual-junction solar cell. The maximum efficiency of 31.2 % is obtained around 300x. Right: A mounted 0.13 cm² concentrator cell.

dual-junction cells under strong forward bias current. Details of this studies will be published elsewhere [22]. On the other hand, these cells have been used for more than 4 years in concentrator modules and no degradation has been observed. Thus more research work is necessary to prove the long-term reliability.

Triple-junction solar cells

As already shown in figure 1, the main interest for the lattice mismatched material lies in the application of monolithic triple-junction concentrator solar cells using Ge as the substrate. Growing the lattice mismatched materials on an elemental semiconductor may lead to additional problems like the formation of antiside defects. Moreover, the thermal expansion coefficient of Ge is different to the other applied materials (at 300K: $\alpha_{Ga(0.83)In(0.17)As} = 5.55 \cdot 10^{-6}\ K^{-1}$, $\alpha_{Ga(0.35)In(0.65)P} = 5.17 \cdot 10^{-6}\ K^{-1}$and $\alpha_{Ge} = 5.9 \cdot 10^{-6}\ K^{-1}$ [23]) which influences the final residual stress in the epitaxial layers. Fortunately, the formation of antiphase domain can be avoided if misoriented wafers are used for the heteroeptixial growth. This is confirmed in our TEM analysis where we do not find any evidence for this defect.

A step-graded $Ga_xIn_{1-x}As$ buffer on Ge-substrates (see table I, buffer 4) was investigated. A plan view and a cross section TEM image of the same sample is shown in figure 8. The generated misfit dislocation network is clearly visible in the plan view image. The cross-section image shows that the dislocations are preferentially located at the interfaces of the step-graded layers with some dislocations connecting adjacent interfaces. However, the two uppermost layers of the step-graded buffer are virtually free of dislocations indicating lack of strain relaxation in these layers. No threading dislocations were found in the active layers of the $Ga_{0.83}In_{0.17}As$ middle and $Ga_{0.35}In_{0.65}P$ top cell.

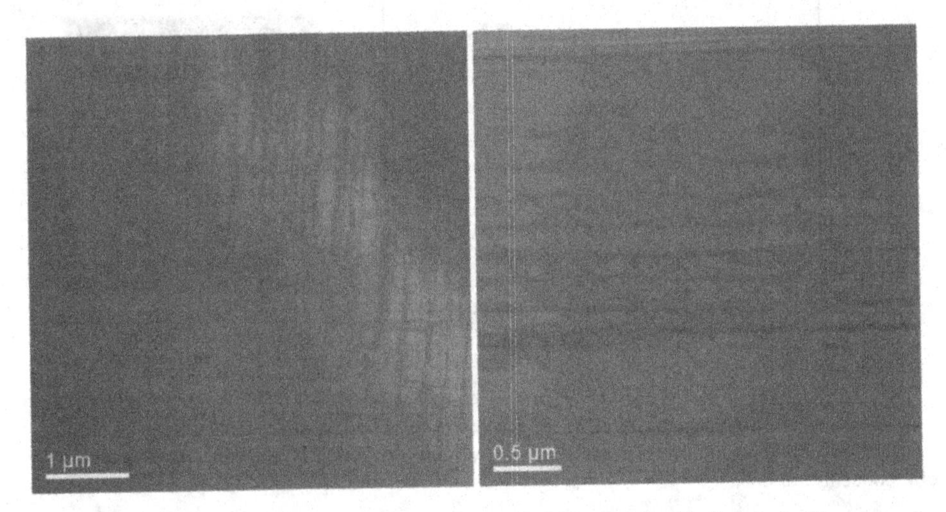

Figure 8: Plan view (left) and cross section (right) TEM image of a $Ga_xIn_{1-x}As$ buffer grown in Ge substrate.

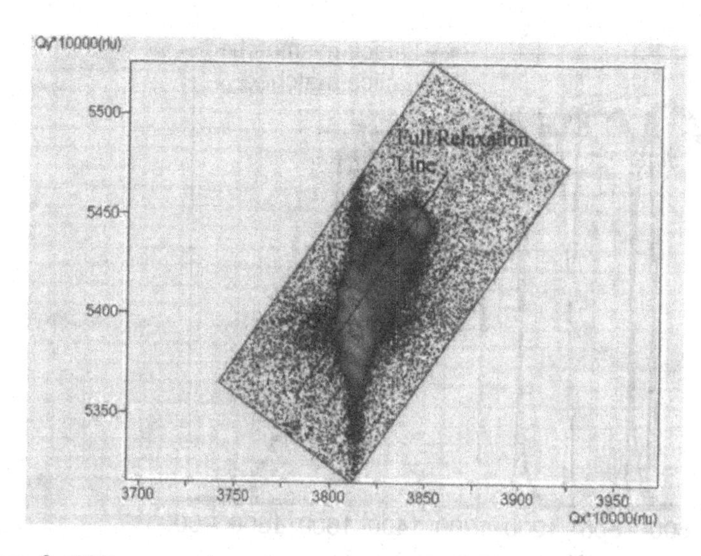

Figure 9: (224) x-ray reciprocal space map of a $Ga_{0.35}In_{0.65}P/Ga_{0.83}In_{0.17}As/Ge$ triple-junction solar cell structure. The relaxation within the buffer is nearly achieved. On the other hand, the lattice matching of the active solar cell layers was not fully obtained.

In agreement with the TEM results x-ray reciprocal mapping (figure 9) shows that strain relaxation is almost completed in the lower part of the step-graded buffer whereas no further relaxation occurs in the top part of the buffer. As result the $Ga_{0.83}In_{0.17}As$ middle and $Ga_{0.35}In_{0.65}P$ top cell show residual tetragonal distortions. A small FWHM-value of 45 arcsec for the (004)-$Ga_{0.83}In_{0.17}As$ reflection was determined. Thus, TEM as well as x-ray analysis show that a step-graded $Ga_xIn_{1-x}As$ buffer on Ge leads to high crystal perfection in the active layer of the solar cells.

The first triple-junction solar cells to be fabricated showed a good performance. Figure 10 shows the external quantum efficiency of a $Ga_{0.35}In_{0.65}P/Ga_{0.83}In_{0.17}As/Ge$ lattice mismatched triple-junction solar cell compared to a lattice matched triple-junction solar cell. Again, as for the dual-junction cell, excellent performance is obtained. Using the EQE-data and the AM1.5direct-new (=AOD) spectrum one can calculate the current generation in the subcells. Table II shows the results of this exercise. Regarding the lattice mismatched cell, the top- and middle cells are well matched. However, due to the lowered band gaps of these sub cells the Ge sub cell is now limiting the current of the triple cell.

Table II: Calculated current densities of the sub cells assuming the AM1.5g and AM1.5direct with low optical aerosol value (AOD) and the EQE shown in figure 10.

	lattice matched			lattice mismatched		
	Ga(0.51)In(0.49)P	Ga(0.99)In(0.01)As	Ge	Ga(0.65)In(0.35)P	Ga(0.83)In(0.17)As	Ge
current density [mA/cm²] AM1.5g	13.06	14.48	18.69	17.70	15.70	13.58
current density [mA/cm²] AM1.5direct (AOD)	12.47	15.06	20.11	16.69	16.40	14.75

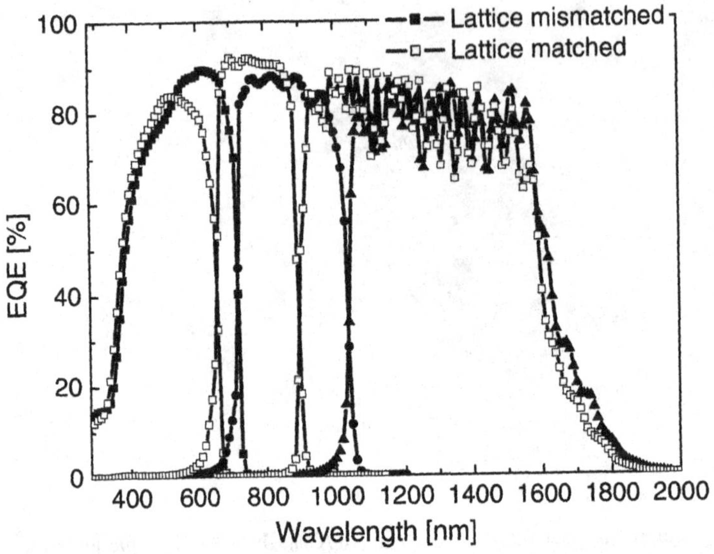

Figure 10: External quantum efficiency measurement of a $Ga_{0.35}In_{0.65}P/Ga_{0.83}In_{0.17}As/Ge$ lattice mismatched triple solar cell.

This is the first time that the bottom cell has limited the performance of the triple cell. In lattice matched cells the Ge sub cell was only considered as a voltage booster. No consideration was given to improving the overall performance of the Ge sub cell. In the case of the lattice mismatched approach it has now been demonstrated that the Ge sub cell limits the triple-junction cell performance. Therefore in the future research will be focussed on increasing the band edge performance of the Ge sub cell.

So far, an one-sun efficiency of 29.6 % (AOD) and 28.4 % (AM1.5g) has been determined for the $Ga_{0.35}In_{0.65}P/Ga_{0.83}In_{0.17}As/Ge$ triple-junction solar. This proves the electrical quality of the grown lattice mismatched solar cell structure.

CONCLUSIONS

Lattice mismatched epitaxial growth of $Ga_{0.83}In_{0.17}As$ and $Ga_{0.35}In_{0.65}P$ on GaAs and Ge misoriented substrates was investigated. Both materials are of great benefit to high efficiency monolithic triple-junction solar cells. Step- and linear-graded buffer structures were examined in the presented research. The buffer characterization was performed with TEM and x-ray analysis. No significant difference between epi-layer material quality grown on step- and linear-graded $Ga_xIn_{1-x}As$ buffers was found. On the other hand, $Ga_yIn_{1-y}P$ linear-graded buffers lead to degraded active solar cell material. Single-, dual- and triple-junction solar cell were fabricated and exhibited very good performance on $Ga_xIn_{1-x}As$ buffers. Thus, the lattice mismatched monolithic triple-junction solar cell seems to be the most promising path to increasing the efficiency of multi-junction solar cells in the near future.

ACKNOWLEDGEMENTS

The authors wish to thank Prof. T.Y. Seong for performing Transmission Electron Microscopy on lattice-mismatched solar cell structures (figure 3-5). Furthermore we acknowledge extremely valuable discussion with Dr. E. Spiecker and Prof. Jäger, University of Kiel about TEM analysis. For the continuous excellent work we would like to thank the members of the group "III-V solar cells and epitaxy" from Fraunhofer ISE. The work was partly financed by the European Commission within the project FULLSPECTRUM Contract number (SES6-CT-2003-502620) and by the Federal Ministry for the Environment, Nature Conservation and Nuclear Safety (BMU) under contract no. 03285554 F. The authors are responsible for the content of this paper.

REFERENCES

1 R. R. King, C. M. Fetzer, P. C. Colter, K. M. Edmondson, J. H. Ermer, H. L. Cotal, H. Yoon, A. P. Stavrides, G. Kinsey, D. D. Krut, and N. H. Karam, Proc. of 29th IEEE Photovoltaic Specialists Conference, 776-781 (2002). .

2 R. R. King, C. M. Fetzer, K. M. Edmondson, D. Law, P. C. Colter, H. L. Cotal, R. A. Sherif, H. Yoon, T. Isshiki, D. D. Krut, G. S. Kinsey, J. H. Ermer, S. Kurtz, T. Moriarty, J. Kiehl, K. Emery, W. K. Metzger, R. K. Ahrenkiel, and N. H. Karam, *Metamorphic III-V Materials, Sublattice Disorder, and Multijunction Solar Cells Approaches with over 37% Efficiency*, In Proc. 19th European PVSC, 3587-3593 (2004).

3 M. Yamaguchi, T. Takamoto, T. Agui, M. Kaneiwa, K. Araki, M. Kondo, H. Uozumi, M. Hiramatus, Y. Miyazaki, T. Egami, Y. Kemmoku, and N. J. Ekins-Daukes, *Japanese Activities of R&D on III-V Concentrator Solar Cells and Modules*, In Proc. 19 European PVSEC 2014-2017 (2004).

4 G. Letay and A. W. Bett, In Proc. 17th European Photovoltaic Solar Energy Conference, 178-180 (2001).

5 K. Emery, D. Myers, and S. Kurtz, In Proc. 29th IEEE PVSC 840-843 (2002).

6 S. R. Kurtz, J. F. Klem, A. A. Allerman, R. M. Sieg, C. H. Seager, and E. D. Jones, *Appl. Phys. Lett.* **80,** 1379-1381 (2002).

7 C. Baur, A. W. Bett, F. Dimroth, S. v. Riesen, B. Kunert, M. Traversa, K. Volz, and W. Stolz, in Proc. WCPEC-3,. 667-670 (2003).

8 F. Dimroth, P. Lanyi, U. Schubert, and A. W. Bett, *Journal of Electronic Materials* **29** 42-46 (2000).

9 F. Dimroth, U. Schubert, and A. W. Bett, *IEEE Electron Device Letters* **21** 209-211 (2000).

10 F. Dimroth, P. Lanyi, M. Meusel, U. Schubert, and A. W. Bett In Proc. 16th European Photovoltaic Solar Energy Conference, 106-109 (2000).

11 S. Sinharoy, M. O. Patton, T. M. Valko, and V. G. Weizer, *Progr. Photovolt.* **10** 427-432. (2002)

12 D. Gonzalez, D. Araujo, G. Aragon, and R. Garcia, *Appl. Phys. Lett.* **72** 1875-7 (1998).

13 F. Romanato, E. Napolitani, A. Carnera, A. V. Drigo, L. Lazzarini, G. Salviati, C. Ferrari, A. Bosacchi, and S. Franchi, *J. Appl. Phys.* **86** 4748-55 (1999).

14 M. T. Bulsara, C. Leitz, and E. A. Fitzgerald, *Appl. Phys. Lett.* **72** 1608-10 (1998).

15 D. Araujo, D. Gonzalez, R. Garcia, A. Sacedon, and E. Calleja, *Appl. Phys. Lett.* **67** 3632-3634 (1995).

16 R. R. King, M. Haddad, T. Isshiki, P. Colter, J. Ermer, H. Yoon, D. E. Joslin, and N. H. Karam, In Proc. Proceedings of 28th IEEE Photovoltaic Specialists Conference 982-985 (2000).

17 B. K. Tanner, P. J. Parbrook, C. R. Whitehouse, A. M. Keir, A. D. Johnson, J. Jones, D. Wallis, L. M. Smith, B. Lunn, and J. H. C. Hogg, *Appl. Phys. Lett.* **77** 2156-2158 (2000).

18 O. Yastrubchak, T. Wosiniski, J. Z. Domagla, E. LUsakowska, T. Figielski, B. Pecz, and A. L. Toth, *Journal of Physics: Condensed Matter* **16** S1-S8 (2004).

19 E. Bourhis and G. Patriarche, *Philosophical Magazin Letters* **84** 373-381 (2004).

20 M. A. Green, K. A. Emery, D. L. King, S. Igari, and W. Warta, *Progr. Photovolt.* **12** 365-372 (2004).

21 J. M. Gordon, E. A. Katz, D. Feuermann, and M. Huleihil, *Applied Physics Letters* **84** 3642-3644 (2004).

22 S. v. Riesen, and A. W. Bett, *Degradation study of III-V solar cell for concentrator applications*, accepted for publication in *Progr. Photovolt* (2005)

23 http://www.ioffe.ru/SVA/NSM/Semicond/index.html

Mater. Res. Soc. Symp. Proc. Vol. 836 © 2005 Materials Research Society

Mid-10^5 cm^{-2} threading dislocation density in optimized high-Ge content relaxed graded SiGe on Si for III-V solar on Si

David M. Isaacson, Carl L. Dohrman, Arthur J. Pitera, Saurabh Gupta, and Eugene A. Fitzgerald

Department of Materials Science and Engineering
Massachusetts Institute of Technology
Cambridge, Massachusetts 02139

ABSTRACT

We present a framework for obtaining high quality relaxed graded SiGe buffers on Si for III-V integration. By avoiding dislocation nucleation in $Si_{1-x}Ge_x$ layers of x>0.96, we have achieved a relaxed $Si_{0.04}Ge_{0.96}$ platform on Si(001) offcut 2° that has a threading dislocation density of 7.4×10^5 cm^{-2}. This 2° offcut orientation was determined to be the minimum necessary for APB-free growth of GaAs. Furthermore, we found that we could compositionally grade the Ge content in the high-Ge portion of the buffer at up to 17 %Ge μm^{-1} with no penalty to the dislocation density. The reduction in both threading dislocation density and buffer thickness exhibited by our method is an especially significant development for relatively thick minority-carrier devices which use III-V materials such as multi-junction solar cells.

INTRODUCTION

The relaxed graded SiGe platform ($\nabla_x[Si_{1-x}Ge_x]/Si$) [1] has proven itself enormously successful for the subsequent growth of III-V devices such as lasers [2,3], waveguides [4], and high-efficiency solar cells [5,6]. The key feature of the relaxed graded SiGe buffer approach is its effectiveness at minimizing deleterious issues related to lattice mismatch, thereby enabling the fabrication of scaleable $Ge/\nabla_x[Si_{1-x}Ge_x]/Si$ structures with dislocation densities on the order of 10^6 cm^{-2} [7]. Despite these successes, with respect to III-V integration on SiGe buffer there remain two factors to improve upon: a further reduction in the threading dislocation density as well as a reduction in the high thermal mismatch (i.e. high-Ge content) portion of the SiGe buffer. A reduction in dislocation density to the mid-10^5 cm^{-2} would be particularly useful, as at this level recombination in solar cells is no longer expected to be dominated by dislocations [8]. A closer look at the dislocation dynamics of relaxed graded semiconductors is therefore warranted.

The dependence of the threading dislocation density (TDD), ρ, on the various growth parameters for relaxed graded buffers has been shown to be:

$$\rho = \frac{2R_g R_{gr} \exp\left(\frac{E_{glide}}{kT}\right)}{bBY^m \varepsilon_{eff}^m} \qquad \text{Eq. 1}$$

where R_g and R_{gr} are the growth and grading rates, respectively, m is an exponent with a value generally between 1 and 2, B is a constant, T is the temperature, Y is the elastic modulus, k is Boltzmann's constant, E_{glide} is the activation energy for dislocation glide, and b is the magnitude of the Burgers vector (60° dislocations assumed) [9]. This equation assumes planar surfaces and no dislocation-dislocation interaction between misfit dislocation strain fields and threading dislocations. Roughening surfaces, dislocation interactions, and other such phenomena appear in Eq. 1 through a change in ε_{eff}. Thus, if non-ideal conditions are created in graded layer growth, the threading dislocation dependence vs. typical parameters will have different dependences, as the ε_{eff} will be dependent on the parameter as well [9]. The three parameters which offer the most flexibility to the experimentalist are the growth rate, the grading rate, and the growth temperature. In this study, we attempt to explore these parameters in order to formulate a framework for obtaining the optimum $\nabla_x[Si_{1-x}Ge_x]/Si$ platform for III-V integration.

EXPERIMENTAL DETAILS

All $\nabla_x[Si_{1-x}Ge_x]/Si$ samples in this study were grown via ultra-high vacuum chemical vapor deposition (UHVCVD) at temperatures ranging from 550-900°C using SiH_4 and GeH_4 gas precursors. Typical growth pressures were 25 mT for Ge contents at or below 50% Ge and 1-10 mT for Ge contents above 50%. Unless otherwise stated, all samples in this study were graded at an average rate of 10% Ge μm^{-1}. The experimental parameters of this study are shown in Figure 1 and are modified from previous reports by our group [7]. Increased grading rates were obtained by proportionately reducing the growth time of each layer, while reduced growth rates were obtained by reducing the flow rates of the precursor gases. Details of the MOCVD III-V growth process on virtual Ge substrates are described elsewhere [2].

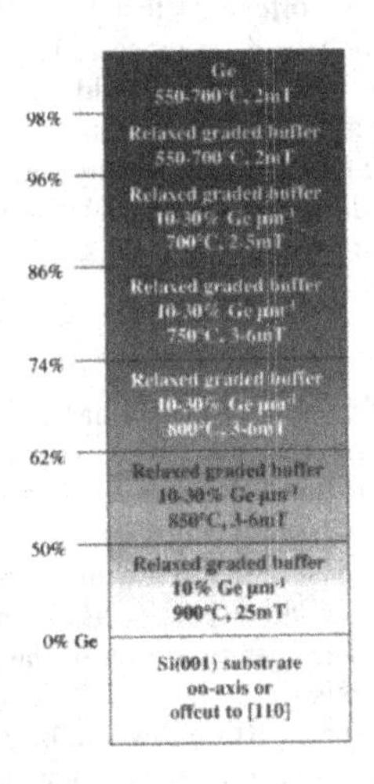

FIGURE 1 - Schematic of the structure and growth conditions used for the samples used in this experiment. Shown at left are the Ge concentrations that delineate each growth regime.

RESULTS AND DISCUSSION

Evolution of threading dislocation density in high-Ge content SiGe buffers

Shown in Figure 2 is a plot of the threading dislocation density as determined by etch-pit density (EPD) measurements as a function of the final Ge concentration in relaxed graded SiGe buffers, using data from previous reports from our group as well [10]. EPD measurements are known to be erroneous in this system and must be carefully correlated with plan view TEM experiments to confirm validity. Three observations can immediately be made from this Figure. First, the threading dislocation density in the 50-96% Ge regime is fairly constant and is less than 10^6 cm^{-2}

for all samples. Second, there is a significant increase in the threading dislocation density at compositions very near to 100% Ge. It is clear that the on-axis samples grown to 96% Ge retain approximately the same value exhibited by the 50% Ge layer, yet the TDD value increases rapidly between 96 and 100% Ge for both substrate types. Finally, there is a clear disparity in the behavior of the TDD values between the on-axis and 6° offcut substrates, with values of 5.6×10^5 cm^{-2} and 9.1×10^5 cm^{-2} at 96% Ge and 3.4×10^6 cm^{-2} and 2.3×10^6 cm^{-2} at 98% Ge for the on-axis and offcut samples, respectively.

As the mechanical strength of $Si_{1-x}Ge_x$ alloy is known to decrease with increasing temperature **[11]**, virtual $Si_{0.02}Ge_{0.98}$ and Ge layers on Si structures were grown at a reduced temperature in an attempt to suppress the hypothesized increase in threading dislocation density from increased dislocation nucleation. For this experiment, the same procedure described above was followed to 96% Ge, i.e. an attempt was made to ensure the most efficient relaxation possible to 96% Ge by growing at the highest possible growth temperature. The growth temperature was then reduced to 550°C, and a 200 nm $Si_{0.02}Ge_{0.98}$ regrowth followed by a 1.5 μm Ge layer was deposited. As shown in Figure 2, the resulting structures had TDD levels of 6.8 and 9.7×10^5 cm^{-2} for the on-axis and 6º offcut samples, respectively. It is important to note that these values are comparable to those for each structure at 96% Ge. This suggests that the reduction in growth temperature for these compositions is directly responsible for arresting the rampant dislocation nucleation in these compositions. Unfortunately the reduced temperature employed to attain these low TDD levels locks in strain, which results in extensive dislocation nucleation when the structure is annealed beyond the growth temperature. This makes virtual Ge ultimately inadequate for the integration of high-quality GaAs layers with Si. As such, we assert that a virtual substrate composition of $Si_{0.04}Ge_{0.96}$ is ideal for the integration of III-V material on Si. However, the trend exhibited with regards to substrate orientation in Figure 2 remains to be explored.

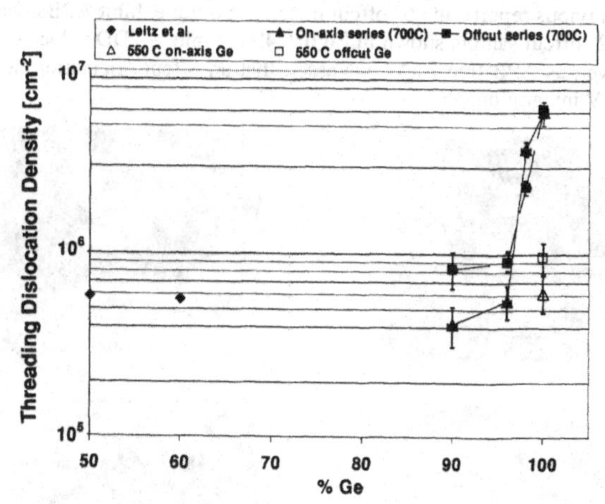

FIGURE 2 – Evolution of the TDD with increasing Ge content.

Effect of offcut orientation on growth of $Si_{0.04}Ge_{0.96}$ on Si(001)

In order to study the effect of offcut orientation on graded buffer growth, a set of 96% SiGe layers were grown on Si(001) substrates offcut various degrees to the nearest [110] direction. The observed TDD values of these structures are presented in Figure 3. From these data we clearly see that it is desirable to use substrates which are as close to exactly on-axis as possible. Unfortunately, however, the degree of offcut cannot be arbitrarily reduced as some degree of offcut is required for the subsequent growth of III-V material that is free of anti-phase boundaries (APBs) [**12**].

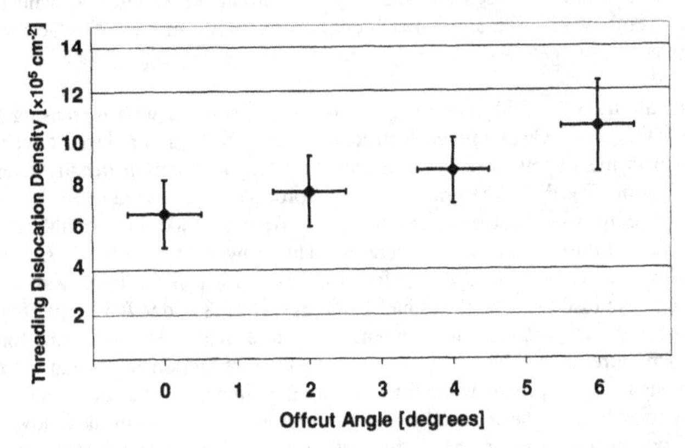

FIGURE 3 - Plot of dislocation density as a function of the degree of substrate offcut to [110].

In order to further study the effect of offcut angle, 1 μm layers of GaAs were deposited on each of the above platforms by MOCVD. The microstructures of these films are presented in Figure 4. As expected from previous reports, the 6° offcut material did not exhibit APBs, while the on-axis sample did. As the 2° offcut sample shown in **Figure 4b** exhibits a TDD value comparable to the on-axis control as well as APB-free GaAs, we assert that an offcut orientation near this value is the optimum for III-V integration.

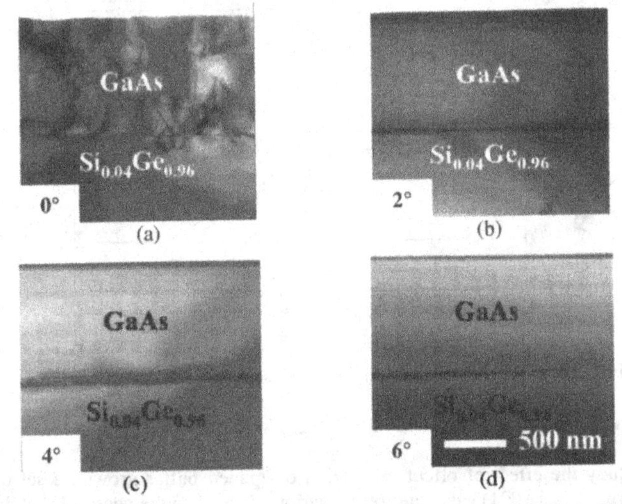

FIGURE 4 – XTEM images of GaAs deposited on graded SiGe buffers grown on Si wafers of various initial degrees of offcut.

The above data demonstrate that the 2° offcut in the initial Si wafer for SiGe graded buffer growth is the optimum orientation; however the effect of grading rate remains to be explored.

Coupled grading-growth rate approach to high-Ge content buffers

For many III-V devices on Si, such as high-efficiency multi-junction solar cells, it would be highly desirable to reduce the thickness of the high-Ge content portion of the SiGe buffer due to the additional thermal mismatch this region adds. To this end, a control grading rate set (closed symbols) was therefore graded at various rates to a final composition of $Si_{0.04}Ge_{0.96}$ on Si(001) offcut 2° to [110] to observe the effect of increasing the grading rate on the TDD. The effect of grading rate on the threading dislocation density for these structures as determined by EPD is shown in Figure 5. As one can clearly see from Figure 5, it is observed that while the TDD is relatively independent of the grading rate up to about 20% Ge μm^{-1}, at some point beyond this the TDD rises appreciably.

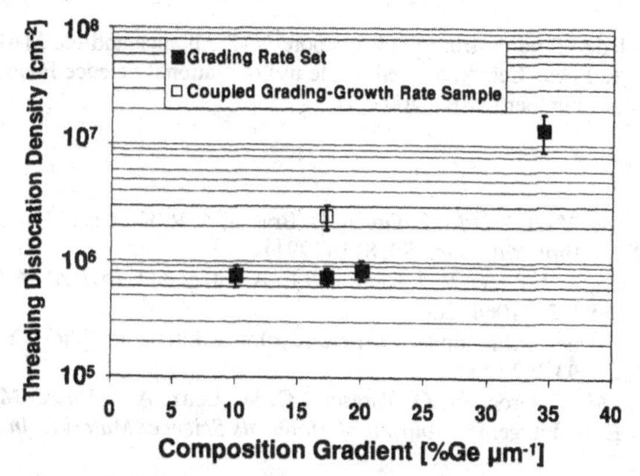

FIGURE 5 - Plot of dislocation density versus composition gradient in $Si_{0.04}Ge_{0.96}$ layers on 2° offcut Si(001).

Also shown in Figure 5 (open symbol) is the TDD value of a sample that was grown at a reduced growth rate, yet in the same amount of time as the 10 %Ge μm^{-1} reference. This coupled grading-growth rate experiment was performed to test Eq. 1, which predicts that one could grade at an increased rate, yet grow proportionately slower and therefore fabricate layers without increasing the TDD. An additional benefit of such an approach is that by grading faster the thickness of the buffer is greatly reduced. This experiment clearly shows that this virtual $Si_{0.04}Ge_{0.96}$ layer demonstrates a significant penalty with regard to threading dislocation density when time and growth rate are exchanged. The increase in threading dislocation density exhibited by this sample is most likely attributable to a combination of an increase in the strain field overlap between the misfit arrays at each interface of the graded buffer [**1**], coupled to an increase in both dislocation blocking and pileup formation, thereby decreasing ε_{eff} in Eq. 1. The lack of any observed dislocation escalation makes a composition gradient of approximately 17 %Ge μm^{-1} the optimum grading rate for the high-Ge content portion of the buffer.

CONCLUSIONS

We have investigated the evolution of the TDD in $\nabla_x[Si_{1-x}Ge_x]/Si$ buffers in the high-Ge composition range, using increased growth temperatures relative to previous reports in attempt to increase dislocation glide kinetics. We found a transition from glide-dominated to nucleation-dominated relaxation for compositions greater than 96% Ge, leading to a drastically higher threading dislocation density. Grading rates for the portion of the graded buffer with high Ge concentration in excess of approximately 20 %Ge μm^{-1} were found to result in significantly increased dislocation densities, as is the case for graded SiGe buffers at lower Ge compositions. [13] Finally, we demonstrated that the dislocation density increases monotonically with substrate offcut, and that Si(001) substrates offcut only 2° to [110] can result in APB-free III-V material. A fusion of the above studies resulted in a relaxed $Si_{0.04}Ge_{0.96}$ buffer on 2° offcut Si(001) graded at 17 %Ge μm^{-1} with a TDD of 7.4×10^5 cm^{-2}.

ACKNOWLEDGMENTS

The authors acknowledge funding from the Singapore-MIT Alliance and the MARCO Materials Structures and Devices Focus Center, as well as the use of National Science Foundation/MRSEC Shared Facilities (award number DMR-9400334).

REFERENCES

1. E. A. Fitzgerald, Y.-H. Xie, M. L. Green, D. Brasen, A. R. Kortan, J. Michel, Y.-J. Mii, and B. E. Weir, *Appl. Phys. Lett.* **59**, 811 (1991).
2. M. E. Groenert, A. J. Pitera, R. J. Ram, and E. A. Fitzgerald, *Journal of Vacuum Science and Technology B* **21**, 1064 (2003).
3. Michael E. Groenert, Christopher W. Leitz, Arthur J. Pitera, and Vicky Yang, *Journal of Applied Physics* **93** 362 (2003).
4. V. K. Yang, M. E. Groenert, G. Taraschi, C. W. Leitz, A. J. Pitera, M. T. Currie, Z. Cheng, and E. A. Fitzgerald, *Journal of Materials Science: Materials in Electronics* **13**, 377 (2002).
5. S.M. Ringel, R.M. Sieg, J.A. Carlin, S. Ting, M. Currie, V. Yang, E.A. Fitzgerald, M. Bulsara, and B.M. Keyes, *Proceedings of the Second World Conference and Exhibition on Photovoltaic Solar Energy Conversion*, July 1998.
6. S.A. Ringel, J.A. Carlin, C.L. Andre, M.K. Hudait, M. Gonzalez, D.M. Wilt, E.B. Clark, P. Jenkins, D. Scheiman, A. Allerman, E.A. Fitzgerald, and C.W. Leitz, *Progress in Photovoltaics: Research and Applications* **10**, 417 (2002).
7. M. T. Currie, S. B. Samavedam, T. A. Langdo, C. W. Leitz, and E. A. Fitzgerald, *Applied Physics Letters* **72**, 1718 (1998).
8. M. Yamaguchi and C. Amano, *J. Appl. Phys.* **58** (9), 3601 (1985).
9. E. A. Fitzgerald, A. Y. Kim, M. T. Currie, T. A. Langdo, G. Taraschi, and M. T. Bulsara, *Materials Science and Engineering B* **67**, 53 (1999).
10. C.W. Leitz, M.T. Currie, A.Y. Kim, J. Lai, E. Robbins, E.A. Fitzgerald and M.T. Bulsara, *Journal of Applied Physics* **90**, 2730 (2001).
11. I. Yonenaga, Journal of Materials Science: Materials in Electronics **10** 329 (1999).
12. R. Fischer, W. T. Masselink, J. Klem, T. Henderson, T. C. McGlinn, M. V. Klein, and H. Morkoc, *J. Appl. Phys.* **58** (1), 374 (1985).
13. E. A. Fitzgerald, Y.-H. Xie, D. Monroe, P. J. Silverman, J. M. Kuo, A. R. Kortan, F. A. Thiel, and B. E. Weir, *J. Vac. Sci. Technol. B* **10**, 1807 (1992).

Mater. Res. Soc. Symp. Proc. Vol. 836 © 2005 Materials Research Society L6.7

1MeV electron irradiation effects of GaAs/Si solar cells

N. Chandrasekaran, T. Soga, Y. Inuzuka, M. Imaizumi [1], H. Taguchi and T. Jimbo.
Department of Environmental Technology and Urban Planning, Nagoya Institute of Technology, Gokiso-cho, Showa-ku, Nagoya 466-8555, Japan.
[1] Japan Aerospace Exploration Agency, Tsukuba, 305-8505, Japan.

ABSTRACT

The characteristics of 1 MeV electron irradiated GaAs solar cells grown on GaAs and Si substrates are studied under dark and AM 0 conditions. The short circuit currents (I_{sc}) for GaAs/GaAs cell and GaAs/Si cell have been decreased at higher fluences. The degradation rate of V_{oc} and P_{max} for GaAs/Si is slower than that of GaAs/GaAs at the fluence $1x10^{16}$ cm^{-2}. This is due to the high radiation resistance of saturation current. It has been due to slow generation of arsnic vacancies related defect (V_{As}) in the GaAs/Si solar cell, which is determined by photoluminescence analyses and deep level transient spectroscopy.

INTRODUCTION

GaAs solar cell has excellent conversion efficiency and good radiation tolerance behavior in radiation environment, so it can be widely used as power generation device in the space applications [1]. In order to reduce the weight, cost effect and to improve the radiation tolerance, GaAs solar cell grown on alternative substrate such as Si [2] has been studied. Also the Si substrates are available in large area and good mechanical strength, which is highly suitable for space applications. But, the efficiency of the GaAs solar cells on Si substrate is inferior to that on GaAs substrate [3]. This inferiority is due to the difference in lattice constant mismatch and thermal expansion coefficient between GaAs and Si. In the space applications, the end-of-life (EOL) efficiency is more important than that of beginning-of- life (BOL) efficiency of the solar cell. A realized report shows that GaAs solar cell on Si substrate was less degraded when compared to that on GaAs substrate [4]. However, till now the degradation mechanisms of GaAs solar cell on Si substrate are not clearly understood because of the insufficient study. There are a few reports on the electron irradiation effects of GaAs solar cell on Si substrates[5]. The 1 MeV electrons have been frequently used to study the defects in a semiconductor. The effects of high-energy electron irradiation on n-GaAs have been the subjects of large number of works [6,7]. Electron irradiation generally produces point defect in semiconductor because of their recoil energy. After electron irradiation in GaAs, the direct formation of anti-sites is clearly explained[8]. In the present study, the high-energy electron (1MeV) irradiation effects on GaAs solar cells and GaAs layer grown by metal organic chemical vapor deposition (MOCVD) on Si substrate and compared with those on GaAs substrate. Electrical characterizations of these samples were carried out deep level defects by deep level transient spectroscopy (DLTS). The optical properties of the samples are studied by photoluminescence (PL) spectroscopy.

EXPERIMENTAL PROCEDURE

In our experiment, AlGaAs-GaAs heretofore solar cell structures were grown on (100) 4° off n-Si substrate by metal organic chemical vapor deposition (MOCVD) technique and employing two-step growth procedure with low temperature buffer layer. Also thermal cycle annealing was performed during the growth of GaAs solar cell on Si substrate for improve the efficiency. For comparison GaAs solar cell is also grown on GaAs substrate for comparison. The cell structure is similar to our previously reported paper [9]. The 3μm thick n-GaAs layers with carrier concentration of $5 \times 10^{16} cm^{-3}$ were grown on n-Si(100) with 4-degree off substrate by MOCVD by two step growth method [10]. Also a GaAs layer of the same thickness was grown on the GaAs substrate under similar conditions for comparison. These samples are used for photoluminescence (PL) study. The circular gold Schottky contacts of area 0.0076 cm^{-2} and thickness of 1500 Å were formed on n-GaAs layer for DLTS study.

After fabrication of the GaAs solar cells on GaAs and Si substrates, 1 MeV electron (fluence $10^{16} cm^{-2}$) from tandem accelerator was used for the irradiation studies on the above samples. The samples were irradiated with fluences of $1 \times 10^{16} cm^{-2}$ at room temperature. Before and after irradiation, the GaAs/GaAs and GaAs/Si solar cell parameters such as short-circuit current density (J_{sc}), open-circuit voltage (V_{oc}), and maximum power (P_{max}) are evaluated from the I-V characteristics of solar cells measured under AM0 condition at 1 sun. The electrically active electron traps were detected by deep level transient spectroscopy in the temperature range from 50K to 400 K using Horiba Impurity Analyzer (DA 1500) and computer controlled system. Low temperature (77K) photoluminescence measurements were made with the before and after irradiated samples mounted in strain free manner on a cold finger. An Ar^+ (514.5nm) laser was provided for an excitation. The luminescence from the sample was analyzed by a scanning spectrometer and through detected by photo multiplier. The spectra were recorded after lock-in amplification.

RESULTS AND DISCUSSION

SOLAR CELL

The open circuit voltage (V_{oc}) is expressed by setting I = 0 and can be written as $V_{oc} = nkT/q \log (I_{sc}/I_o+1)$ [11]. The saturation current is expressed as $I_o = qD_pP_{no}A/L_p$ in the case of p^+n junction, where D_p is diffusion co-efficent of hole in n-type GaAs layer, P_{no} is equilibrium hole density in n - GaAs later, L_p is diffusion length of a hole in n- GaAs layer and A is active area of the solar cell. The short-circuit current can be expressed in terms of total current $I = I_o \{\exp eV/nkT-1\}-I_{sc}$ where I_o is saturation current, k is Boltzman constant, T is temperature and when the applied voltage is equal to zero.

Figure 1 shows the room temperature dark I-V characteristics of before and after electron irradiated GaAs/GaAs solar cell (a) and GaAs/Si solar cell (b), respectively. The saturation current (I_o) of these cells has been analyzed from forward I-V characteristics. Before irradiation, the saturation current density for GaAs/GaAs and GaAs/Si are 1.2×10^{-11} (A/cm^{-2}) and 2.09×10^{-9} (A/cm^{-2}) respectively. After 1 MeV electron with fluence $10^{16} cm^{-2}$, the saturation current for GaAs/GaAs solar cells are increases nearly 3 orders than before irradiated one. But the is no considerable changes in saturation current for GaAs/Si solar cells after $10^{16} cm^{-2}$ fluence. This is due to slow generation of As vacancy defects in GaAs/Si solar cell after electron irradiation.

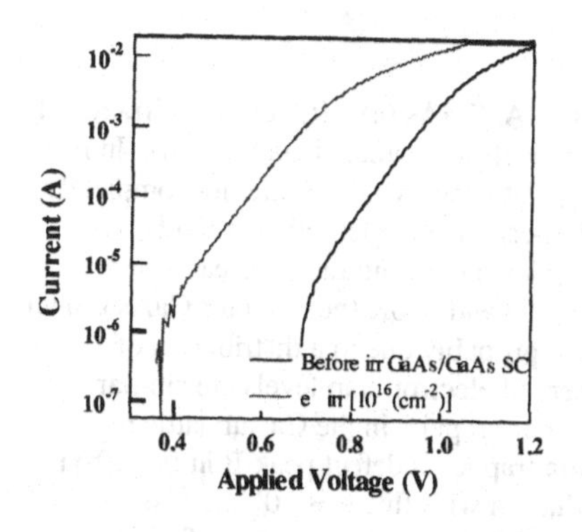

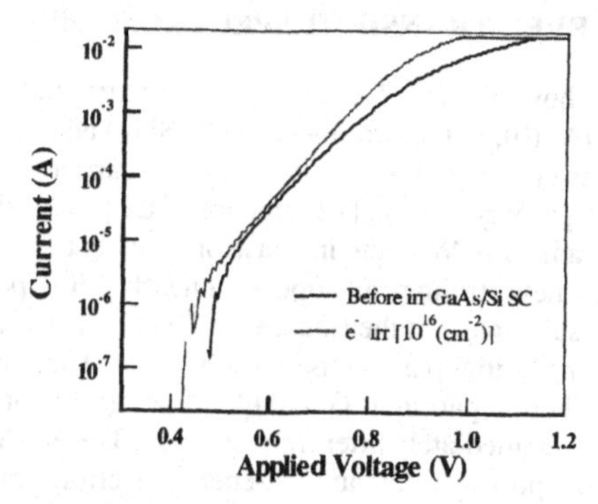

Figure. 1 The current – voltage characteristics of before and after electron irradiated GaAs/GaAs cell (a) and GaAs/Si cell (b) under dark condition.

Table 1. The photovoltaic properties of before and after electron irradiated GaAs solar cells on GaAs and Si substrate under AMO illumination at 1 sun.

SC Structure	fluence(cm^{-2})	J_{sc}(mA/cm^2)	V_{oc}(V)	E_{ff}(%)	FF(%)
GaAs/GaAs	0	24.5	0.94	13.5	79.5
	1×10^{16}	15.3	0.72	5.8	71.3
GaAs/Si	0	23.6	0.87	10.9	72.1
	1×10^{16}	16.1	0.77	6.6	72.4

Table 1 shows short circuit current density (J_{sc}), Open circuit voltage (V_{oc}), fill factor and conversion efficiency of before and after electron irradiated (fluence 10^{16} cm^{-2}) GaAs/GaAs solar cell and GaAs/Si solar cell which was calculated from the cell data's measured under AM 0 condition at sun 1. The rate of degradation of short circuit current density (J_{sc}) for GaAs/GaAs and GaAs/Si are almost equal. This is due to the same degradation rate of minority carrier lifetime for GaAs/Si cell and GaAs/GaAs cell.

The V_{oc} of GaAs/GaAs solar cell decreases rapidly than that of GaAs/Si solar cell with respect to higher electron fluences. The higher value of V_{oc} after 10^{16} cm^{-2} fluence for GaAs/Si solar cell is due to high radiation resistance of saturation current. It has been due to slow generation of As- vacancies related defect [9] in the GaAs/Si solar cell during high-energy electron irradiation. Before irradiation the saturation current of GaAs/Si cell is larger than that of GaAs/GaAs cell because of high dislocation density. However, the degradation rate of saturation current for GaAs/Si cell is smaller than that of GaAs/GaAs cell after 10^{16} cm^{-2} irradiation. The degradation rates of fill factor are almost same in the case of GaAs/GaAs and GaAs/Si cells after electron irradiation. Therefore, the rate of degradation of P_{max} for GaAs/Si cell is lower than GaAs/GaAs cell.

DEEP LEVEL TRANSIENT SPECTROSCOPY

Figure 2 shows the DLTS spectra of before irradiated GaAs/GaAs (a), after electron irradiated GaAs/GaAs (b), before irradiated GaAs/Si (c) and after electron irradiated GaAs/Si (d). In n-type GaAs on GaAs substrate, the majority carrier traps, detected by DLTS, are not completely related to primary defects [12] and only EL2 [13] peak (peak A) is observed at around 400K before irradiation. With the increase of fluence, the deep level concentration increases and moreover, new strong peak appears at nearly 150K(peak B) and 300K (peak C) for GaAs grown on GaAs substrate at the fluences 10^{16} cm^{-2}. These are probably due to a distribution of vacancy-interstitial pairs in the As sub lattice. The observed electron trap levels are similar to V_{As} - As distribution pair in n- GaAs[14]. The vacancy-interstitial pairs in the Ga sub lattice recombine immediately after their creation. The electron trap level defect peak B in Fig 2(b) is similar to reported results in high-energy electron irradiation with fluences 10^{16} cm^{-2} for GaAs/GaAs[15].. On the other hand, another peak (peak D) appeared at around 250 K in addition to EL2 peak (peak E) for before irradiated GaAs/Si sample. This is due to the trap by Si dislocation complex defect [16] during epitaxial growth.

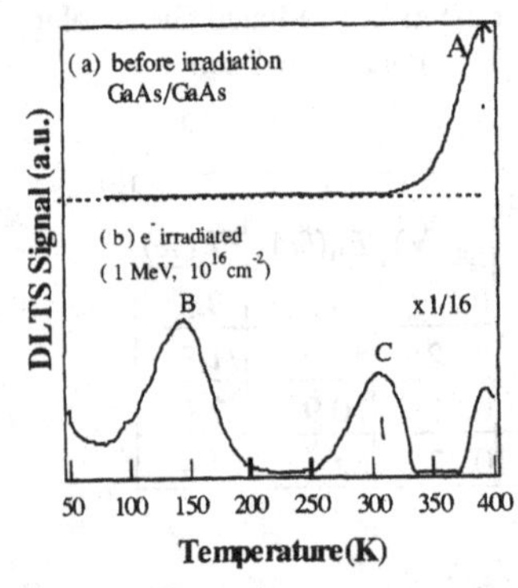

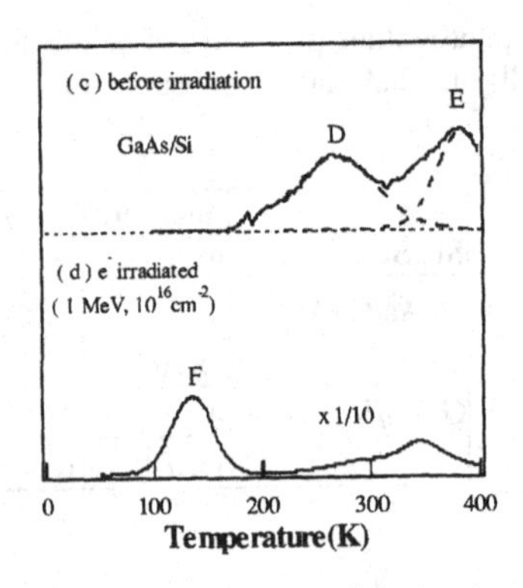

Figure. 2 The DLTS spectra of before irradiated GaAs/GaAs (a), after irradiated GaAs/GaAs (b), before irradiated GaAs/Si (c) and after irradiated GaAs/Si (d).

After irradiation the level related with dislocation complexes slowly starts to disappear for the higher fluences because of arsenic vacancies removal effect on GaAs by electron irradiation. The deep level at around 150K (peak F) appears at fluence 10^{16} cm^{-2} for GaAs grown on Si substrate. For fluences 10^{16} cm^{-2} the number of defect peak is lesser than GaAs layer on GaAs substrates i.e., the deep level corresponding to peak A in GaAs/GaAs is not observed in GaAs/Si.

PHOTOLUMINESCENCE

The photoluminescence (PL) intensity of band-to-band emission for before and after 1MeV electron irradiation with fluences 10^{16} cm^{-2} on GaAs/Si and GaAs/GaAs are shown in Fig

3. Before irradiation, a strong band-to-band emission peak observed at 1.508 eV for GaAs/GaAs. After electron irradiation with fluence 10^{16} cm^{-2}, a new defect peak appeared at 1.466 eV that is similar to in an article after 1 MeV electron irradiation[14]. On the other hand the PL spectra of GaAs/Si are splits in to two peaks, because of the valance band splitting into heavy hole and light hole by tensile stress caused by difference in the thermal expansion coefficients [17]. The PL band-to-band emission peaks are observed even after 10^{16} cm^{-2} fluences, and also the peak related to V_{As} is almost negligible.

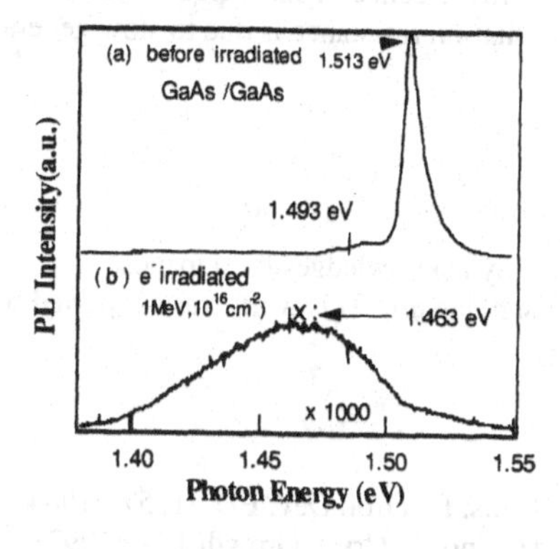

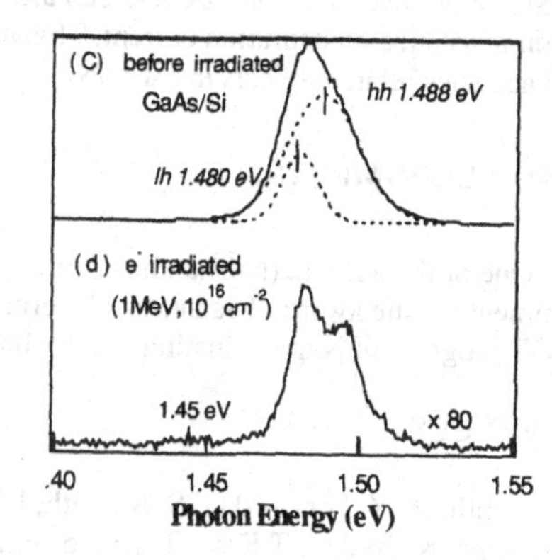

Figure. 3 The 77K PL spectra of before irradiated GaAs/GaAs (a), after irradiated GaAs/GaAs (b), before irradiated GaAs/Si (c) and after irradiated GaAs/Si (d).

The PL intensity of GaAs/Si and GaAs/GaAs decrease gradually with increasing the fluences. The degradation of GaAs on Si is slower than that of GaAs/GaAs. The noticeable point is that for the fluence at 10^{16} cm^{-2}, the PL intensity of GaAs on Si becomes stronger than that of GaAs on GaAs substrate, which is due to As vacancy defect. In general, the reduction of the PL intensity has been attributed to the formation of radiation-induced complexes, which capture conduction band electrons and therefore reduce the number of radiative transitions [18]. For GaAs/Si the decrease of PL intensity at higher electron fluences is slower because the defect creations are controlled by slow generation of As vacancies.

The decrease of PL intensity for GaAs/GaAs at higher electron fluence is correlated to the generation of DLTS defect peak B &C in Fig. 2(a). These defect peaks affects the number radiative of transitions to conduction band, so the intensity of the GaAs/GaAs decreases drastically at fluences10^{16} cm^{-2}. For GaAs/Si, the PL intensity degradation is slow at higher fluences, because the number of defect peak in DLTS spectra is lower and also the intensity of peak F is low which is shown in Fig. 2(b). In this case, the defect peak not much affect the transaction signals to the conduction bands, so the decrease of GaAs/Si intensity is lower when compared with GaAs/GaAs at higher fluences.

The slow generation of *As*-vacancies at higher electron fluences has been controls the saturation current values for GaAs/Si. So the GaAs/Si solar cell properties are less degraded when compared to GaAs/GaAs solar cells after 1 MeV electron irradiation with fluences10^{16} cm^{-2}. The above results seems the GaAs/Si materials has good radiation tolerant behavior, which is due to slow generation of As- vacancies. The slow generation of As vacancies may be due to the point defects in the case of GaAs/Si.

CONCLUSION

1 MeV electron irradiation effects of GaAs solar cells grown on GaAs and Si substrates are studied under dark and AM 0 condition. The rate of degradation of Voc and P_{max} for GaAs/Si cell is lower than GaAs/GaAs cell after 10^{16} cm^{-2} fluences This attributed to the high radiation resistance of saturation current. This high radiation resistance is due to slow generation of V_{As} vacancies related defects in GaAs/Si .

ACKNOWLEDGMENTS

One of the authors (N.Chandrasekarn) great fully acknowledges the Japanese Government for the award of Japanese Government scholarship. This work was supported by 21st COE program of Nagoya Institute of Technology.

REFERENCES

[1] Knechtli, R. Y. Loo, and G. S. Kamath, IEEE Trans. Electron Dev. ED-31, 577(1984).
[2] T. Soga, K. Baskar, T.Kato, T. Jimbo, and M. Umeno, J. Cryst. Growth 174, (1997) 579.
[3] M. Yamaguchi: J. Mater. Res. 6(1991) 376.
[4] M.. Imaizumi, S. Matsuda, M. Yamaguchi, and T. Ohara, Proc., 28 th IEEE Photovoltaic Specialists Conference, Sep 2000, Anchorage, USA. P.776.
[5] Y. Itoh, M. Yamaguchi, T. Nishioka and A. Yamamoto J. Appl. Phys., 61 (1987) 762.
[6] Bourgoin J. C, von Bardeleben H. J, J. Appl Phys **64** , (1988) R65-R91.
[7] N.G. Goswami, R.C Newman, and J.E. Whitehouse Solid State Communications, vol. 40, (1981) 473.
[8] T. Mattile and R. M. Nieminen, Phys. Rev lett., 74, (1995) 2721.
[9] T. Soga, N. Chandrasekaran, M. Imaizumi, Y. Inuzuka, H. Taguchi, T. Jimbo and S . Matsuda. Jpn. J.Appl. Phys. 42 (2003) 1054.
[10] M. Akiyama, Y. Kawarada and K. Kaminishi, J. Crystal Growth 68 (1984) 21.
[11] S.M . Sze, Semiconductor Devices: physics and Technology. John Willey & sons USA (2002).
[12] D. C . Look and J. R. Sizlove, J. Appl. Phys **62**, (1987) 3660.
[13] G. M. Martin, A. Mintonneau, and A. Mircea, Electron Lett. 13, (1977) 191.
[14] D. Stievnard, X. Bboddaert, J. C . Bourgoin, H. J. Bardeleben, Phycal Review B, **41**, (1990) 5271.
[15] H. Yousefi, J. B. Webb, R. Rousin, S. M. Kanna. J. Electron Mater. 24, (1995) 15.
[16] E. K Kim, H.Y. cho, Y. Kim, M.S. Kim, H.S. Kim and S.k Min, J. Appl. Phy`s. **67**, (1990) 2554.
[17] Y. Huang, P. Y. Yu, H. Lee and S. Wang Appl. Phys. Lett., 52, 579 (1988)
[18] S. M. Khanna, A. Houdayer, A. Jorio, C. Carlone, M. Parenteasu, and J. W. Gerdes, Jr., IEEE Trans. Nucl. Sci. 43, (1996) 2601.

II-VI Semiconductors and Transparent Conducting Oxides

Mater. Res. Soc. Symp. Proc. Vol. 836 © 2005 Materials Research Society L7.3

Scanning Tunneling Luminescence of Semiconductors

M.J. Romero
National Renewable Energy Laboratory (NREL), 1617 Cole Boulevard, Golden, CO 80401-3393

ABSTRACT

We review scanning tunneling luminescence (STL) – a photon emission spectroscopy based on scanning tunneling microscopy (STM) – and report on its application to photovoltaics. As part of this exploratory research, we have investigated $CuInSe_2$ thin films, solar cells based on quantum dots, and dilute nitride compounds. STL is very attractive because it is capable of nanometer resolution, which is being demanded by the spectacular advancement of nanoscience and nanotechnology. In addition, STM offers both unipolar and bipolar excitation of the luminescence and, consequently, the transport and recombination of electrons and holes can be investigated independently.

INTRODUCTION

In the last decade, photon emission from semiconductors has been observed in STM, a phenomenon commonly referred to as scanning tunneling luminescence [1]. The opportunity to use tunneling effects for a photon emission spectroscopy with nanometer resolution is obviously of great interest. Nanostructures might be successfully applied in next-generation photovoltaics, and STL will contribute decisively to our understanding of electron transport and recombination in the nanoscale. As proof of principle, luminescence from individual InP [2,3] and InAs [4] quantum dots has been reported.

Improved resolution is not the only benefit of STM. Below the activation energy for secondary electron-hole excitation, STL is excited by diffusion of tunneling electrons and subsequent recombination with available holes for *p*-type semiconductors – the reverse is true for *n*-type. In this case, the polarity of the tunneling current needed for STL is specific to the type of semiconductor. The effect can be exploited to observe local type inversion, which will be established by the need to invert the polarity of the tunneling current in order to detect photon emission.

In this contribution, we report on the development of STL at the National Renewable Energy Laboratory and its application to semiconductors relevant to our program in photovoltaics. As part of this exploratory research, and in order to illustrate the capabilities of this method, we have investigated grain boundaries in $CuInSe_2$ (CIS) thin films, InGaAs quantum dots (QDs) for third-generation photovoltaics, and GaPN epilayers grown on silicon substrates for tandem solar cells.

EXPERIMENTAL DETAILS

In STL, and for semiconductors, the quantum efficiency of the radiative recombination is 10^{-4} to 10^{-3} for each tunneling electron. Therefore, the most critical aspects in the development of this STM are achieving excellent photon collection and detection sensitivity. We have adapted an STM to be operated in conjunction with cathodoluminescence (CL). One of the advantages of this approach is that the tip can be accurately positioned into the focal point of the parabolic mirror of the CL detection system, improving dramatically the photosensitivity in STL (see Fig. 1). Eclipsing of the luminescence by the STM is largely minimized in our setup. Emission spectra are acquired either by a Roper Scientific Silicon EEV 1340×400 CCD or an InGaAs 512×1 multichannel detector, depending on the spectral range of interest. In the case of photon mapping, a GaAs photomultiplier or a Ge detector are available.

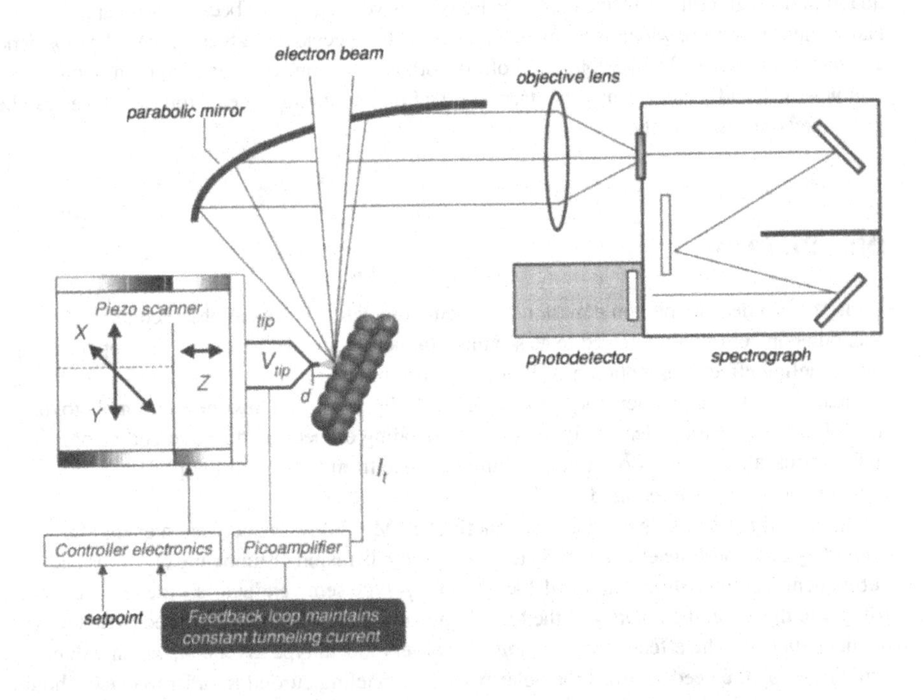

Fig. 1. Schematics of the STM for STL. The tip is positioned into the focal point of the parabolic mirror using cathodoluminescence. When this alignment is completed, the electron beam is turned off and the STM operation resumes. The detection system for cathodoluminescence is then used for detection of the STL.

STM observations are performed in constant current mode. An image of the tunneling current is acquired simultaneously to confirm that variations in photon intensity cannot be correlated with instabilities of the tunneling current. We chose electrochemically etched PtIr tips, provided by Molecular Imaging, for these measurements. For photon mapping, acquisition times by line are 2 to 5 seconds and, therefore, thermal drift on STM images is prevented. STL spectroscopy is performed scanning the tip on a selected area of an image for 100 seconds or more, and some drift is then unavoidable.

RESULTS AND DISCUSSION

Grain boundaries in $CuInSe_2$ (CIS) thin films.

This issue is attracting considerable attention in our community because such grain boundaries are believed to contribute effectively to photoconversion efficiency [5]. A better understanding of the observed anomalous behavior is needed to further advance efficiency and deliver higher performance of terrestrial photovoltaics. The surface of CIS shows more pronounced stoichiometric deviations and, consequently, very distinct electronic behavior when compared to grain interiors [6]. Because the *p–n* junction is located near this surface when the solar cell is completed [7], grain boundaries that intersect it are of special interest. Recently, we studied such grain boundaries by CL [8] and scanning Kelvin probe microscopy (SKPM) [9]. Using CL, we found that the photon energy of the emission spectrum near grain boundaries does not depend on the excitation density, in contrast to grain interiors. However, CL does not provide the needed spatial resolution to investigate individual grain boundaries without interference of carriers excited in grain interiors. On the other hand, SKPM demonstrated the existence of a built-in potential on grain boundaries. Here, we present the results of our STL observations from CIS thin films that produced a recent record solar cell of 14.5% efficiency [10].

Figure 2 shows the bias dependence of the photon intensity in STL. The STM tip scans the CIS surface in constant current mode with the tunneling current set to $I_t = 50$ nA. We clearly identify a threshold voltage at about 5 to 6 volts. Below this threshold, STL is excited by tunneling of *electrons* and subsequent recombination with available *holes* in CIS. Excitation is thus *unipolar*. Therefore, the photon intensity should be essentially proportional to the density of majority carriers seen by the tip in the semiconductor. Above the threshold, secondary electrons and holes are generated by impact ionization of hot electrons. In this case, STL becomes fundamentally similar to cathodoluminescence (or photoluminescence) because excitation is *bipolar*.

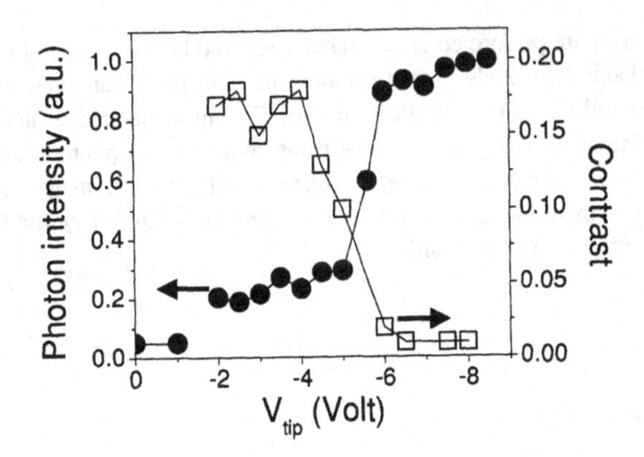

Fig. 2. Bias dependence of the photon intensity and contrast of grain boundaries and dislocations in CIS thin films, measured under the excitation provided by tunneling electrons in STM. $I_t = 50$ nA. $T = 120$ K. The contrast is defined as the difference in photon intensity between the grain boundaries and grain interiors, normalized to the second.

The diffusion length of *hot tunneling electrons* in CIS determines the spatial resolution of STL. To answer this very common question, Fig. 3 shows the emission spectrum measured simultaneously by STM and CL. CL is primarily excited in grain interiors and, as reported before [8], the spectrum is dramatically sensitive to the external excitation (represented by the electron-beam current, I_b), with an absolute shift in photon energy of +30 meV. Above saturation, which is observed at approximately $I_b = 1$ nA, the photon energy of the CL spectrum becomes constant. The photon energy of the STL spectrum, in contrast, is not affected by V and I_t settings used for STM observations. This suggests that STL is excited under saturation; that is, the density of minority carriers induced by tunneling electrons becomes comparable to majority carriers. In addition, the emission observed in STM is considerably higher in energy when compared to CL. Because there is considerable evidence of an unusually increased bandgap at the surface of CIS thin films, and absolutely no correspondence between CL and STL, we believe that the STM emission is confined to the surface layer, which is tenths of nanometers in depth and more deficient in Cu than grain interiors, as revealed by secondary-ion mass spectrometry (SIMS) measurements.

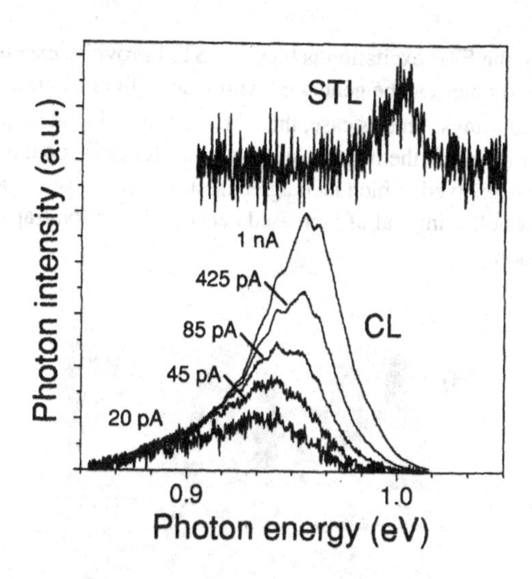

Fig. 3. STL emission spectrum from CIS thin films at V = - 8 V, I_t = 25 nA, and T = 120 K. Acquisition time = 100 s. The effect of excitation density (electron-beam current, I_b) on the CL spectrum is shown for comparison. Because the spectra are not shown to scale, the readout noise of the detector can be used for estimating the relative intensities.

Photon mapping is performed in constant current mode, with the synchronization of the Ge detector (which offers high photosensitivity in the infrared) with the STM tip while scanning the CIS surface. Figure 4 shows the STM and corresponding STL images, acquired simultaneously, at V = - 3 V and I_t = 50 nA. Extended linear defects (grain boundaries and dislocations) are revealed by their reduced emission in STM (Fig. 4b). Grain boundaries do not necessarily follow the topography, mainly because we are actually seeing a {112} reconstructed surface. Therefore, we can only partially correlate surface morphology with the location of these defects.

These images were acquired below the threshold voltage and, therefore, the excitation is unipolar, as discussed above. In that case, reduced STM emission at grain boundaries and dislocations can be explained by either a reduced density of holes or increased density of electron traps. We attempt to solve the puzzle by investigating how the voltage applied to the STM tip affects the contrast of these defects in STL. As seen in Fig. 2, the contrast is practically nullified above the threshold voltage, and no grain boundaries or dislocations are seen by STL; they just disappear.

Figure 4c shows the corresponding photon map with V = - 8 V and I_t = 25 nA. To demonstrate that we indeed observe emission at this voltage, we withdrew the STM tip twice during the scanning and, consequently, the emission dropped (see Fig. 4c).

Above the threshold, the STL excitation is bipolar. STM provides excellent resolution, and it is likely that electrons and holes can be generated within an individual grain boundary without interfering with its surroundings. In this case, the STL emission should be proportional to $V \times I_t$, which is constant. Assuming that the trap density is similar for grain boundaries and grain interiors, those cannot be resolved, which is in agreement with our observations. This suggests that a reduced density of holes, instead of increased density of electron traps, is responsible for the reduced STM emission.

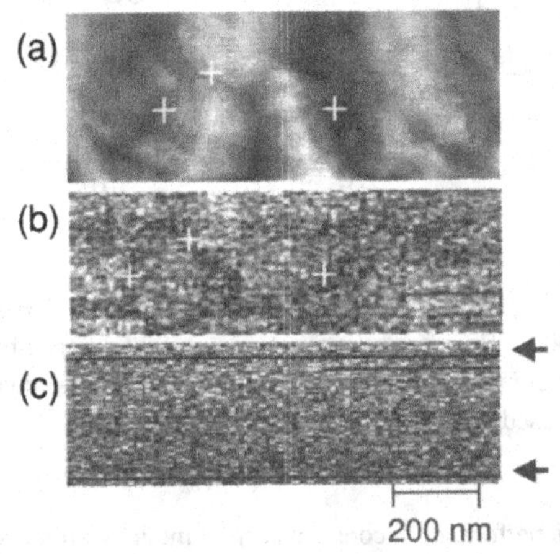

Fig. 4. (a) A constant current mode STM image and corresponding photon maps (raw data), which are acquired simultaneously while the tip is scanning the CIS surface: (b) $V = -3$ V, $I_t = 50$ nA, $T = 120$ K. (c) $V = -8$ V, $I_t = 25$ nA, $T = 120$ K. Time/line = 2.5 s. P-gain = 9, I-gain = 10. The arrows on Fig. 4c correspond to the withdrawal of the STM tip during the scanning.

In summary, when STL is excited by recombination of *tunneling electrons* with available *holes* (majority carriers in CIS), grain boundaries and dislocations near the CIS surface are revealed by their reduced emission. We have investigated how the voltage applied to the STM tip affects the photon emission, and concluded that a reduced density of holes is responsible for the observed behavior.

Self-assembled quantum dots for third-generation photovoltaics.

Self-assembled quantum dots have the potential to demonstrate the novel approach to the very high efficiency solar cells suggested by Luque and Marti [11], based on additional photon-induced transitions at intermediate levels. To succeed, a very strict control and uniformity in the nanoscale are required. We have investigated self-assembled InGaAs QDs grown by metalorganic vapor phase epitaxy (MOVPE) on (001) and (311)B *n*-type GaAs substrates. The surface reconstruction of the second promotes the alignment of QDs along certain orientations, as shown on the STM image (Fig. 5a). The quantum dots are not directly imaged, because a 1-nm, n^+ GaAs top layer was deposited to improve hole transfer for STL measurements.

Figure 5b shows the photon map obtained at $V = 4$ V and $I_t = 60$ nA. Under these observation conditions, holes are transferred directly to QDs, recombining with available electrons (*n*-type), and individual dots are resolved – or at least, some correspondence can be established between the STL and STM images. Figure 5c shows the STL spectrum obtained when the tip is positioned on the A and B QDs selected from the STL image. The CL spectrum is shown for comparison. Cathodoluminescence is excited from the InGaAs QDs and the GaAs barriers. In contrast, no emission from the GaAs is seen on the STL spectrum, corroborating that tunneling holes recombine at QDs. We investigated these QDs simultaneously by CL and STL to assess the degree of uniformity in both the micro- and nanometer scales and found substantial fluctuations in the photon energy associated with QD transitions in both scales. This is a very interesting aspect of self-assembled structures that should be addressed to improve QD uniformity for photovoltaic applications.

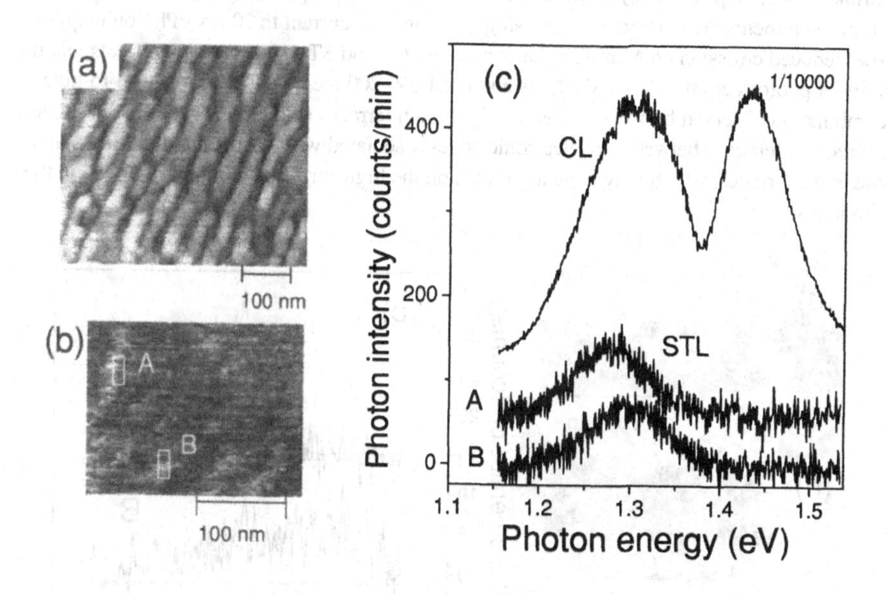

Fig. 5. (a) A constant current mode STM image of InGaAs QDs ($V = 4$ V, $I_t = 1$ nA, $T = 300$ K, Z range = 5.5 nm) and (b) corresponding STL image (raw data), which is acquired by increasing I_t to 60 nA. (c) STL emission spectrum obtained from locations A and B on the STL image. Acquisition time = 100 s. The CL spectrum is shown for comparison.

When the voltage applied to the STM tip exceeds 5 V, relaxation of *hot tunneling holes* leads to electron-hole excitation and the correspondence between the STM images and photon maps is lost. In this case, carriers are not confined to QDs and recombine in the GaAs barriers.

GaPN epilayers grown on Si substrates for tandem solar cells.

Dilute nitride compounds based on GaAsP are very promising because the combination of GaAsPN and silicon can potentially achieve a nearly optimal efficiency of 37% while avoiding issues of severe lattice mismatching found in 4-junction cells [12], or mismatched GaInP/GaInAs/Ge [13].

Nitrogen is an isoelectronic impurity in GaAsP, and how it is incorporated during the epitaxy determines the electronic properties of the compound. To answer these questions, we are investigating the surface reconstruction in dilute nitride compounds by STM and STL. STM images of *n*-type GaPN epilayers are acquired in constant current mode, with $V = 8$ V and $I_t = 1$ nA (see Fig. 6a). These images show the reconstruction of the GaPN surface, with multimonolayer steps (MmLs) of about 1 to 2 nm in height and periodicity of 50 to 100 nm. STL measurements are performed increasing the tunneling current to 20 nA. Photon mapping shows reduced emission on MmLs, as shown in Fig. 6b, and STL spectroscopy reflects that the energy of photons emitted from MmLs is about 2.1 eV, 200 meV below the electron-to-hole recombination observed between individual MmLs on terraces (Fig. 6c). These results suggest important differences between the electronic states associated with nitrogen at the MmLs and those in the terraces, which may have an impact on the high surface recombination seen in these compounds.

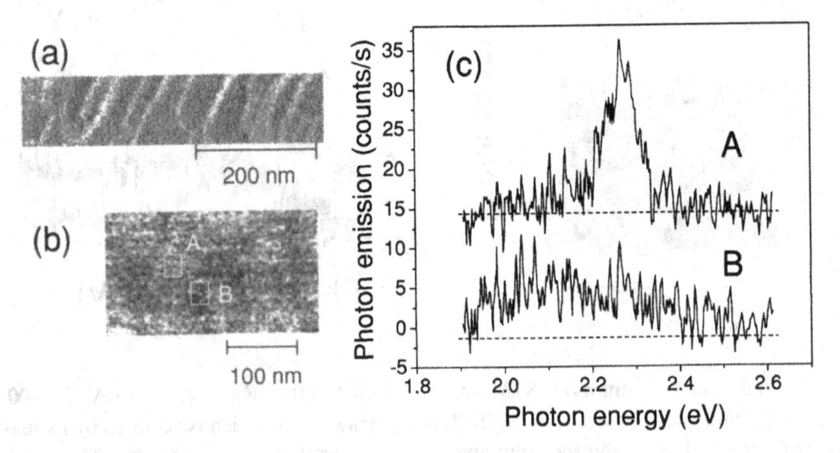

Fig. 6. (a) STM image of the GaPN epilayer, epitaxially grown on silicon. $V = 8$ V, $I_t = 1$ nA, $T = 300$ K. To enhance the contrast of the surface steps, the derivative of Z is shown

instead. STL image (raw data), which is acquired by increasing I_t to 20 nA. (c) STL emission spectrum obtained from locations A and B on the STL image. Acquisition time = 20 s.

CONCLUSIONS

Scanning tunneling luminescence, which takes advantage of the extreme resolution of STM, offers new opportunities in research that are not accessible by other methods. Because of the low quantum efficiency, the most critical aspect for STL is achieving excellent collection of photons and high photon detection sensitivity. In our STM, the tip can be accurately positioned in the focal point of the parabolic mirror of the cathodoluminescence equipment, monitoring the transition of the tip from under focus to above focus. We have exemplified the potential of STL by applying this method to individual grain boundaries, nanostructures, and surface steps.

ACKNOWLEDGMENTS

This work was supported by the U.S. Department of Energy (DOE) under Contract No. DE-AC36-99GO10337.

REFERENCES

1. S.F. Alvarado, Ph. Renaud, D.L. Abraham, A. Ch. Schönenberger, D.J. Arent, and H.P. Meier, *J. Vac. Sci. Technol. B* **9**, 409 (1991).
2. U. Hakanson, M.K.-J. Johansson, J. Persson, J. Johansson, M.-E. Pistol, L. Montelius, and L. Samuelson, *Appl. Phys. Lett.* **80**, 494 (2002); U. Hakanson, M.K.-J. Johansson, M. Holm, C. Pryor, L. Samuelson, W. Seifert, and M.-E. Pistol, *Appl. Phys. Lett.* **81**, 4443 (2003).
3. U. Hakanson, T. Sass, M.K.-J. Johansson, M.-E. Pistol, and L. Samuelson, *Phys. Rev. B* **66**, 235308 (2002).
4. T. Tsuruoka, Y. Ohizumi, and S. Ushioda, *Appl. Phys. Lett.* **82**, 3257 (2003).
5. C.-S. Jiang, R. Noufi, K. Ramanathan, J.A. AbuShama, H.R. Moutinho, and M.M. Al-Jassim, *Appl. Phys. Lett.* **85**, 2625, (2004).
6. Y. Yan, K.M. Jones, J. Abushama, M. Young, S. Asher, M.M. Al-Jassim, and R. Noufi, *Appl. Phys. Lett.* **81**, 1008 (2002).
7. C.-S. Jiang, F.S. Hasoon, H.R. Moutinho, H.A. Al-Thani, M.J. Romero, and M.M. Al-Jassim, *Appl. Phys. Lett.* **82**, 127 (2003).
8. M.J. Romero, K. Ramanathan, M.A. Contreras, M.M. Al-Jassim, R. Noufi, and P. Sheldon, *Appl. Phys. Lett.* **83**, 4770 (2003).
9. C.-S. Jiang, R. Noufi, J.A. AbuShama, K. Ramanathan, H.R. Moutinho, J. Pankow, and M.M. Al-Jassim, *Appl. Phys. Lett.* **84**, 3477 (2004).

10. J. AbuShama, S. Johnston, T. Moriarty, G. Teeter, K. Ramanathan, and R. Noufi, *Prog. Photovolt. Res. Appl.* **12**, 39 (2004).
11. A. Luque and A. Marti, *Phys. Rev. Lett.* **78**, 5014 (1997).
12. S.R. Kurtz, D. Myers, and J.M. Olson, *Proceedings of the 26th IEEE Photovoltaic Specialists Conference*, pp. 875-878 (1997).
13. A.W. Bett, R. Adelhelm, C. Agert, R. Beckert, F. Dimroth, and U. Schubert, *Solar Energy Materials and Solar Cells* **66**, 541 (2001).

Mater. Res. Soc. Symp. Proc. Vol. 836 © 2005 Materials Research Society L7.4

Expanded Experimental Space for Luminescence Studies of Thin Film CdS/CdTe Solar Cells

Scott Feldman, Tim Ohno, Victor Kaydanov, and Reuben Collins
Physics Department, Colorado School of Mines
Golden, CO 80401

ABSTRACT

We have explored a large range of experimental space for photoluminescence (PL) and electroluminescence (EL) measurements of CdS/CdTe solar cells. This space includes changes in temperature, injection intensity (laser power for PL, current for EL), electrical bias for PL, and laser energy for PL. Measurements were resolved both spectrally and spatially (2D for EL, 1D for PL). Combination of EL and PL measurements revealed that most spatial inhomogeneity was the result of non-uniform current transport rather than local variations in recombination rate. The greatest spectral resolution was obtained with low temperature EL at low injection rates. High injection EL as well as high forward biased PL suppressed band-edge emission at low temperatures. Spectral structure was found to be greater in EL than in PL. These effects likely originated from preferential current transport along grain boundaries and/or certain grains.

INTRODUCTION

CdS/CdTe photovoltaics represents one of the leading technologies for manufacture of polycrystalline thin film solar cells. State of the art CdTe devices have reached efficiencies of 16.5% [1]. Yet much is still not understood about the basic properties of the material and device structure, limiting further advances in conversion efficiency.

One factor that hinders cell performance is the non-uniform nature of the devices [2]. Though there are many techniques capable of probing on the scale of individual grains, such microscopic techniques are difficult and time-consuming to perform, disallowing the study of large sample sets. However, several simpler techniques have been employed to quantify meso-scale (larger than grain size, smaller than cell size) non-uniformities including laser beam induced current [3], photoluminescence (PL) mapping [4], and electroluminescence (EL) [5].

Besides studying spatial non-uniformities, EL and PL can also be resolved spectrally to study defects. In this paper, we use EL and PL with simultaneous spatial and spectral resolution. We utilize a large range of experimental variation including injection level, temperature, bias (light for EL, electrical for PL), and laser energy for PL.

EL and PL are techniques that are complimentary to each other. Both methods utilize the detection of radiative recombination of excess carriers. However, the method of carrier injection differs. With EL, carriers are injected electrically with forward bias. With PL, carriers are injected optically with a laser. The electrical nature of EL makes it very sensitive to transport properties of the material. Because of the direct optical injection, PL is less sensitive to electrical transport. Combination of both techniques allows one to isolate effects due to transport. In order to observe EL, one needs a complete device whereas PL is also applicable to films and single crystals. However, measurements on CdTe crystals or films are not necessarily analogous to measurements of CdTe within a device because of addition processing steps such as $CdCl_2$ treatment and back contact application. In this study only complete devices were used.

In previous work, using intentionally engineered non-uniformities (in the form of patterned Cu doping), we have shown that inhomogeneity in EL intensity results mostly from variations in transport, rather than variations in recombination [6]. Comparatively, PL intensity varied little with location. Subtle spectral changes were noted between Cu-rich and undoped regions in both EL and PL. Not addressed was the difference in spectral structure between EL and PL. This difference between EL and PL spectra has also been observed in other samples and is the focus of this paper.

EXPERIMENTAL DETAILS

For combined EL/PL measurements, the solar cell was first mounted in a cryostat. To obtain spatial images of EL intensity, the cell was forward biased at constant current conditions and the resultant EL was focused onto the entrance slit of an Acton 300i spectrometer with a Princeton Instruments Spec-10:100BR CCD array detector. The spectrometer had two gratings and a mirror mounted on a rotating turret. Using the mirror, and opening the slit to 2 mm, spatial images of EL were obtained with resolution ~30 microns. For spectral resolution, a grating is rotated into position and the slit width is reduced. The signal that passes through the spectrometer is dispersed onto the CCD detector. Because the detector is a 2-dimensional array, spectrographic information is obtained along the horizontal axis of the CCD, while spatial information is obtained along the vertical axis. With PL, the path of the laser beam to the sample was several meters resulting in a wide beam that was roughly uniform over the area of the cell imaged through the slit. Except where indicated, a 1.96 eV HeNe laser was used to induce PL. Apart from bias-dependent PL, PL was measured under open-circuit conditions (V_{oc}).

DISCUSSION

EL and PL spectra were found to have the greatest intensity and most amount of structure at lower temperatures, thus most of the our effort was concentrated on low temperature measurements. As mentioned above, EL and PL spectra can vary significantly from each other. One way in which this effect manifests itself is a spectral variation by location that is present in EL but not in PL as can be seen in Fig. 1. Injection level variation revealed some key differences between EL and PL, shown in Fig. 2. As injection level increased, EL and PL intensity both increased linearly. We note that this observation is in contrast to Compaan *et al.* who have observed a super-quadratic increase in EL intensity with current density [7]. In EL, there are two major trends in the spectra as injection level increases. First, structure becomes less prevalent as injection level increases. Second, the band edge features diminish with increasing injection level. By contrast, PL spectra show no change in structure with varying injection and show increased band edge emission with increased injection.

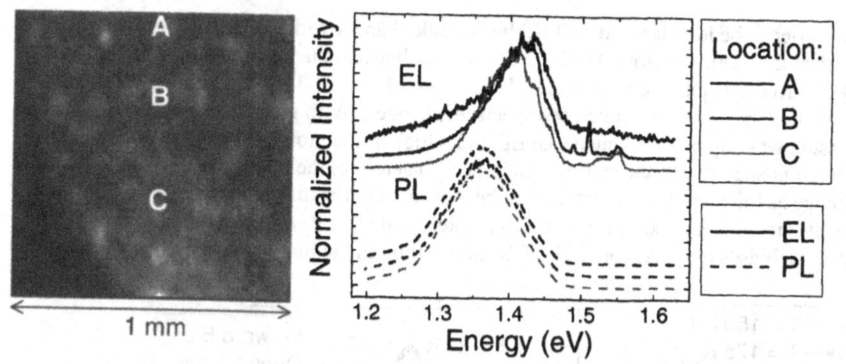

Figure 1. (color online) EL image and EL and PL spectra taken from three different points on the cell. The EL spectrum varies by location but the PL spectrum does not. T = 83 K. EL injection = 0.13 mA/cm^2. PL injection = 25 mW/cm^2

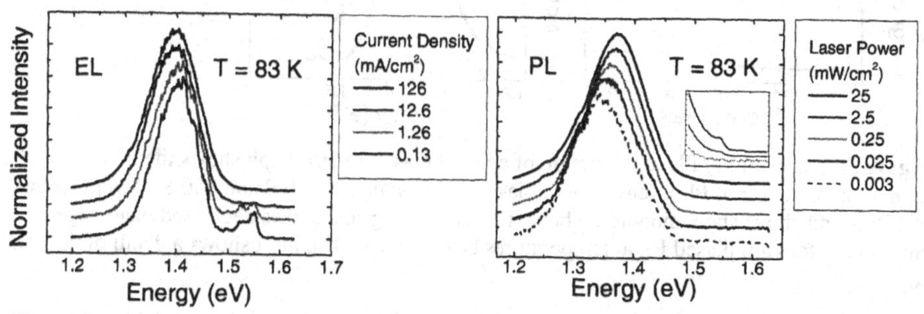

Figure 2. (color online) Normalized EL and PL spectra vs. injection level. EL shows an increase in structure and band-edge emission with decreasing injection. In contrast, PL shows no change in structure and an increase in band edge emission (detailed in the inset figure) with injection.

To find the origin of these dissimilarities, the PL conditions were varied to better resemble the conditions under which EL is measured. One difference between EL and PL is the applied bias. Though some forward bias will result from the laser in PL, EL is under much higher forward bias for even moderate injection. At higher temperatures (175 K), forward bias (FB) increased PL intensity whereas reverse bias (RB) suppressed it with little change in spectrum. These results agree with Ref. 7. FB decreases the size of the depletion region, allowing injected carriers more time to recombine while RB increases the depletion region, sweeping carriers away before they can recombine. As the temperature is decreased to 150 K, the RB behavior remained qualitatively the same whereas FB behavior changed. +1V FB greatly increased PL intensity with respect to V_{oc} but +2V FB yielded decreased emission with respect to +1V (though still slightly greater than V_{oc}). Again, the spectrum changed little with applied bias. This non-monotonic dependence of PL intensity on voltage can be explained by considering effects of the back contact Schottky barrier (present in CdTe cells due to its high work function). When the size of the main junction is attenuated under FB, the Schottky barrier is under RB and therefore grows in size. There is no reason to expect the temperature dependence of the main junction and

Schottky barrier to be identical. At 150 K, the Schottky barrier width apparently grows faster with bias than the main junction. At +2V FB, the back barrier is large enough to separate light-inducted electron-hole pairs and suppress PL emission. Below 100 K, PL intensity was nearly constant with applied bias. However, the spectrum changed. With greater than +1V FB, the band edge emission was suppressed, much like EL under higher injection conditions. Still, no change was observed in spectral structure of the main peak. Therefore, the opposite trend of the EL and PL band edge as injection level is varied can be attributed to the difference in bias conditions. These results are summarized in Fig. 3 below. Note, PL intensity was much greater than EL intensity at the biases used. Accordingly, EL contribution to the biased PL signal was negligible.

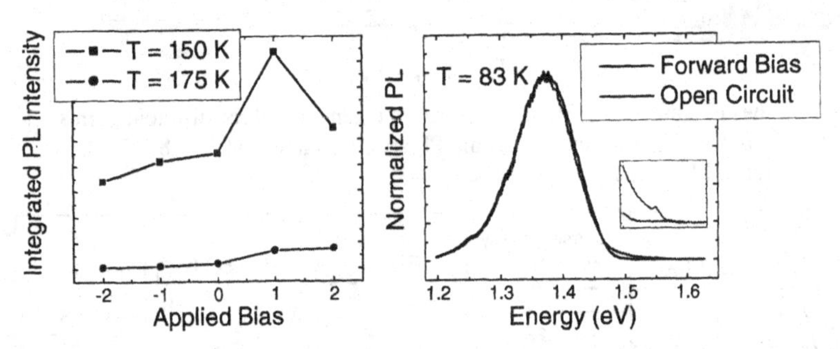

Figure 3. (color online) PL as a function of applied bias. The left graph shows the non-monotonic behavior of PL intensity with bias at temperatures of ~100 and 150 K. Temperatures of 175 K and above show monotonic behavior. The right graph shows decreased band edge emission in forward biased PL at temperatures below 100 K. The inset shows a detail of the band edge.

Besides voltage conditions, EL conditions could be dissimilar from PL because of the location of recombination. Besides lateral variation as shown in Fig. 2, EL and PL could originate from different depths. PL should mainly originate from approximately the same depth at which the laser is absorbed, 99% within the first micron of CdTe for photons of 1.96 eV. The depth of EL recombination can be approximated by taking the product of free carriers [8]. Modeling places the peak EL at the main junction, as in PL, but penetrating deeper into the CdTe than the PL. By changing to a lower energy laser, however, the laser absorption depth (and therefore the PL emission) can be changed to resemble the depth of the EL emission. A tunable Ti-sapphire laser was used for this purpose, providing a range of laser energies down to 1.57 eV. This energy was very close to the CdTe band gap at 83 K and thus should provide electron-hole pairs throughout the depth of the device. No change in the PL spectrum was observed with the change in laser energy indicating that the difference in EL and PL spectra is not due to a difference in the depth of recombination.

We could not achieve an increase in spectral structure by changing the bias conditions of PL or the depth of PL. Therefore, the increased structure of EL likely results from recombination that is laterally separated from that of PL. This hypothesis is further supported by the EL spectral variations observed by altering location. The fact that PL spectra do not vary by location implies that the variations occur on a scale smaller than can be resolved with our equipment (less than 10 microns). This scale is approximately that of CdTe grains. There is much evidence for increased conduction along grain boundaries and/or certain grains. These results include increased current injected with a scanning tunneling microscope along grain boundaries and certain grains [9], depletion of grain boundaries as measured by scanning capacitance measurements [10], and increased photocurrent collection along grain boundaries [11,12]. Because EL is selective of areas of lowest resistance, perhaps it originates only from grain boundaries and/or certain grains under low injection conditions. If these areas are more defective than the bulk, this reasoning explains the increased spectral structure EL possesses at small currents. The lack of spectral structure of PL is also explained because the injection is uniform throughout grain boundaries and the bulk.

CONCLUSIONS

This study explored the wide variation of conditions available for EL and PL measurements of CdTe. The differences between EL and PL spectra were explored. The greatest spectral structure was obtained with low injection EL at low temperature. The opposite behavior of band-edge features with injection level between EL and PL was explained with the different electrical bias conditions of EL and PL. The increased spectral structure of EL may be attributed to the sensitivity of EL to current transport that may preferentially select defective areas of the device. For this reason, EL may be superior to PL for defect studies of CdTe solar cells.

ACKNOWLEDGEMENTS

This material was based on work supported by the Department of Energy under Grant No. ADJ-2-30630-05 and the National Science Foundation under grant No. DMR-013945.

REFERENCES

1. Martin A. Green, Keith Emery, David L. King, Sanekazu Igari, and Wilhelm Warta, *Prog. Photovolt: Res. Appl.* **11,** 39 (2003).
2. V.G. Karpov, A.D. Compaan, and Diana Shvydka, Appl. Phys. Lett. **80**, 4256 (2002).
3. J. Hiltner, *Investigation of Spatial Variations in Collection Efficiency of Solar Cells* (Ph.D. thesis, Colorado State University, Fort Collins, CO), 2001.
4. D. Shvydka, A.D. Compaan, and V.G. Karpov, *J. Appl. Phys.*, **91**, 9059 (2002)
5. Scott Feldman, Fred Seymour, Tim Ohno, Victor Kaydanov, and Reuben Collins, *Mat. Res. Soc. Symp. Proc.* **763,** B5.10 (2003).
6. S.D. Feldman, R.T. Collins, V. Kaydanov, and T.R. Ohno, Appl. Phys. Lett. 85, 1529 (2004)
7. A.D. Compaan, D. Shvydka, K. J Price, A. Vasko, and V.G. Karpov, *Proceedings of the NCPV Program Review Meeting* (2001).
8. P.R. Thornton, *The Physics of Electroluminescent Devices* (E. & F. N. Spon Limited, London, 1967) p. 18.
9. P. Sutter, E. Sutter, and T.R. Ohno, Appl. Phys. Lett. 84, 2100 (2004).
10. Iris Visoly-Fisher, Sidney R. Cohen, and David Cahen, Appl. Phys. Lett. 82, 556 (2003).
11. M.K. Herndon, A. Gupta, V. Kaydanov, and R.T. Collins, Appl. Phys. Lett. 75, 3503 (1999).
12. S. Smith, P. Zhang, T. Gessert, and A. Mascarenhas, Appl. Phys. Lett. 85, 3854 (2004).

Mater. Res. Soc. Symp. Proc. Vol. 836 © 2005 Materials Research Society L7.5

Preparation and characterization of monolithic HgCdTe/CdTe tandem cells

S. L. Wang*, J. Drayton, V. Parikh, A. Vasko, A. Gupta, and A. D. Compaan
Department of Physics & Astronomy, University of Toledo,
Toledo, OH, 43606, USA

ABSTRACT

A prototype monolithic HgCdTe/CdTe superstrate tandem cell has been fabricated by RF sputtering, comprising a CdTe/CdS top cell, a ZnTe:N/ZnO:Al interconnect junction and a HgCdTe/CdS bottom cell. The $Hg_{1-x}Cd_xTe$ film as the bottom absorption layer was deposited by RF sputtering with 70% or 85% Cd content in the $Hg_{1-x}Cd_xTe$ magnetron target. $Hg_{1-x}Cd_xTe$ films with band gap from 0.98 eV to 1.45 eV were obtained by controlling the deposition temperature. $CdCl_2$ thermal treatments were used to improve the $Hg_{1-x}Cd_xTe$ film electrical properties. A nitrogen-doped ZnTe film combined with an aluminium (Al) doped ZnO film formed a good interconnect junction. Results of $V_{oc} = 0.99$ V and $J_{sc} = 2.1$ mA/cm^2 were obtained in the best such tandem cell at one sun (AM1.5).

INTRODUCTION

25% conversion efficiency is a goal for polycrystalline thin films in the high performance PV program of the US Department of Energy [1, 2]. The predicted achievable efficiency at AM1.5 is 28.2% for top and bottom cell band gaps of 1.72 eV and 1.14 eV, respectively. As a single-junction solar cell material, the best sputtered CdTe cell is 14% [3] and has reached 16.5% for close-spaced sublimation [4]. Large-scale manufacturing of low-cost CdTe cells has been achieved [5], therefore, CdTe-based alloys such as CdZnTe or CdMnTe, which can be adjusted to 1.74 eV band gap, are attractive candidates for top cell materials [6, 7]. In addition to these II-VI alloys with wide band gaps, alloys of CdTe with HgTe form possible narrow-band-gap bottom cell candidates. $Hg_{1-x}Cd_xTe$ is flexible enough to tailor the band gap from -0.15 eV to 1.5 eV.

Although polycrystalline CdZnTe and CdMnTe can fit the optical requirement for such tandem cells, their poor electronic properties remain obstacles to tandem cell applications. Compared to CdZnTe and CdMnTe, the electronic properties of HgCdTe can be easily controlled by proper annealing. A sputtered, single heterojunction HgCdTe/CdS cell showed typically $V_{oc} \sim 0.39$ V and $J_{sc} \sim 13$ mA/cm^2. This has encouraged us to fabricate HgCdTe/CdTe tandem cell as a first step toward approaching a high efficiency, all II-VI tandem cell. In this paper, we investigate processing of a HgCdTe/CdTe two-terminal tandem cell, particularly studying the characteristics of RF sputtered $Hg_{1-x}Cd_xTe$ films and HgCdTe/CdS single junctions.

EXPERIMENTAL DETAILS

We have chosen to maintain a separate deposition system for the Hg-based alloys and also a separate system for sputtering of the ZnTe/ZnO layers. Therefore the whole tandem structure was deposited in three separate RF sputtering systems. The deposition sequence started with a cleaned SnO_2:F-coated TEC-15 glass substrate loaded into a dual magnetron sputtering system for CdS and CdTe deposition; then vapor $CdCl_2$ treatment was carried out in dry air at 390 °C.

After $CdCl_2$ treatment, ZnTe:N and ZnO:Al layers were deposited on the sample in a similar dual magnetron ZnO/ZnTe sputtering system. After this interconnect junction was fabricated, a CdS layer was deposited again in the CdS/CdTe chamber. Finally the sample was loaded into the HgCdTe sputtering system for HgCdTe layer deposition. After a second vapor $CdCl_2$ treatment of the HgCdTe layer, a Cu(3 nm)/Au(20 nm) metal contact was evaporated to serve as back contact.

For the $Hg_{1-x}Cd_xTe$ alloy film a modified bell jar sputtering system with a 2-inch magnetron sputtering gun was used. The target was a cold-pressed mixture of high purity HgTe (5N) and CdTe (5N) with the CdTe content of either 70% or 85%. The growth temperature was monitored by an optical interference technique through the glass substrate [8]. Pure argon gas flowed through the chamber during the deposition, the deposition pressure was controlled in the range of 5 to 30 millitorr and the RF power was from 20 to 40 W. A one inch diameter quartz tube was used to carry out the vapor $CdCl_2$ treatment on $Hg_{1-x}Cd_xTe$ films, dry air was induced during heating and temperature was controlled in the range of 360 °C to 400 °C. The films were analyzed by infrared (IR) transmission, energy dispersive x-ray spectroscopy (EDS) and x-ray diffraction (XRD).

RESULTS AND DISCUSSION

Characterization of HgCdTe films and the HgCdTe/CdS single junction

The composition x of $Hg_{1-x}Cd_xTe$ affects optical and electronic properties through the relationship between the composition and the band gap E_g. According to the Hanson equation [9],

$$E_g = -0.302 + 1.93\,x - 0.81\,x^2 + 0.832\,x^3 + (5.35e10^{-4}\,(1-2x))T, \quad (1)$$

for $x = 0.77 \pm 0.05$, the band gap of HgCdTe would be 1.0 ± 0.1 eV. We used EDS and IR transmission spectra to determine the film composition. For example, we did EDS measurement on two samples which deposited at room temperature, there was no supplemental Hg introduced during the deposition. We assumed that the films were stoichiometric, and found that the compositions of two sample were x = 0.79 and 0.76, respectively. Using Equation (1), we estimated the band gaps of these two samples were 1.03 eV and 0.98 eV, respectively.

Figure 1 shows the effect of substrate temperature on the composition of four $Hg_{1-x}Cd_xTe$ films. The films were deposited in sequence from the same target at different temperature. The band gap was determined from IR transmission and composition was deduced using Equation (1). The films deposited at higher temperature show a larger band gap due to more loss of Hg during deposition. The quality of $Hg_{1-x}Cd_xTe$ films deposited on glass is comparable to CdTe films on glass. As-deposited samples show high resistance ($1 \sim 5\times10^{10}$ Ω/sq.) at one sun; such samples do not make good devices, so a post annealing process was used to lower the resistance. We carried out heat treatment of $Hg_{1-x}Cd_xTe$ samples in a 1 inch diameter tube. The annealing temperature was controlled within ± 5 °C in the range of 360 °C ~ 400 °C with different ambients, N_2, air, with or without $CdCl_2$. Figure 2 shows the X-ray diffraction of as-deposited and annealed samples. After annealing at 370 °C or 390 °C in dry air, the films still exhibit a (111) orientation similar to the as-deposited samples. The low intensity of other peaks suggests a few randomly oriented grains appear after annealing.

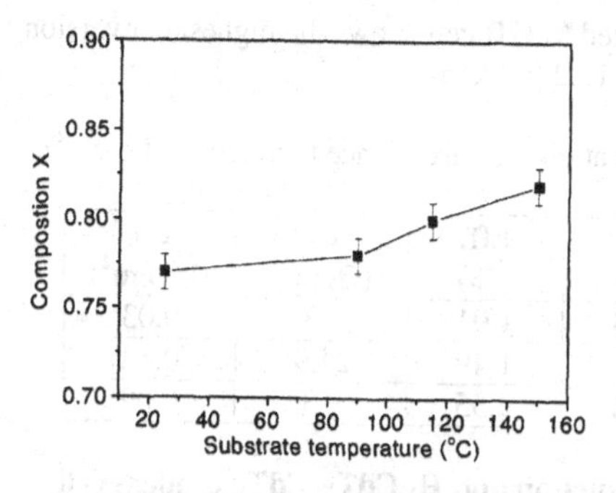

Figure 1. Substrate temperature dependence of composition of $Hg_{1-x}Cd_xTe$ films.

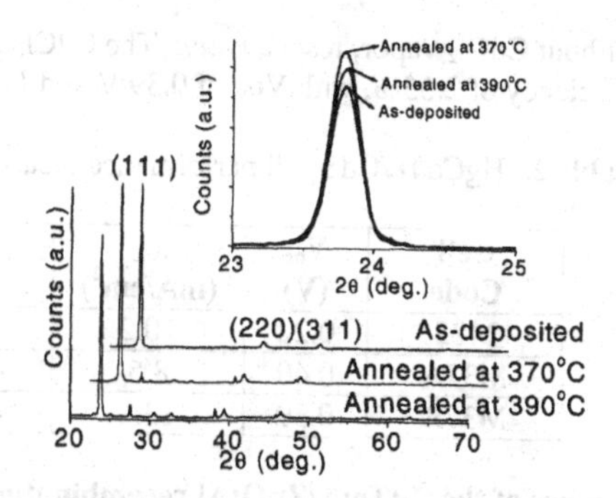

Figure 2. X-ray diffraction of as-deposited and annealed (in air with $CdCl_2$) $Hg_{1-x}Cd_xTe$ samples.

Table 1 shows the film sheet resistance under different annealing conditions. The annealing temperature cannot be raised to more than 400 °C without damaging the film. Sample M24a was annealed in Ar at 250 °C for 20 hrs; its sheet resistance was about 9.5×10^8 Ω/sq at AM1.5, higher than that of other samples which were annealed in air. Sample M19a was annealed in air at 390 °C without $CdCl_2$ vapor, while sample M22b was annealed with $CdCl_2$ vapor, both samples showed lower sheet resistance; however, sample M22b showed much lower sheet resistance than sample M19a, indicating that annealing in air with $CdCl_2$ vapor improves the sample conductivity.

Table 1. Sheet resistance (by transfer length method [10]) of samples annealed at different conditions.

Sample code	Annealed ambience	Temperature (°C)	Time (minute)	R_{Sheet} at dark (Ω/sq)	R_{Sheet} at AM1.5 (Ω/sq)
M15b	As-grown	N/A	N/A	8.9×10^{10}	1.2×10^{10}
M19a	Air	390	15	5.7×10^8	1.7×10^8
M18a	$CdCl_2$, Air	390	1	4.2×10^8	1.1×10^8
M22b	$CdCl_2$, Air	390	5	9.0×10^7	3.9×10^7
M24a	Ar	250	1200	3.9×10^9	9.5×10^8

Several cells were fabricated with the structure of Tec15/CdS/HgCdTe and a Cu (3 nm)/ Au (20 nm) layer was evaporated as a back contact at room temperature followed by diffusion at 150 °C. Since we successfully deposited p-ZnTe by doping with N using the reactive sputtering technique [11], we also made some efforts to make p-HgCdTe by using reactive sputtering in 5% N_2/Ar ambient. Table 2 shows the performance of several of these solar cells. The M22 cell was made using pure Ar as the deposition gas, while the others were made from 5% N_2 /Ar as the deposition gas. The postdeposition annealing process was the same for all samples with the temperature at 390 °C in dry air. Cells M22 and M37A were annealed in $CdCl_2$ vapor and without $CdCl_2$ vapor, respectively. The $CdCl_2$ treated M37B cell shows the highest conversion efficiency of 2.55%, with Voc of 0.39 V and Jsc of 13.2 mA/cm^2.

Table 2. HgCdTe/CdS cell performance measured at one sun irradiance (cell area: 0.06 cm^2).

Cell Code	V_{oc} (V)	J_{sc} (mA/cm^2)	FF (%)	Eff. (%)	R_s (Ωcm^2)	R_{sh} (kΩcm^2)
M22	0.22	10.2	38.3	1.01	6.3	0.03
M37A	0.40	8.5	44.2	1.49	24.9	0.12
M37B	0.39	13.2	49.6	2.55	9.3	0.09

Studies of the ZnTe:N/ZnO:Al recombination junction and HgCdTe/CdTe tandem cell

A good interconnect junction in a monolithic tandem cell should introduce negligible resistance. We made several interconnect junctions which were composed of 0.16 μm ZnTe:N and 0.25 μm ZnO:Al layers. The typical doping levels of ZnTe:N and ZnO:Al were 3 ~ 5×10^{18} cm^{-3} and ~1×10^{20} cm^{-3}. Figure 3 is a sketch of this interconnect junction with an area of 0.25 cm^{-2}. The effective contact resistance for a small applied bias was measured to be ~ 1.75 Ω-cm^2. A true tunnel junction needs a doping level of 10^{20} cm^{-3} or higher to achieve tunnelling in the narrow depletion layers, however, the highly defective structures of amorphous and polycrystalline material can produce defect-assisted tunnelling or recombination junctions that can yield acceptable performance [12]. Defect-assisted tunnelling and recombination may be occurring here.

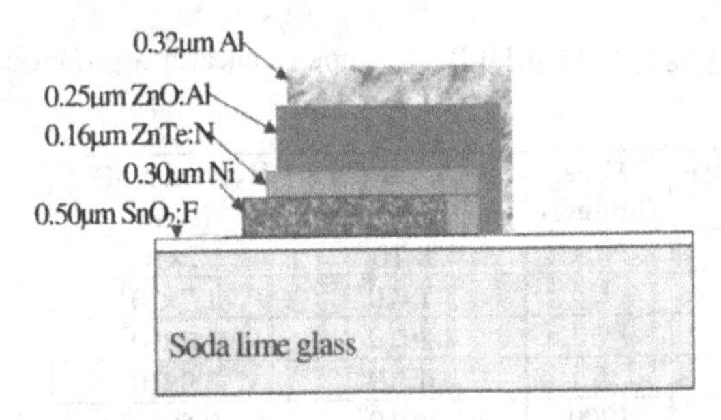

Figure 3. A schematic diagram of ZnTe:N/ZnO:Al interconnect junction.

To investigate the possibility of making monolithic tandem cells with sputtered HgCdTe materials, a prototype two-terminal tandem cell structure was fabricated. The top cell consisted of CdTe (1.8 μm)/CdS (0.13 μm) layers and the bottom cell of HgCdTe (1.5 μm)/CdS (0.13 μm) with a HgCdTe band gap around 1.1 ~ 1.15 eV. 0.16 μm ZnTe:N and 0.25 μm ZnO:Al layers were deposited to form an interconnect junction. A Cu (3 nm)/Au (20 nm) back contact was evaporated as the last step forming a cell area of 0.06 cm^2.

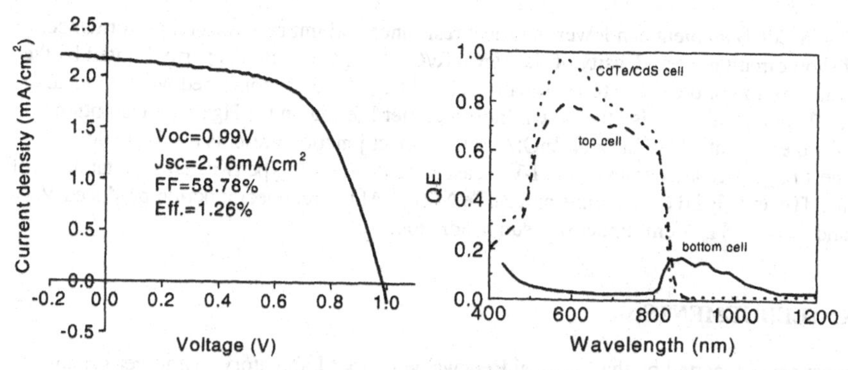

Figure 4. J-V curve of a CdTe/HgCdTe tandem cell at one sun illumination.

Figure 5. Quantum efficiency of CdTe/HgCdTe tandem cell and Au back contact CdTe/CdS cell.

Figure 4 displays the J-V curve of one such tandem cell which was measured at NREL. The V_{oc} is 0.99 V, J_{sc} is 2.16 mA/cm^2, and efficiency is 1.26 %. Figure 5 shows the quantum efficiency of a similar CdTe / HgCdTe based tandem cell and a CdTe / CdS reference cell with Au back contact. Normally we use 3 ~ 4 nm of Cu before the Au contact; however, here we use a reference cell QE from a cell prepared no Cu, since the tandem cell also used no Cu at the CdTe back contact. From Figure 5, we notice that the CdTe top cell current is nearly as high as that of the reference cell, whereas the HgCdTe bottom cell has low QE. A CdTe cell with ZnTe:N based back contact (ZnTe:N/Ni) produced a V_{oc} ~ 0.71 V and we expect that the ZnTe:N/ZnO:Al back contact would not have a V_{oc} greater than this. Thus the bottom HgCdTe cell probably contributes over 0.2V to the V_{oc} of tandem cell. We have obtained the transmission of different thickness CdTe films in CdTe/CdS structures, about 50% of light in the range of 830 ~ 1100 nm passes through the CdTe/CdS structure, and assuming the light is absorbed by the HgCdTe layer completely, it would generate about 6 mA/cm^2 of current density. The low J_{sc} of 2.16 mA/cm^2 of whole structure is mainly due to low photon current generated in HgCdTe bottom cell. Low response to long wavelengths (~1050 nm) in QE measurement indicates poor carrier collection which points to limited minority carrier lifetimes in the HgCdTe absorption. Figure 5 also shows the QE of a sputtered CdTe/CdS single junction cell with Au back contact. Note that the QE of the top cell of the tandem is not much different from a cell prepared with no Cu as part of the back contact, confirming that the top cell structure does not degrade substantially during the fabrication of HgCdTe bottom cell. Although the conversion efficiency is only 1.26%, we believe that with optimizing of the HgCdTe bottom cell fabrication and increasing the top cell optical transmission the whole tandem cell performance will be improved.

CONCLUSIONS

We have successfully deposited $Hg_{1-x}Cd_xTe$ films on glass by RF sputtering. The properties of the $Hg_{1-x}Cd_xTe$ films have been investigated by IR transmission, EDS and XRD. The composition of the $Hg_{1-x}Cd_xTe$ films is affected by the deposition temperature. Films with the 1.0 eV band gap required for tandem cells were obtained at low deposition temperature.

Annealing in $CdCl_2$ ambient can lower the sheet resistance and improve the cell performance. The low short circuit current density of the HgCdTe/CdS single junction is partly limited by the optical transmission of the CdTe top cell with its band gap of 1.5 eV compared with the ideal of 1.72 eV. There is evidence also that low minority carrier lifetime in the HgCdTe absorption layer limits the current. The ZnTe:N/ZnO:Al interconnect junction presents a suitably low interconnect resistance as seen from the I-V measurement. A prototype two-terminal tandem cell based on CdTe and HgCdTe materials and ZnTe:N/ZnO:Al interconnect junction produced V_{oc} = 0.99 V and J_{sc} = 2.16 mA/cm^2 under one sun irradiation.

ACKNOWLEDGEMENTS

This project was supported by the National Renewable Energy Laboratory - High Performance PV Project (Martha Symko-Davies).

REFERENCES

1. T. J. Coutts, K. A. Emery and J. S. Ward, *Prog. Photovolt: Res. Appl.* **10**,195-203, (2002)
2. T. J. Coutts, J.S. Ward, D. L. Young, K. A. Emery, T. A. Gessert and R. Noufi, *Prog. Photovolt: Res. Appl.* **11**,359-375, (2003)
3. A. Gupta and A. D. Compaan, *Applied Physics Letters*, 85(4) 684 (2004)
4. X. Wu. J. C. Keane, R. G. Dhere, C. DeHart, D. S. Albin, A. Duda, T. A. Gessert, S. Asher, D. H. Levi, and P. Sheldon, *Proc. 17th Europen PVSEC*, pp.995, (2001)
5. D. Rose, R. Powell, U. Jayamaha, M. Maltby, D. Giolando, A. McMaster, K. Komanyos, G. Faykosh, J. Klopping, and G. Dorer, *Proc. 28th IEEE Photovoltaic Specialists Conference 2000*, pp. 428, (2000).
6. S. H. Lee, A. Gupta, A. D. Compaan, *Phys. Stat. Sol. (c)* **1**, pp.1042, (2004)
7. B. E. McCandless, *Proc. 29th IEEE Photovoltaic Specialists Conference 2002*, pp. 488, (2002).
8. Z. Feng, C. N. Tabory, and A. D. Compaan, *1st World conference of photovoltaic energy conversion (1994)*, pp.350, (1994)
9. E. D. Palik, *Handbook of Optical Constants of Solids, II (Academic Press)*, p.655, (1990)
10. D. K. Schroder, *Semiconductor material and device characterization (John Wiley & Sons, Inc.)* p109, (1990)
11. A. D. Compaan, J. Drayton, V. Parikh, A. Gupta, Y. Yu, C, Taylor, T. Osborn and R.G. Bohn, *Proceedings of 3rd World Conference on Photovoltaic Energy Conversion,* 2P-A8-3 (2003)
12. J. Drayton, V. Parikh, G. Rich, A. Gupta, T. Osborn, R. G. Bohn, A. D. Compann, B. E. McCandless, P. D. Paulson, *Mat.Res.Soc.Symp.Proc.* **763**, 353, (2003)

Silicon Thin Films

Mater. Res. Soc. Symp. Proc. Vol. 836 © 2005 Materials Research Society

Effects of Grain Boundaries in Amorphous/Multicrystalline Silicon Heterojunction Photovoltaic Cells

M. Farrokh Baroughi and S. Sivoththaman
Department of Electrical & Computer Engineering, University of Waterloo
200 University Avenue West, Waterloo, Ontario N2L 3G1, Canada.

ABSTRACT

Spectral response and dark current-voltage characteristics of heterojunctions are used to investigate grain boundary degradation in photovoltaic properties of a-Si/mc-Si heterojunction solar cells. Measured spectral response inside the grain and on the grain boundary shows small but consistent QE degradation due to minority carrier recombination at the grain boundaries. No consistent difference is observed in dark current-voltage characteristics because of large diode area and periphery leakage current in the employed heterojunction diodes. Comparing measurement results and results from device modeling using the simulation software Medici, a recombination velocity of 4900 cm/sec is found at the grain boundaries of employed multicrystalline silicon wafer. The modeling and experimental results can also be used to define an effective grain area that serves as a measure of grain boundary recombination and the influence of grain size.

INTRODUCTION

Cost reduction is an important issue in the fabrication of silicon (Si) photovoltaic (PV) cells, where the material cost accounts for nearly half of the overall cost. Materials like multicrystalline silicon (mc-Si), silicon ribbons etc., offer a cost effective option for Si PV cells compared to single crystalline Si [1]. In most of those materials however, the presence of large number of grains (mm to cm scale), grain boundaries (GBs), and crystallographic defects necessitates defect passivation. Defect passivation by atomic hydrogen is a very efficient method. However, this imposes a temperature (T) limit for any post-passivation processes such as pn junction diffusion at high-T. Implementation of amorphous Si (a-Si)/crystalline Si heterojunctions (HJ) in place of diffused homojunctions in defective Si can keep the process temperature low thereby preserving the defect passivation [2,3].

mc-Si material normally comprises large grains in the range of mm^2 to cm^2. mc-Si production by casting and Edge-defined Film-fed Growth (EFG) techniques result in very large grain sizes in cm^2 range while Electro Magnetic Casting (EMC) and Silicon Sheet from Powder (SSP) techniques results in small grain sizes in mm^2 range[4]. Grain boundaries include lots of trap centers that can act as generation-recombination centers. Therefore, grain boundaries can potentially degrade short circuit current by recombining photogenerated carriers; and also fill factor & open-circuit voltage by increasing leakage current of the heterojunction diode in low forward bias region. GB effects in solar cells become very important for small grain mc-Si wafers [5-7]. In this work we have assessed the effects of grain boundaries on spectral response and dark current-voltage (IV) characteristics of a-Si/mc-Si HJ solar cells. HJ diodes and solar cells are fabricated and characterized to reveal GB degradation effects. Two-dimensional device simulation using Medici is employed to validate experiment results.

EXPERIMENTAL

Device fabrication

Followed by wafer cleaning and HF dip, Al layer is deposited on back side of the wafer and annealed at 475°C to obtain suitable ohmic rear contact. Followed by ultrasonic treatment in IPA, native oxide of the wafers is removed by 1% HF solution. HJs are fabricated by depositing a 20nm (n^+) a-Si layer on mc-Si wafer using PECVD technique. Process parameters for film deposition are: substrate temperature 250°C, chamber pressure during deposition 150mTorr, RF signal frequency 13.56MHz and RF power density 10mW/cm^2. For fabricating HJ diodes, Al layer is deposited and patterned. (n^+) a-Si is finely mesa etched by well tuned RIE using Al as a hard mask, figure 1(a). For fabrication of PV cells, followed by (n^+) a-Si deposition, a 65-70nm Indium Tin Oxide (ITO) layer is deposited in ITO sputtering zone of the cluster tool system. ITO layer is sputtered at 180°C using an In_2O_3+SnO_2 (90/10) target with 15 sccm Ar flow at 5mTorr without adding oxygen. For patterning of front Al layer of 1cm^2 PV cells a Cr photo-mask with 8% metal coverage has been used, figure 1(b).

Characterization setup

Dark IV characteristics of HJ diodes with 250μm x 100μm area are measured using a Cascade probe station equipped with Keithley 4200 characterization system. External Quantum Efficiency (EQE) of 1cm^2 HJ PV cells is measured inside grains and on the grain boundaries. Characterization setup includes a conventional lock-in based spectral response setup equipped with optical monochromator and lenses to achieve small and narrow monochrome light spot. This small and narrow monochrome beam with about 100μm width and 1mm length is focused right on the GB using a fine micromanipulator keeping the boundary at the center of the beam. A section of a 1cm^2 PV cell in figure 2 shows grain boundaries, metal grids and the size of narrow light spot covering a part of GB. A reference silicon photodetector, with tabulated quantum efficiency (QE) throughout the wavelength window, is used for calibration of the PV cell spectral response.

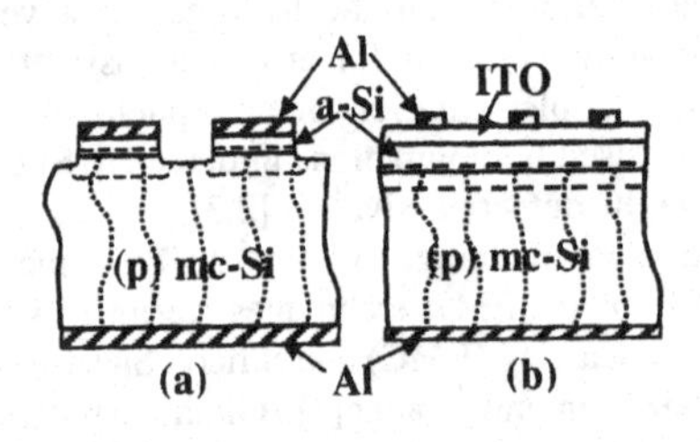

Figure 1: Cross section of (a) Mesa etched (n^+) a-Si/ (p) mc-Si HJ diode (b) Heterojunction solar cell

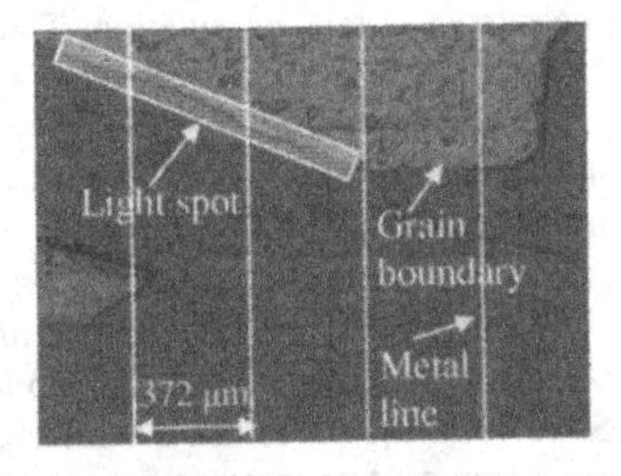

Figure 2: Fabricated HJ PV cell, the structure in figure 1(b), showing the alignment of the beam on GB

MEASUREMENT RESULTS

Dark IV characterization

Dark IV characteristics of 250μm x 100μm mesa etched HJ diodes, structure in figure 1(a), are measured on- and off-GB. Measurement results do not show IV degradation due to GB. Figure 3 shows dark IV characteristics of 2 adjacent HJ diodes, one on-grain and the other off-GB. As shown, IV characteristics are too close to be differentiated. Meanwhile, measurements on different samples do not result in consistent results because some diodes on GB show slightly higher diode current while some other GB diodes yield less dark current.

The ratio of GB length inside the diode area to diode periphery in fabricated HJ diodes vary between 0 and 0.38. This is the main drawback of our measurement because mesa etched diodes leave lots of dangling bonds at the diode periphery increasing diode leakage current especially at low forward bias regime. Similarly, the ratio of GB length inside the diode to the diode area varies between 0/cm and 108/cm. Since the diode area is relatively large, the GB effect in comparison with the diode current is negligible. Therefore, measurement of GB effect on dark IV characteristics of HJ diode requires very narrow but long mesa etched HJ diode, with very well passivated periphery, aligned on GB. In the current work we do not use measured dark IV characteristics for GB characterization but we model GB effect on dark IV characteristics of HJ solar cells.

External Quantum Efficiency

EQE analysis is employed to measure photogenerated charge collection efficiency on- and off-GB.

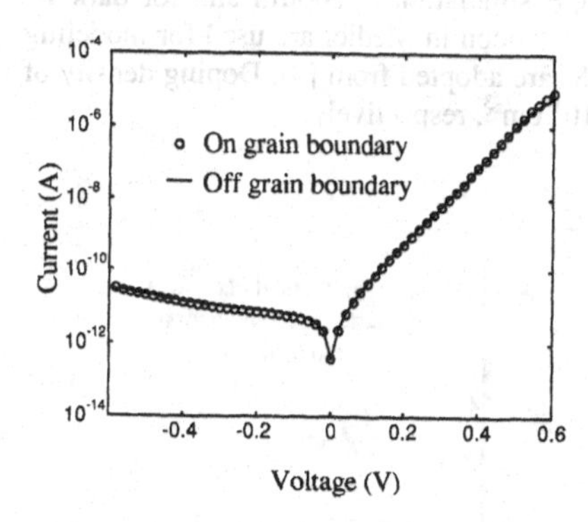

Figure 3: Dark IV of (n^+) a-Si/(p) mc-Si HJ diodes located on- and off- GB. The curves are almost identical because of peripheral effects in the diodes. [structure in figure 1(a)]

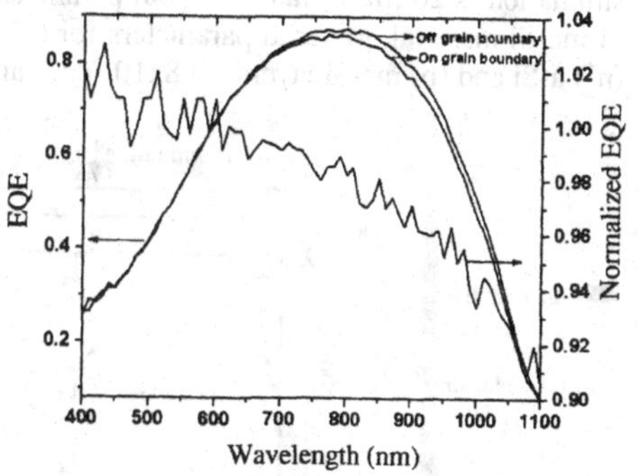

Figure 4: EQE and normalized EQE of HJ solar cell (figure 1(b)) on and off GB

EQE measurements on different PV cells show consistent degradation in EQE due to recombination in GB. EQE degradation, measured in different samples at 900nm, is between 3% and 5% for light spot size of 100μm x 1mm covering GB (figure 2). Figure 4 shows EQE of HJ PV cell on- and off-GB. Normalized EQE in figure 4 is defined as the ratio of EQE on GB to EQE off GB. Measurement results point out two main issues: (i) quantum efficiency on GB is less than QE off GB, and (ii) QE degradation due to GB is wavelength dependent and low energy photons (photons with long wavelengths) result in more EQE degradation. We discuss these issues in modeling section.

MODELING OF GRAIN BOUNDARY

Basically, GB in multicrystalline material includes a very thin amorphous tissue decorated with lots of crystallographic defects and lots of metallic, carbon and oxygen impurities. Some of these defects result in trap levels inside band gap that give rise to carrier generation and recombination. Generally, precise modeling of GB effect on EQE degradation requires information about density of states due to different defect sources within the band gap. Because of compositional and geometrical complexity of grain boundaries, providing such information is almost impossible. Therefore, we use a simplified GB representation. We treat GB as a hypothetical sheet with certain recombination velocity for electrons and holes [8]. Spectral response degradation originates from the fact that photogenerated minority carriers close to the GB can easily diffuse towards GB and recombine in trap states. Meanwhile, in the vicinity of GB, the injected electrons from (n^+) emitter into (p) mc-Si experience a smaller diffusion length due to recombination in GB. Smaller diffusion length tends to increase leakage current of the heterojunction diode.

For all of the simulations we use Medici device simulator. The simulated device structure is shown in figure 5. Width of the device for EQE simulation is 200μm and for dark IV simulation is 20μm. Default material parameters for silicon in Medici are used for modeling of mc-Si material. Material parameters for (n^+) a-Si are adopted from [9]. Doping density of (n^+) a-Si and (p) mc-Si layers are $8\times10^{18}cm^{-3}$ and $10^{16}cm^{-3}$, respectively.

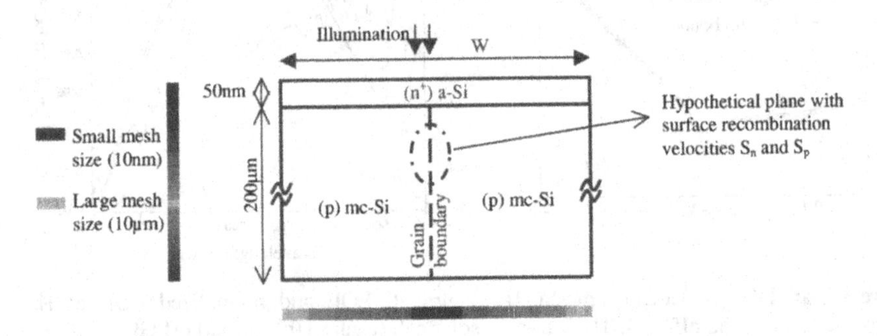

Figure 5: Device structure defined in Medici for dark IV and spectral response simulations; W=20μm is used for dark IV modeling and W=200μm is used for IQE modeling.

Effect of dark IV characteristics

Dark IV of HJ solar cell has a great impact in fill factor and hence in efficiency of the solar cell. GB defects increase both leakage current and ideality factor of the HJ diode resulting in less fill factor and conversion efficiency. The impact of surface recombination velocity (S) on dark IV characteristics of HJ diode is shown in figure6. Three main results can be obtained from this analysis: (i) as we expect, dark current of the diode increase with increasing S, (ii) significant deviation from reference IV characteristics of HJ diode is obtained for S>1000 cm/sec for the device with 20μm width, and (iii) recombination in GB affects low forward bias region more than high forward bias region because of limited minority carrier sinking capacity of GB. We do not see these effects in our cells but we expect a narrow long heterojunction diode well aligned on the GB with well passivated walls is able to reveal these effects.

Effect on spectral response

Spectral response directly shows the photogenerated carrier collection efficiency at the HJ. Figure 7 shows normalized Internal Quantum Efficiency (IQE) degradation due to increasing surface recombination velocity of the GB.

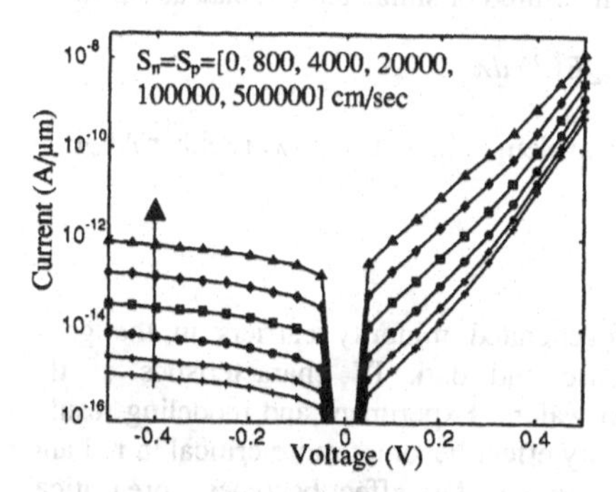

Figure 6: Impact of GB on dark IV of HJ diodes

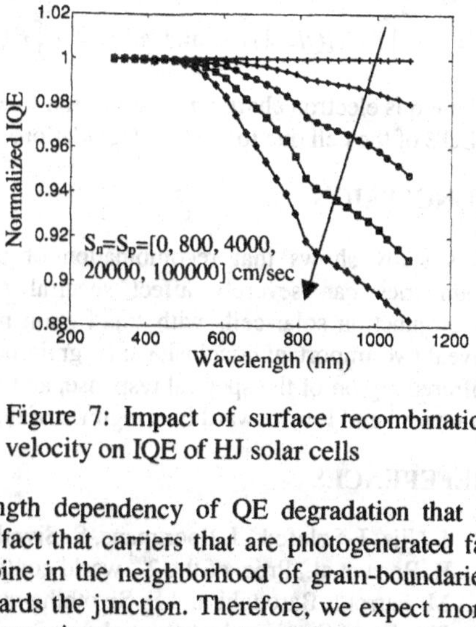

Figure 7: Impact of surface recombination velocity on IQE of HJ solar cells

Simulation results show the same wavelength dependency of QE degradation that is achieved in measurements. This is due to the fact that carriers that are photogenerated far away from junction are more likely to recombine in the neighborhood of grain-boundaries since they need to diffuse longer distances towards the junction. Therefore, we expect more QE degradation at larger wavelength values. Comparing measurement and simulation results we see that normalized EQE in figure 4 is in good agreement with normalized IQE in figure 7 with S_n=S_p=4000 cm/sec. More precise simulation show that the best simulation curve that fits measurement results in figure 4 is obtained for S_n=S_p=4900 cm/sec.

"EFFECTIVE GRAIN AREA" DUE TO PRESENCE OF GRAIN BOUNDARIES

Defining an effective grain area (A_{eff}) can be useful as a measure of GB recombination activity, and the influence of periphery/area ratio of the grains. Knowing QE degradation either from measurements or 2D modeling, effective GB area can be modeled. Assume external or internal QE degradation is available for a grain with periphery of P_G and area of A_G. Measurement and modeling is performed in a window of width W located inside the grain (figure 8). We can assume that the GB degradation reduces effective grain area by reducing width of grain by ΔW. ΔW can be expressed by,

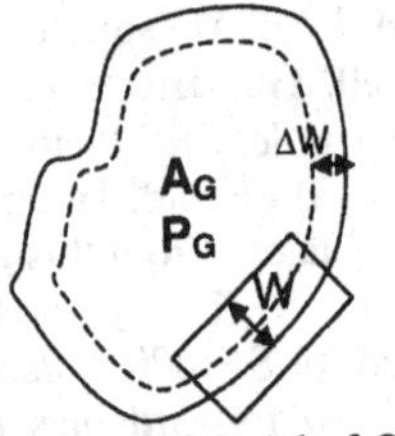

Figure 8. Model of GB

$$\Delta W = \frac{\Delta J_{SC}}{J_{SC}} W$$

J_{SC} is short circuit current density of the solar cells without GB and ΔJ_{SC} is degradation in J_{SC} due to GB. Effective grain area is approximated by $A_G - P_G \Delta W$. Therefore,

$$\frac{A_{eff}}{A_G} = 1 - \frac{P_G W}{A_G} \frac{\Delta J_{SC}}{J_{SC}}$$

J_{SC} and ΔJ_{SC} in this model can be calculated from measured or simulated QE data as below.

$$J_{SC} = q \int_{\lambda_{min}}^{\lambda_{max}} F(\lambda) QE(\lambda)\, d\lambda \text{ and } \Delta J_{SC} = q \int_{\lambda_{min}}^{\lambda_{max}} F(\lambda) \Delta QE(\lambda)\, d\lambda$$

where q is electron charge, F(λ) is photon flux at wavelength λ, and ΔQE(λ) is either ΔIQE or ΔEQE of the cell due to the GB degradation.

CONCLUSION

This study shows that recombination of photogenerated minority carriers in the grain boundaries can severely affect spectral response and dark IV characteristics of the heterojunction solar cells with small grain mc-Si wafers. Experiment and modeling results reveal two important conclusions: (i) grain boundary effect becomes more critical in red and infrared region of the spectral response, and (ii) grain boundary effect becomes more critical in reverse and low forward bias regimes of the dark IV characteristics.

REFERENCES

1. J. Nijs, J. Szlufcik, J. Poortmans, S. Sivoththaman, R. Mertens, ITED, **46,** p.10 (1999)
2. R. Rosa et al., Proc. of the 2nd world conference on PV energy conversion, p.2440 (1998)
3. M. Farrokh Baroughi and S. Sivoththaman, Proc. of 27th International Conference on the Physics of Semiconductors, under review
4. B. Qun et al., Proc. of 3rd World Conference on PV Energy Conversion, p.1282 (2003)
5. T. F. Ciszek et al., Proc. of 23rd IEEE Photovoltaic Specialists Conference p.101 (1993)
6. Edmiston, et al., Journal of Applied Physics, **80,** p.6783 (1996)
7. H. Nouri et al., Thin Solid Films 451-452, p.312 (2004)
8. C. Donolato, Semiconductor Science Technology, **15,** p.15 (2000)
9. L. Jiang, Mat. Res. Soc. Symp. Proc. **609**, p. A18.3.1 (2001)

Mater. Res. Soc. Symp. Proc. Vol. 836 © 2005 Materials Research Society L8.2

Improved Efficiency in Hydrogenated Amorphous Silicon Solar Cells Irradiated by Excimer Laser

A. A. Damitha T. Adikaari, S. Ravi P. Silva, Michael J. Kearney and John M. Shannon
Nano-Electronics Centre,
Advanced Technology Institute, University of Surrey,
Guildford, GU2 7XH, United Kingdom.

ABSTRACT

Excimer laser crystallisation is used to fabricate nanocrystalline thin film silicon Schottky barrier solar cells, in a superstrate configuration with indium tin oxide as the front contact and chromium as the back contact. 150 nm thick intrinsic absorber layers are used for the solar cells, and was crystallised using an excimer laser with different laser energy densities. These layers were characterised using Raman spectroscopy and optical absorption before device fabrication. External quantum efficiencies of the devices were calculated from the spectral response data of the devices. A maximum efficiency of 70 % is observed for low energy irradiation, which is significant for very thin absorber layers. Device operation is discussed with proposed band structures for the devices and supplementary measurements.

INTRODUCTION

Excimer laser (EL) crystallisation has been the preferred method for nanocrystalline thin film formation from hydrogenated amorphous silicon (a-Si:H) thin film transistors [1-4]. Pulsed laser energy melts and solidifies thin a-Si:H films within nanosecond time scales, systematically evolving hydrogen and forming nanocrystalline silicon (nc-Si:H). The short wavelength, short pulse duration, ultra violet energy of the excimer lasers is absorbed in the a-Si:H within a few nanometres, with a minimum percentage of heat reaching the substrate. This enables the use of cheap substrates such as glass, which cannot withstand conventional annealing temperatures for large area electronic applications. Although the application has primarily been focused on thin film transistors, its adaptability for photovoltaics has attracted considerable interest over time [5-7]. Significant emphasis has been placed on investigating layered nanocrystalline silicon formation from partially melting a-Si:H using the excimer laser [8,9]. Partial melting occurs when the laser energy density is low, but sufficient to melt a layer of the film, leaving a continuous solid layer underneath. This process results in a stratified structure with large crystallites from the laser irradiated surface, followed by a fine-grained silicon layer. A third unconverted a-Si:H layer is expected at the bottom, depending on how high the laser energy density used, but its properties may be affected due to conducted heat. In this report, we investigate the utilisation of these very thin, stratified nanocrystalline silicon films as absorber layers in Schottky barrier solar cells. The intrinsic nanocrystalline silicon was characterised using Raman spectroscopy and optical absorption measurements before device fabrication. External quantum efficiency (EQE) of the cells was calculated from spectral response of devices and they will be discussed in detail explaining the comparatively high efficiency obtained at low film thickness.

EXPERIMENTAL DETAIL

150 nm thick a-Si:H films were deposited by 13.56 MHz plasma-enhanced chemical vapour deposition at 250ºC. 100 nm thick indium tin oxide (ITO) covered 0.7 mm thick Corning 1737 glass was used as the substrates. A KrF Lambda Physik excimer laser (LPX 210i) operating at 248 nm with 25 ns full width half maximum pulse duration was used to crystallise a-Si:H films. The films were scanned with a 4 mm wide 10 mm in length semi-Gaussian beam profile with the scanning speed at 2.5 mm/s with a pulse repetition rate of 50 Hz, in vacuum (base pressure of 0.0133 Pa) over a total scan area which could be extended to 100x100 mm^2. Scanning of samples was carried out along the Gaussian direction with an overlap of 5% between two scan lines. This arrangement results in a 'multiple pulse irradiation' scheme with a pulse density of 200 at a given spot. The laser energy densities are quoted for a single pulse. A series of laser energy densities were used to crystallise the samples from 40 to 160 $mJcm^{-2}$, in steps of 20 $mJcm^{-2}$. Due to pulse-to-pulse variation of excimer laser energy, the accuracy of energy density measurements is expected to be close to 10% [10].

Chromium back contacts were made to the films, after depositing a heavily phosphorus doped 50 nm thick layer of n-type a-Si:H (n^+ a-Si:H), during the post crystallisation device fabrication process. 500 μm diameter circular chromium contacts were evaporated through a shadow mask and n-type a-Si:H layers surrounding chromium contacts were etched using a plasma etch, in order to reduce leakage currents to a minimum.

Optical transmission measurements of the samples were done with a Camspec M330 UV-visible spectrophotometer, and Raman spectroscopy with a Renishaw 2000 Raman microscope with 782 nm excitation. Current–voltage measurements of the samples were performed using a Kiethley 487 Pico ammeter/Voltage source. EQEs of the devices were calculated from the measurement of spectral response of the devices with a Centronics OSD5.8-7 reference photodiode using a standard lock-in system without a bias light.

RESULTS & DISCUSSION

Material characteristics

Raman spectra for the EL crystallised silicon show a mixed phase silicon spectrum, with increasing crystalline intensities with increased laser energy density [11]. The crystalline volume fractions of the samples were calculated from the integrated intensities of the Raman peaks, with Gaussian fits for amorphous peaks and Lorentzian fits for crystalline peaks. Calculations were performed as proposed by Tsu et al. [12] with the ratio of the backscattering cross sections of amorphous and crystalline phases taken to be 0.8. The technique yields a rough estimate of crystalline to amorphous ratio in the film, where the amorphous contribution results mainly from the unconverted a-Si:H at the bottom layer and some grain boundary contributions. Figure 1 shows the variation of crystalline volume fraction and Tauc gap of EL crystallised silicon with series of laser energy densities. A maximum crystalline volume fraction of 73% resulted at 160 $mJcm^{-2}$. Crystalline volume gradually increases with the increase of EL energy density, which in turn increases the grain boundary density of the film. The amorphous contribution for these films is thought to be from the grain boundaries as well as the unconverted a-Si:H bottom layer.

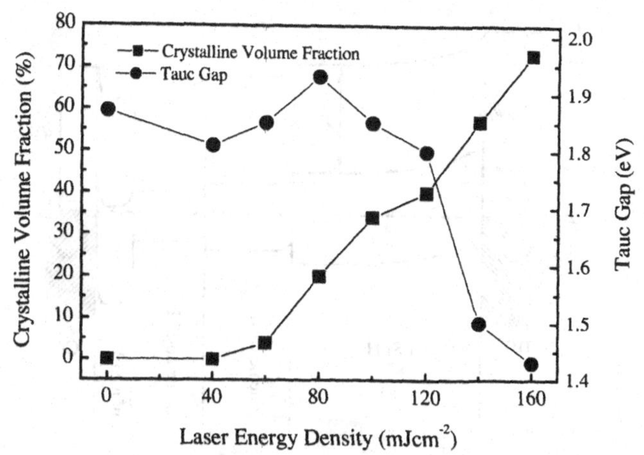

Figure 1. Variation of crystalline volume fraction and Tauc gap of EL crystallised silicon with laser energy density.

The Tauc gap of EL crystallised silicon was estimated by optical absorption measurements and is shown in Figure 1. Due to the stratified nature of the resulting films, the estimated optical gap is a representation of collective absorption of all three different nanocrystalline/amorphous layers. It can be seen that the influence of the amorphous fraction is minimal at laser energy densities above 140 mJcm^{-2}, which reduces the optical gap from initially higher values to lower 'composite' bansgaps.

It should be noted that the presence of thin oxide layers both between EL crystallised silicon and n$^+$a-Si:H layer, and between n$^+$a-Si:H and chromium back contact is possible. Although crystallisation is done in vacuum, it is expected that there is enough oxygen, for the films to react in the liquid phase. Oxide formation between n$^+$ a-Si:H and the chromium back contact, which occurs between n$^+$a-Si:H deposition and metal deposition, is less prominent. The oxide layers are assumed to be transparent to carriers, if present.

Electronic Characteristics

The electronic band structure of the devices was formulated based on the above findings. Figure 2 shows the proposed band structure for the devices under discussion. Depending on the crystallisation energy density, the depth of each type of nanocrystalline layer changes. At higher energy densities, the large grained layer will extend close to the ITO back contact, whereas at lower energy densities, most of the film will be amorphous. The large grained nc-Si:H layer at the laser irradiated surface is assumed to have a lower optical gap than the fine grained nc-Si:H layer sandwiched between this and the unconverted a-Si:H. The fine-grained nc-Si:H layer is expected to have a lower optical gap to a-Si:H. With the work function of ITO approximately 4.7 eV [13] and electron affinity of a-Si:H approximately 4 eV, it is suggested that the a-Si:H makes a Schottky contact to the ITO front contact [14]. The heavily phosphorous doped a-Si:H layer makes a very thin barrier to the Cr back contact, so that carriers tunnel through, resulting in a low

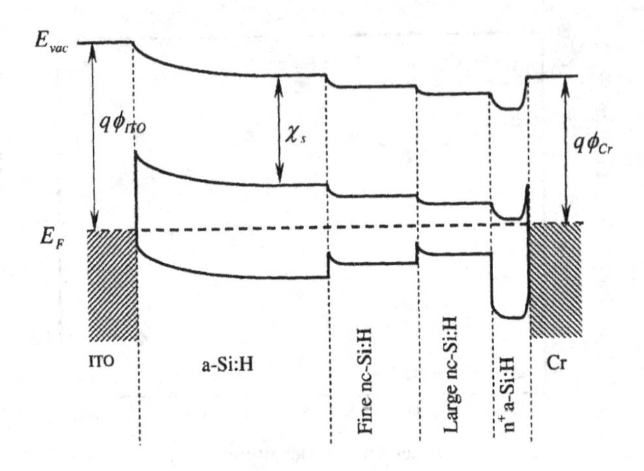

Figure 2. Band structure of EL crystallised silicon Schottky barrier cells indicating work functions of ITO (4.7eV) and Cr (4.5 eV) and electron affinity of a-Si:H (4.0 eV).

resistance back contact. The cascaded band arrangement is expected to aid electron transport upon photo-excitation. From current-voltage measurements of cells, it was observed that the rectifying behaviour of the device is lost after 120 mJ/cm^2 crystallisation energy density. This is thought to be due to considerable structural changes during laser crystallisation. Crystalline volume fractions of these cells are comparatively higher, suggesting increased grain boundary contributions, which increase the conductivity of the films, resulting in near ohmic contacts. Activation energy measurements conducted on these devices confirm this observation, with barrier heights for the ITO/a-Si:H interface calculated to be 0.7 eV for devices made below 120 mJ/cm^2. From the barrier height calculations, it is concluded that the ITO/a-Si:H interface is not significantly affected by the EL treatment up to 100 mJcm^{-2} and the collection mechanisms are similar for all the cells fabricated below this energy.

The EQE of each cell was calculated from the spectral response of the devices within the 300-800 nm wavelengths. Figure 3 shows the EQE plots for devices made with different laser energy densities. The EQE curves show similar spectral response at almost all crystallisation energy densities except for 140 and 160 mJcm^{-2}, where the collection mechanism is insignificantly weak. Three distinct peaks of EQE can be identified for the devices at 330 nm (3.75 eV), 380 nm (3.26 eV), and 475 nm (2.61 eV) wavelengths. It is suggested that the 3. 75 eV peak is due to ITO, where the band gap is around 3.5 eV. [15] The other two peaks are suggested to arise from extended state transitions of the disordered absorber layer. The best EQE corresponds to cells fabricated after crystallising at 40 mJcm^{-2}, with a maximum of 70%, at 475 nm (2.6 eV) of the spectrum. For non-laser treated a-Si:H cells, when the EQE was calculated, it was 30 % at the same wavelength. At 60 and 80 mJcm^{-2} the EQEs are higher than for the amorphous case, however, significantly lower than the 40 mJcm^{-2} devices. With the very thin absorber layers utilised for the cells, this EQE response is significant, compared to previously reported quantum efficiencies of a-Si:H Schottky barrier solar cells. [16]

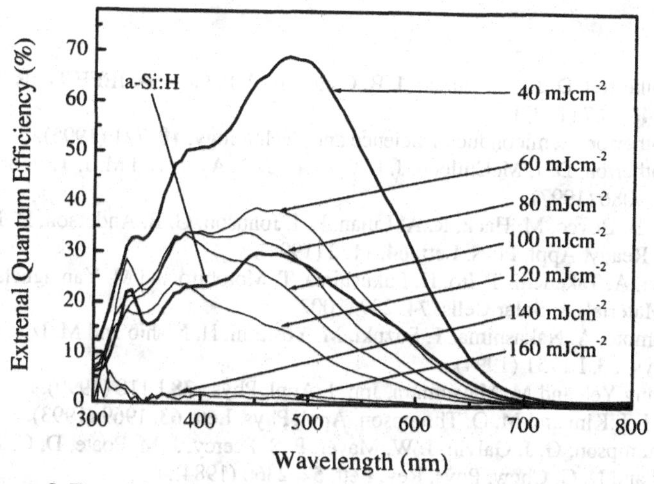

Figure 3. External Quantum efficiency of EL crystallised Schottky barrier cells.

From Figure 1, it is evident that the devices resulting in the highest EQEs show no significant crystalline volume. The whole absorber layer seems to be amorphous, with no apparent changes in optical gap. However, an ultra thin layer of nano crystallites is expected to form at 40 mJcm^{-2} irradiation. This ultra thin layer is expected to have a lower optical gap, higher mobility of carriers with a lower H content compared to rest of the absorber layer. Hence, it is proposed that the efficiency improvements are associated with the interface modification between the absorber layer and the back contact assembly.

CONCLUSION

Very thin Schottky barrier solar cells were fabricated using EL crystallised silicon and ITO on glass superstrates. At 150 nm absorber layer thickness, a maximum EQE of 70% is observed, whereas a-Si:H cells fabricated under the same conditions show 30% EQE. It is concluded that an ultra thin nanocrystalline layer at the back contact assembly from excimer laser treatment improved collection efficiency of photo-generated carriers.

ACKNOWLEDGEMENTS

The authors would like to thank Philips research laboratories, Redhill, Surrey, UK, for the supply of a-Si:H samples and post laser treatment fabrications and G.Y. Chen for Raman measurements. This work is supported by the Portfolio Partnership Award from EPSRC.

REFERENCES

1. S. D. Brotherton, D. J. McCulloch, J. B. Clegg and J. P. Gowers, IEEE Trans. Electron Devices, **40**, 407 (1993).
2. S. D. Brotherton, Semiconductor Science and Technology, **10**, 721 (1995).
3. S. D. Brotherton, D. J. McCulloch, J. P. Gowers, J. R. Ayres and M. J. Trainor, J. Appl. Phys. **82**, 4086 (1997).
4. P. Mei, J. B. Boyce, M. Hack, R. A. Lujan, R. I. Johnson, G. B. Anderson, D. K. Fork and S. E. Ready, Appl. Phys. Lett. **64**, 1132 (1994).
5. H. Azuma, A. Takeuchi, T. Ito, H. Fukushima, T. Motohiro and M. Yamaguchi Solar Energy Materials & Solar Cells, **74**, 289 (2002).
6. K. Yamamoto, A. Nakashima, T. Suzuki, M. Yoshimi, H. Nishio and M. Izumina, Jpn. J. Appl. Phys. **33**, L1751 (1994).
7. Wen-Chang Yeh and M. Matsumura, Jpn. J. Appl. Phys., **38**,L110 (1999).
8. J.S. Im, H. J. Kim and M. O. Thompson, Appl. Phys. Lett. **63**, 1969 (1993).
9. M. O. Thompson, G. J. Galvin, J. W. Mayer, P. S. Peercy, J. M. Poate, D. C. Jacobson, A. G. Cullis and N. G. Chew, Phys. Rev. Lett. **52**, 2360 (1984).
10. A. T. Voustas, Applied surface science, **208-209**, 250 (2003).
11. C. Smit, R. A. C. M. M. van Swaaij, H. Donker and A. M. H. N. Petit, W. M. M. Kessels, and M. C. M. van de Sanden, J. Appl. Phys., **94**, 3582 (2003).
12. R. Tsu, J. Gonzalez-Hernandez, S. S. Chao, S. C. Lee, and K. Tanaka, Appl. Phys. Lett. **40**, 534 (1982).
13. Y. Park, V. Choong, Y. Gao, B. R. Hsieh, and C. W.Tang, Appl. Phys. Lett. **68**, 2699 (1996).
14. Jerzi Kanicki, Appl. Phys. Lett. **53**, 1943 (1988).
15. H. Kim, C. M. Gilmore, A. Piqué, J. S. Horwitz, H. Mattoussi, H. Murata, Z. H. Kafafi, and D. B. Chrisey, J. Appl. Phys., 86, 6451 (1999).
16. D. Gutkowicz-Krusin, C. R. Wronski, and T. Tiedje, Appl. Phys. Lett. 38, 87 (1981).

Mater. Res. Soc. Symp. Proc. Vol. 836 © 2005 Materials Research Society L8.3

Harvesting Betavoltaic and Photovoltaic Energy with Three Dimensional Porous Silicon Diodes

Wei Sun[1], Nazir P. Kherani[2], Karl D. Hirschman[3], Larry L. Gadeken[4], Philippe M. Fauchet[1,5]
[1]Department of Biomedical Engineering, University of Rochester, Rochester, New York 14627, USA
[2]Department of Electrical and Computer Engineering, University of Toronto,
Toronto, Ontario M5S 3G4, Canada
[3]Microelectronic Engineering Department, Rochester Institute of Technology,
Rochester, New York 14623, USA
[4]BetaBatt, Inc., 12819 Westleigh Drive, Houston, Texas 77077, USA
[5]Department of Electrical and Computer Engineering, University of Rochester,
Rochester, New York 14627, USA

ABSTRACT

Conventional two-dimensional *p-n* diodes can be used for betavoltaic and photovoltaic energy conversion, but the device efficiency is limited by the planar geometry. We propose and demonstrate a novel three-dimensional diode geometry based on porous silicon. The 3D pore array provides two very important features: (1) the storage of the radioisotope energy source and (2) its extreme proximity to the *p-n* junction on each pore wall. The particle energy losses are thereby minimized prior to entering the conversion layer. In betavoltaics, our 3D betavoltaic device efficiency is 10 times that of a similar planar device. In photovoltaics, photons play the role of beta particles and photon trapping inside the pores enhances the conversion efficiency. Further fabrication and geometry optimization can result in practical, high performance devices.

INTRODUCTION

As hydrocarbon resources dwindle and environmental concerns increase, utilization of abundant nuclear energy and renewable solar energy becomes more and more attractive. A material that can convert radioactive or photonic energy directly to electricity is desired for a practical energy conversion device. Following the discovery in 1954 [1] that *p-n* junctions can generate electric current from beta and alpha particles emitted from radioactive materials, research studies have been carried out to investigate various materials and techniques for the construction of efficient and practical nuclear batteries [2-4].

In betavoltaics, the charged particles emitted by unstable nuclei lose their kinetic energy by ionizing the surrounding material. Power is generated by the separation of electron-hole pairs under the influence of the built-in electric field of the junction. The classic planar geometry requires volumetric separation of the energy source and conversion layer that results in inherent limits to the conversion efficiency.

We have developed a direct energy conversion (DEC) chip that intimately combines the energy reservoir and the conversion layer using the 3D character of porous silicon (PSi). PSi is a material containing a network of pores formed by electrochemical anodization of silicon substrate. The resulting large internal surface area retains most of the characteristics associated with planar silicon [5-6]. Recent research has established many physical and chemical properties of this complex material [5-8]. It also has been demonstrated that PSi components can be integrated into microelectronic circuits for practical devices [9]. Further, PSi has been investigated for photovoltaics, however only as an antireflection and surface passivation layer [10].

EXPERIMENTAL DETAILS

3D PSi diode fabrication

Figure 1 illustrates the process of fabricating 3D PSi diodes. Boron doped p<100> silicon wafers with a resistivity of 20-30 Ω-cm were employed as substrates. The macroporous silicon with pore dimension of 0.8-2 μm was generated using an anodization process in a custom-designed etching cell. An electrolyte of 4 wt. % hydrofluoric acid (HF) in dimethylformamide (DMF) was used for the anodization. The wafers were etched using a current density of $2 mA/cm^2$ for 180 minutes to produce a PSi layer thickness of approximately 50 μm.

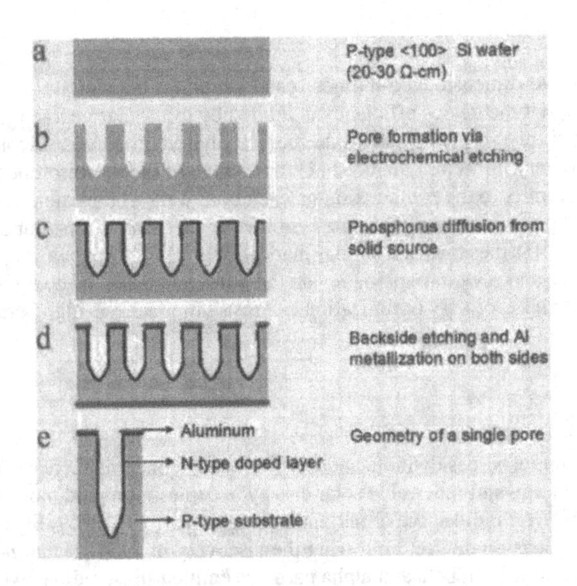

Figure 1. 3D PSi diode fabrication procedure.

After standard RCA cleaning was carried out to eliminate metal and organic species contamination, p-*n* junctions were introduced into the pore surfaces of the PSi layer by solid source diffusion. The solid P_2O_5 disk source was activated in a 1000 °C, nitrogen atmosphere for 18 hours. The silicon wafers were soaked for 9 minutes in the same conditions to make a shallow *n*+ layer (~200 nm). Subsequently, the *n*+ layer formed on the backside of silicon wafer was removed by SF_6 plasma etching. After the samples were dipped in buffered oxide etchant, 50 nm and 200 nm thick aluminum layers were evaporated on the porous side and backside respectively under high vacuum (< 10^{-7} Torr). The samples were sintered at 420 °C for 15 minutes. The wafers were then cleaved into 1 in. × 1 in. (2.54 cm × 2.54 cm) square chips and rinsed in acetone, isopropyl alcohol and deionized water for further tests.

For comparison, planar silicon diodes were constructed in the same manner except that the pore formation step was skipped.

Geometry Characterization

The pore geometry of the PSi layer was visualized using high-resolution Scanning Electron Microscopy (SEM). SEM viewgraphs are shown in figure 2. The average area around each pore and average pore diameter were estimated by measuring the distribution of pore diameters in a 15 x 22.5 μm^2 area. The average pore diameter, d_{avg}, was $0.837 \pm 0.125 \mu m$ and the average pore depth, h_{avg}, was $43.1 \pm 1.18\ \mu m$ estimated from the SEM top view and cross-sectional view, respectively. The fraction of the surface containing the pores was 0.31 and the remaining planar surface fraction was 0.69.

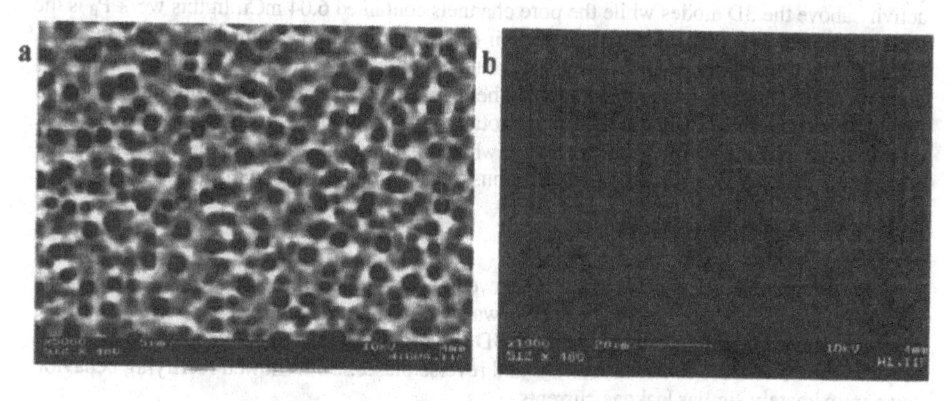

Figure 2. SEM images of device geometry.

Since the targeted phosphorus concentrations were in the range $N_d = 10^{17} \sim 10^{20}\ cm^{-3}$, the depletion width on the "p" side was estimated to be 1.32-1.49 μm, while the depletion width on the "n" side was negligible.

Tritium Loading

Both porous silicon and planar silicon *p-n* diodes were exposed to tritium within the confines of the Chip Test Fixtures (CTFs). Each assembled CTF consisted of two steel plates bolted together which served to clamp a copper fill pipe, an indium wire seal ring, and a diode chip tightly together. An insulator electrically isolated the front and backsides of the test chip from the top and bottom plates. The front-side *p-n* junctions faced the fill pipe. Each completed CTF was installed on a high vacuum system and standard techniques used to confirm vacuum leak tightness of less than 10^{-8} atm·cm^3/s. Thereafter, 232 mCi of tritium gas at a pressure slightly above 1 atm was loaded into each CTF. The CTFs were separated from the Tritium Handling System by cold-weld pinching of the Cu fill pipe at the mid point.

The analysis of the pore geometry together with the area enclosed by the indium seal in each CTF gave the estimated number of pores exposed to tritium as 103 million. The corresponding total internal surface area was 116 cm^2 and the pore volume was 2.43×10^{-3} cm^3.

Device Performance Measurements

The working performance of the *p-n* diodes fabricated in this project was investigated through *I-V* measurements. Samples were placed within a desktop dark box connected to an HP 4145B Semiconductor Parameter Analyzer. Photo response measurements used illumination either from fluorescent room lighting or a 25 W collimated light source.

Data analysis of both betavoltaic and photovoltaic devices is very similar other than the source of energy and generation of excess carriers. The energy conversion efficiency, $\eta = P_{ex}*FF/P_{\beta}$. was determined from the *I-V* curves of the devices [3,11]. The available tritium beta power per unit radioactivity is 33.7 μW per curie (Ci). In each CTF, there was 106 mCi of tritium activity above the 3D diodes while the pore channels contained 6.04 mCi. In this work P_{β} is the total available power contained in the reservoir (pore) volume. We assumed that the planar surface of the PSi diodes performed the betavoltaic energy conversion identically to the 2D planar diodes. Thus, the power developed in the pores could be estimated by subtracting the power developed in the planar area from the total power developed from the porous silicon diode. Hence, this data provides the extent to which the solid angle of 4π contributes to the energy conversion process in the pores, and thus the potential utilization of the available power.

RESULTS AND DISCUSSION

The *I-V* characteristics of our samples showed good *p-n* diode properties and demonstrated significant photo response before tritium gas was loaded. Figure 3a shows the overall diode character of both planar and PSi chips. Both 3D PSi samples and planar silicon controls show very similar *I-V* characteristics in forward and reverse biases. Both showed rectifying behavior and approximately similar leakage currents.

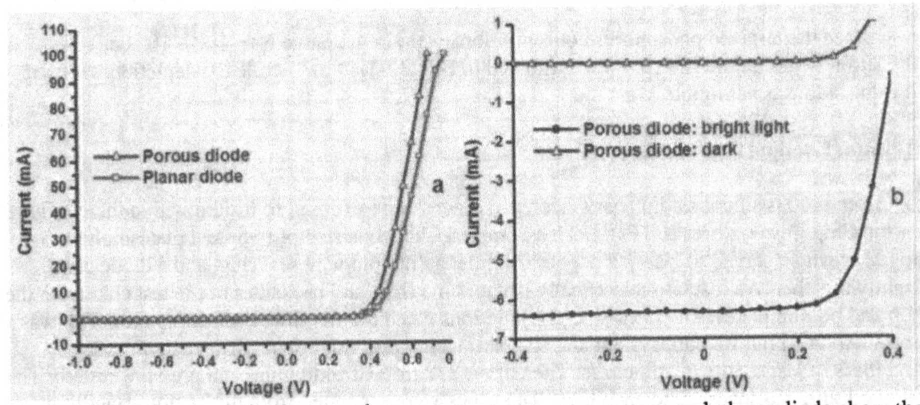

Figure 3. *I-V* characteristics measured at room temperature. *a*, porous and planar diode show the same *I-V* characteristics in the dark. *b*, striking photovoltaic effect in the porous diode.

Under illumination, the 50 nm aluminum front contact blocked light from reaching the *p-n* junctions on planar chips and the only response observed was a very small offset due to light entering at the clean cleaved edges. By contrast, the PSi chips showed a very pronounced photo response contributed by photo conversion in the *p-n* junctions along the pore walls. As shown in figure 3b, the reverse current significantly increased when the device was exposed to a collimated light source.

Representative *I-V* responses for a porous and a planar diode each installed within a CTF are shown in figure 4 before and after loading with tritium gas. Although the "knee" apparent in figure 3a for the "bare" samples was less distinct, evidently due to unexpected resistive effects in the CTFs, the overall diode character was preserved. The presence of tritium displaces the *I-V* response curve into the 4th quadrant indicating that electrical power is being generated.

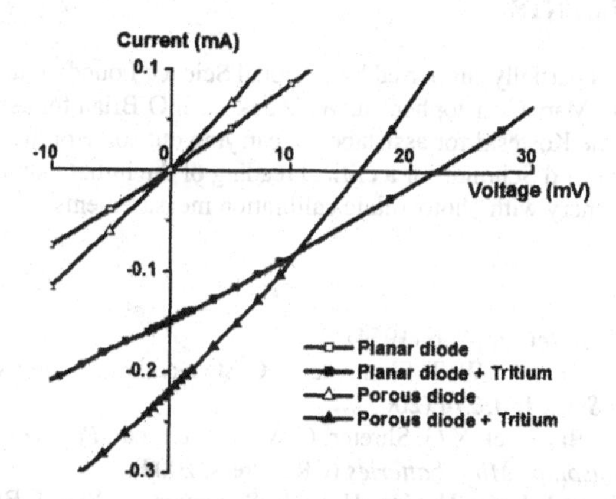

Figure 4. *I-V* characteristics measured with diodes in the CTFs.

The average observed betavoltaic energy conversion efficiencies were 0.023% for 2D diodes and 0.22% for the pore channels of the porous diodes, respectively. This order of magnitude increase in efficiency in the pore channels corresponds to the almost unity probability that tritium decay electrons enter the betavoltaic conversion layers (*p-n* junctions) in the pore walls. Given that the length to diameter ratio of the pores is ~60:1, whatever the direction of flight, the average path length prior to striking the pore walls is no more than 2 μm. Moreover, the tiny pore diameter eliminates self absorption in the tritium gas. Although it is estimated that ~⅓ of the incident betas may be elastically backscattered [3], this loss is negligible in the pores owing to the close proximity of "adjacent" betavoltaic junctions.

Preliminary investigation of the photovoltaic characteristics of the first 3D porous *p-n* diodes was performed with the assistance of the National Renewable Energy Laboratory. Assuming that only the pore channels of the 3D diodes contribute to the photo response, the NREL measurements show that the photovoltaic conversion efficiency was 8%. With appropriate geometry modifications and diode performance improvements, it is reasonable to expect the resulting photo conversion efficiencies can be more than twice this value.

CONCLUSIONS

We have demonstrated that 3D porous silicon diodes can be constructed with standard semiconductor industry fabrication techniques. This geometry increases the betavoltaic power generation efficiency 10 times compared to planar diode devices. It is expected that an increase in the radioisotope energy density in the pores using high density tritiated compounds along with improvements in device quality will yield a far more efficient, long life betavoltaic device. Qualitative evaluation of the photovoltaic response of 3D diodes suggests that this geometry may provide significant benefits for solar cells, photo detectors and other devices for which light-semiconductor interactions are important.

ACKNOWLEDGEMENTS

This work was partially supported by National Science Foundation. We benefited from communications with Marc Christophersen. We thank Sean O'Brian for assistance in solid source diffusion, Tome Kosteski for assistance in carrying out some of the device measurements, Wade Adams and Howard Schmidt for a critical reading of the initial manuscript, as well as the assistance of Keith Emery with photovoltaic calibration measurements.

References

1. P. Rappaport, *Phys. Rev.* **93**, 246 (1954).
2. T. Kosteski. N. P. Kherani, W. T. Shmayda, S. Costea, and S. Zukotynski, *IEE Proc. – Circuits Devices Syst.*, **150**, 274 (2003).
3. K.E. Bower, Y.A. Barbanel, Y.G. Shreter, G.W. Bohnert, ed., *Polymers, Phosphors, and Voltaic for Radioisotope Microbatteries* (CRC Press, 2002).
4. A. Lal, R. M. Bilbao Y León, H. Guo, H. Li, S. Santanam, R. Yao, J. Blanchard, and D. Henderson, *A nuclear microbattery for MEMS devices* (Proc. 9th International Conference on Nuclear Engineering, Nice, **2001**).
5. L. Canham, ed., *Properties of porous silicon,* (IEE/INSEPC, 1997).
6. O. Bisi, S. Ossicini, L. Pavesi, *Surf. Sci. Rep.* **38**, 1(2000).
7. V. Lehmann, S. Ronnebeck, *J. Electrochem. Soc.* **146**, 2968 (1999).
8. M. Christophersen, J. Carstensen, A. Feuerhake, H. Foll, *Mat. Sci. Eng. B* **2000**, *69-70*,194.
9. K. D. Hirschman, L. Tsybeskov, S. P. Duttagupta, P. M. Fauchet, *Nature* **384**, 338 (1996).
10. C. C. Striemer, P. M. Fauchet, *Appl. Phys. Lett.* **81**, 2980 (2002).
11. S. M. Sze, *Physics of Semiconductor Devices* (Wiley, ed. 2, 1981).

Mater. Res. Soc. Symp. Proc. Vol. 836 © 2005 Materials Research Society L8.9

Density of States in Tritiated Amorphous Silicon Measured Using CPM

Simone Pisana†, Stefan Costea, Tome Kosteski, Nazir P. Kherani, Stefan Zukotynski and Walter T. Shmayda††
Department of Electrical & Computer Engineering, University of Toronto
Toronto, ON M5S 3G4, Canada
† Present address: Department of Engineering, University of Cambridge, Trumpington Street, Cambridge, CB2 1PZ, UK
†† Laboratory for Laser Energetics, University of Rochester, Rochester, NY 14623-1299, USA

ABSTRACT

The constant photocurrent method has been used to obtain the density of occupied electronic states of tritiated amorphous silicon thin films. The analyses showed a peak of defects located 1.24 eV below the conduction band edge, suggesting that the main type of defect present in the films was a doubly occupied dangling bond. The concentration of defect states increases as a result of tritium decay by about two orders of magnitude over a period of 500 hours. The defect density in the tritiated amorphous silicon samples could be reduced by thermal annealing, after which it increased once more.

INTRODUCTION

Hydrogenated amorphous silicon (a-Si:H) has been studied extensively since the pioneering work of Spear and Le Comber in the seventies [1]. Amorphous silicon today is an integral material for a variety of large area optoelectronic and electronic applications. Yet, amorphous silicon continues to be of research interest because of its metastable character and associated influence on the optoelectronic properties of the material [2]. Under illumination, a-Si:H experiences an increase in its defect density [3] and accordingly deterioration in material properties. The dynamics of defect states have been studied widely and various models describing the same have been proposed [3,4].

Tritium is a radioactive isotope of hydrogen that can be easily incorporated in the a-Si:H network during film deposition [5]. The decay of tritium ($T \rightarrow {}^3He^+ + \beta^-$, $t_{1/2} = 12.3$ yrs) leads to the creation of dangling bonds as well as the decrease in total hydrogen content. This has been used to study dangling bonds in a-Si:H [6]. Energetic beta particles released in the process of tritium decay create excess carriers. Excess carriers are separated when generated in the intrinsic layer of a *p-i-n* junction device. This configuration has been used to demonstrate a self-powered a-Si:H based betavoltaic device [7].

In this work, tritium decay is used to generate dangling bonds at a predictable rate in tritiated amorphous silicon (a-Si:H:T) and thus provide a tool for the study of defect reactions and to investigate the susceptibility of material properties to the creation of defects.

EXPERIMENT

a-Si:H:T was deposited using the dc saddle-field plasma deposition technique [8]. Tritiated films were deposited on Corning 7059 glass and crystalline silicon wafer substrates at a pressure of 150 mTorr. The samples were placed on a substrate holder and elevated to temperatures of 200°C and 250°C. The gas mixture composition was 5:4:1 of silane:hydrogen:tritium, at a total flow rate of 10 sccm. High purity tritium is generated by heating a uranium tritide bed. [5]. Tritium is introduced into the deposition chamber via a pin hole which is calibrated for flow as a function of tritium pressure. The samples' deposition parameters and film characteristics are summarized in Table I.

Table I – Sample Deposition Conditions and Film Characteristics

Sample #	Substrate Temperature [°C]	Pressure [mTorr]	SiH_4 Flow Rate [sccm]	H_2 Flow Rate [sccm]	T_2 Flow Rate [sccm]	Anode Current [mA]	Film Thickness [nm]	Tauc Gap [eV]
561	250	150	5	4	1	20	200	1.815
805	200	150	5	4	1	14	260	1.946

Tritium effusion was used to determine the bonded tritium content [7]. Effusion was carried out at a temperature of 800°C in an argon atmosphere at a pressure of 900 mTorr. Tritium was released directly into an ionization chamber which measured the tritium beta ionization current and hence the released tritium. Tritium content of the samples examined in this study was of the order of 1.5 at.%.

The constant photocurrent method (CPM) [9] was employed to measure the electronic density of states (DOS) from the valence band edge to the mid-gap at room temperature. The apparatus included a 100 Watt halogen lamp, a CVI Spectral Products CM110 monochromator followed by a crystalline silicon/CG 610 order sorting filter assembly, and a Si/InGaAs sandwiched pair photodetector. The photocurrent was measured with a Keithley 6517 electrometer and the CPM feedback loop was computer controlled. For the CPM measurements, coplanar aluminum ohmic contacts were evaporated on top of the a-Si:H:T thin film which was deposited on glass.

RESULTS AND DISCUSSION

CPM measurements were carried out 2.5 years after the samples were prepared. The samples were stored under dark and dry conditions at room temperature. After the initial measurements, the samples were annealed at 120°C for 1 or 2 hours and then measured again. Sample 805 was repeatedly measured to observe its aging behavior. Figures 1 and 2 show the dispersion of the optical absorption and the density of states plots derived thereof. The interruption in the DOS curves originates from the calculations used to obtain the DOS from the absorption data [10].

The large absorption and large mid-gap defect state density are indicative of a large number of defects present in the unannealed aged material. This is consistent with the long period of time during which the tritium decayed. The defect density and absorption can be reduced by thermal annealing and is found to be proportional to the annealing period. The total number of defect states, obtained by integrating the peak in the DOS curve from near the valence band edge to near the mid-gap, was found to change as a result of annealing as follows ('DOS before annealing' to 'DOS after annealing'): 6.8×10^{19} to 5.7×10^{18} cm^{-3} for sample 561 and 5.1×10^{19} to 1.4×10^{17} cm^{-3} for sample 805. Following thermal annealing the defects in the a-Si:H:T network increase again as shown in Figure 3, eventually saturating within about 3 weeks at levels similar to those found after 2.5 years of aging. The post-anneal DOS aging curve evolution appears to follow the radioactive decay of tritium albeit larger in magnitude.

The position of the peak of the defect states in the band gap suggests that the majority of the defect states are doubly occupied dangling bonds (D^-) [11,12,13]. The presence of the doubly occupied dangling bond defects at such concentration levels is supported by the relatively low values of activation energy measured for the samples, ~0.6 eV, and by the presence of excess electrons originating from the tritium decay [14]. In fact, on average each energetic beta particle produced by each tritium decay produces ~1500 electron-hole pairs.

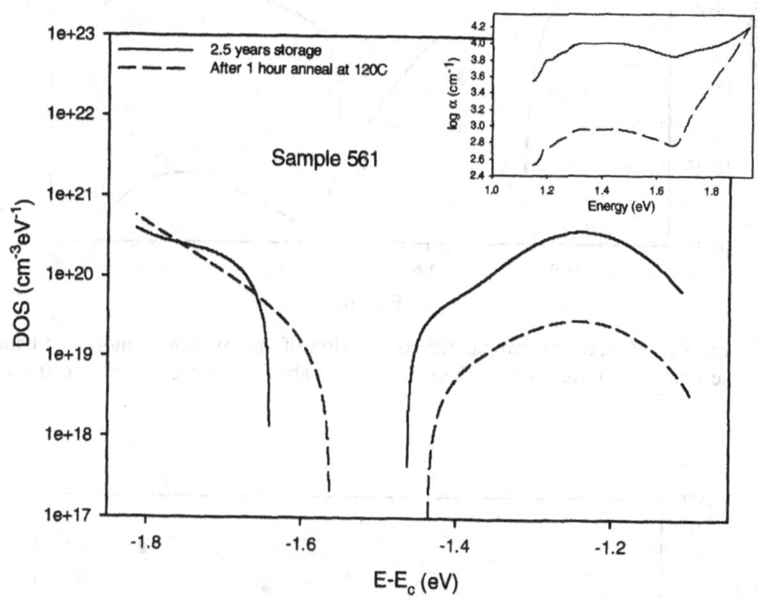

Figure 1 – Density of states of sample 561 as function of energy from conduction band (0 eV represents the conduction band edge). Inset shows the absorption coefficient spectrum derived via CPM.

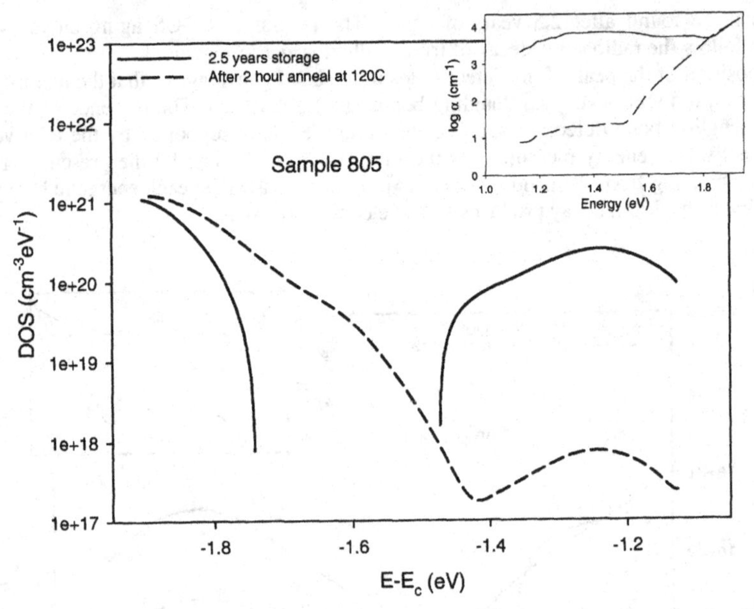

Figure 2 – Density of states of sample 805 as function of energy from conduction band (0 eV represents the conduction band edge). Inset shows the absorption coefficient spectrum derived via CPM.

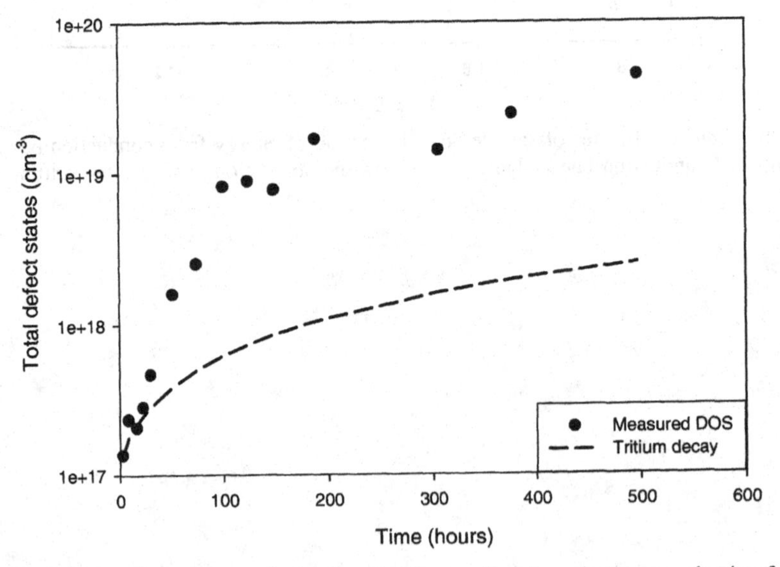

Figure 3 – Evolution of the integrated defect density as a function of post-anneal aging for sample 805.

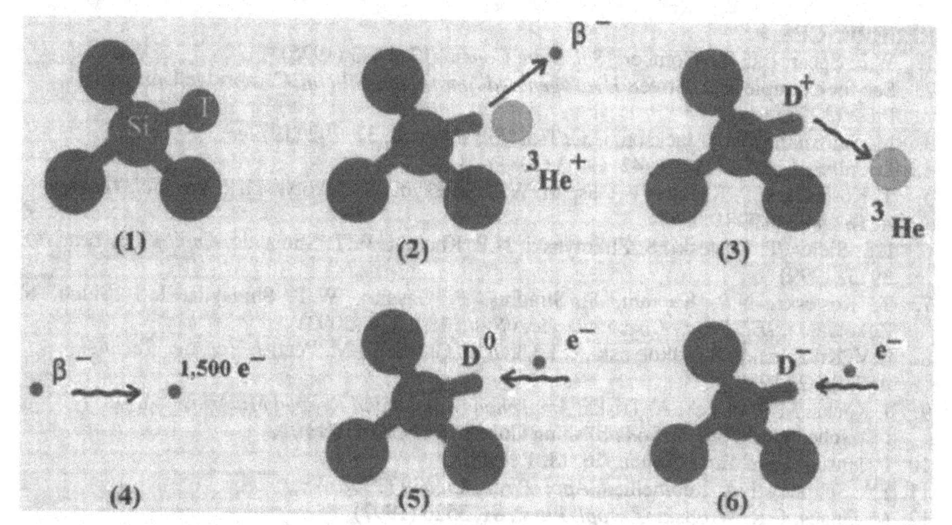

Figure 4 – Proposed D^- defect creation mechanism (the sequence of events is indicated by the number in the brackets).

In view of the large presence of D^- states, we propose the following defect creation mechanism for a-Si:H:T with reference to the illustration in Figure 4. The typical arrangement found in the amorphous network near a T bonding site is as shown in Figure 4-1 with a center Si atom bonded to three Si atoms and one T atom. Upon tritium decay, the T atom transmutates into a positively charged He ion while an energetic beta particle is emitted (Figure 4-2). The He ion, given its high electron affinity, takes an electron from the previous bonding site thereby now becoming chemically inert and leaves behind a D^+ dangling bond (Figure 4-3). The beta particle on the average excites about 1,500 electrons through inelastic scattering (Figure 4-4). One of these electrons has a likelihood of being trapped by the D^+ state, changing it to a neutrally charged D^0 state (Figure 4-5). Given the high concentration of excess electrons, it is likely that a second electron also may be trapped at the site, changing it to a D^- state (Figure 4-6). At thermal equilibrium there will be a mix of defect state occupancies.

CONCLUSIONS

CPM measurements of a-Si:H:T provide evidence of a high concentration of doubly occupied dangling bonds. The associated defect state density increases with time as a result of tritium decay. This defect density can be reduced by thermal annealing. The measured change in defect density varies by more than two orders of magnitude over a period of three weeks. The high concentration of negatively charged dangling bonds is attributed to the abundance of excess electrons generated during tritium decay.

REFERENCES

1. W.E. Spear, P.G. Le Comber, *Sol. State Comm.*, **17**, 1193 (1975)
2. See for example R.A. Street, *Hydrogenated Amorphous Silicon*, Cambridge University Press (1991)
3. M. Stutzmann, W.B. Jackson, C.C. Tsai, *Phys. Rev. B*, **32**, 292 (1977)
4. D. Adler, *J. Phys. (Paris)*, **42**, C4, 3 (1981)
5. T. Kosteski, N.P. Kherani, F. Gaspari, W.T. Shmayda, S. Zukotynski, *J. Vac. Sci. Technol. A*, **162**, 893 (1998)
6. L.S. Sidhu, T. Kosteski, S. Zukotynski, N.P. Kherani, W.T. Shmayda, *Appl. Phys. Lett.*, **74**, 3975 (1999)
7. T. Kosteski, N.P. Kherani, P. Stradins, F. Gaspari, W.T. Shmayda, L.S. Sidhu, S. Zukotynski, *IEE Proc.- Circuits Devices Syst.*, **150**, 274 (2003)
8. R.V. Kruzelecky, S. Zukotynski, C.I. Ukah, F. Gaspari, J.M. Perz, *J. Vac. Sci. Technol. A*, **7**, 2632 (1989)
9. J. Kocka, M. Vanecek, A. Triska, *Amorphous Silicon and Related Materials*, ed. H. Fritzsche, World Scientific Publishing Co., pp. 297-327 (1988)
10. P. Jensen, *Solid State Comm.*, **76**, 1301 (1990)
11. J.A. Schmidt, F.A. Rubinelli, *J. Appl. Phys.*, **83**, 339 (1998)
12. M. Gunes, C.R. Wronski, *J. Appl. Phys.*, **81**, 3526 (1997)
13. M. Gunes, C.R. Wronski, T.J. McMahon, *J. Appl. Phys.*, **76**, 2260 (1994)
14. S. Costea, F. Gaspari, T. Kosteski, S. Zukotynski, N.P. Kherani, W.T. Shmayda, *Mat. Res. Soc. Symp. – Proc.*, **609**, A2741 (2000)

AUTHOR INDEX

Adikaari, A.A. Damitha T., 279
Aernouts, Tom, 81
Ahn, Seung Eon, 113
Ahn, Young Rack, 107
Akimoto, D., 197
Anderson, Wayne A., 203
Andre, Carrie L., 211
Assender, Hazel E., 43

Baba, T., 191
Bae, Chang Hyun, 113
Bailey, Sheila G., 55
Balkus Jr., Kenneth J., 137
Banerjee, D., 23
Baroughi, M. Farrokh, 273
Baur, C., 223
Bett, A.W., 223
Bindu, K., 173
Bornais, Bryan, 89
Brand, J.I., 185
Brett, Michael J., 131
Briggs, G. Andrew D., 43
Burlakov, Victor M., 43

Campos, J., 167
Caruso, A.N., 185
Castracane, J., 61
Castro, Stephanie L., 55
Chan, Wai Kin, 149
Chandrasekaran, N., 241
Chen, G., 23
Chochos, Christos L., 143
Choi, Michael, 3
Chun, Young-Gab, 119, 125
Collins, Reuben, 259
Compaan, A.D., 179, 265
Costea, Stefan, 291
Cousins, Michael A., 161

Dames, C., 23
Dimroth, F., 223
Djurišić, Aleksandra B., 149
Dohrman, Carl L., 235
Drayton, J., 265

Economopoulos, Solon, 143
Efstathiadis, Harry, 49
Eichhorn, S. Holger, 89
Euler, W.B., 61
Evans, Chris M., 55

Fauchet, Philippe M., 285
Feldman, Scott, 259
Ferraris, John, 69
Fitzgerald, Eugene A., 211, 235
Forbes, Ian, 155
Fox, Nicholas, 89
Friend, Richard H., 93

Gadeken, Larry L., 285
Georg, Andreas, 35
Georg, Anneke, 35
Gish, Douglas A., 131
Govaris, Giannis K., 143
Grätzel, Michael, 11
Gregoriou, Vasilis G., 143
Grimes, Craig A., 29
Grovenor, Christopher R.M., 43
Gunn, A.S., 185
Gupta, A., 179, 265
Gupta, Saurabh, 235

Ha, Jeong Sook, 113
Haldar, Pradeep, 49
Hammond, Paula T., 17
Henry, Bernard M., 43
Heremans, Paul, 81
Hirahara, H., 191
Hirschman, Karl D., 285
Huang, J.Y., 23
Huang, Susan, 49
Hynes, Kathleen M., 155

Iida, T., 191, 197
Imaizumi, M., 241
Inoue, Kanzan, 69
Inuzuka, Y., 241
Isaacson, David M., 211, 235

Jensen, Martin O., 131
Ji, Chunhai, 203
Jimbo, T., 241

Jo, Seong Mu, 107

Kallitsis, Joannis K., 143
Kano, Mitsuru, 43
Kaydanov, Victor, 259
Kearney, Michael J., 279
Ketipearachchi, Udaya S., 161
Kherani, Nazir P., 285, 291
Kiema, Gregory K., 131
Kim, Dong Yong, 101, 107
Kim, Gyu Tae, 113
Kim, Hyunsuk, 113
Kim, Joon-Dong, 203
Kim, Kang-Jin, 101
Kim, Ki-Hyun, 119, 125
Kim, Sangsig, 113
Kirov, Kiril R., 43
Kosteski, Tome, 291
Kunwar, S., 23
Kwong, Chung Yin, 149

Landi, Brian J., 49, 55
Lane, David W., 155, 161
Lee, Hee-Gyoun, 49
Lee, Jong Soo, 113
Lee, S.H., 179
Lee, Sergey B., 69
Levitsky, I.A., 61
Losovyj, Y.B., 185
Lowman, Geoffrey M., 17
Lueck, Matthew, 211
Lunca-Popa, P., 185
Lutkenhaus, Jodie L., 17

Macias-Guzman, Minedys, 137
Madakasira, Pallavi C., 69
Man, Kitty Ka Yan, 149
Mandal, Krishna C., 3
Messina, S., 167
Miles, Robert W., 155
Mor, Gopal K., 29
Müllen, Klaus, 93

Nair, M.T.S., 167, 173
Nair, P.K., 167, 173
Nanba, H., 197
Noblitt, Caleb, 3
Nose, A., 197

Ohno, Tim, 259
Opara-Krasovec, Ursa, 35

Painter, Jonathan D., 155, 161
Parikh, V., 265
Park, Byung-Ok, 119, 125
Park, Seung Min, 113
Pasquier, Aurelien Du, 75
Paulose, Maggie, 29
Peña, Y., 167
Pisana, Simone, 291
Pitera, Arthur J., 211, 235
Poortmans, Jef, 81
Poudel, B., 23

Raffaelle, Ryne P., 49, 55
Rauh, R. David, 3
Ren, Z.F., 23
Ringel, Steven A., 211
Rodríguez-Lazcano, Y., 167
Rogers, Keith D., 155, 161
Romero, M.J., 249
Ruf, Herbert J., 55

Saito, K., 197
Sakuragi, G., 197
Sakuragi, S., 191, 197
Schmidtke, Johanna P., 93
Schmidt-Mende, Lukas, 11
Schöne, J., 223
Shankar, Karthik, 29
Shannon, John M., 279
Shimazaki, T., 197
Shmayda, Walter T., 291
Silva, S. Ravi P., 279
Sivoththaman, S., 273
Soga, T., 241
Song, Mi Yeon, 101, 107
Sugiyama, Y., 191
Sun, Wei, 285

Taguchi, H., 241
Takanashi, Y., 191, 197

Tokranova, N., 61
Tokuhisa, Hiroaki, 17
Topic, Marko, 35
Tsukahara, Yusuke, 43

Ulbricht, Ross, 69

Vanlaeke, Peter, 81
Varghese, Oomman K., 29
Vasko, A., 265

Wang, D.Z., 23
Wang, S.L., 179, 265
Wang, W.Z., 23
Wentzel, Paul, 75
Wilt, David M., 211
Wong, Hei Ling, 149

Xie, Zhibin, 43
Xiong, Chunrong, 137
Xu, B., 61

Yiannoulis, Panagiotis, 143
Yoon, Kyung-Hoon, 119, 125

Zakeeruddin, Shaik M., 11
Zakhidov, Anvar A., 69
Zhou, Miaoxin, 69
Zukotynski, Stefan, 291

SUBJECT INDEX

absorber films, 167
amorphous silicon solar cells, 279
anatase nanotubes, 23
anodization, 29
antimony chalcogenide, 167

betavoltaics, 285
bulk heterojunction solar cell, 81

CdMnTe, 179
CdS, 161
CdTe, 259
chemical
    bath deposition, 161, 173
    deposition, 167
CIGS, $CuInGaSe_2$, 119
columnar, 89
concentrator solar cells, 223
constant photocurrent method, 291
copolymer, 143
copper, 49
    phthalocyanine, 61
crystal growth, 191
Cu-In, 155
$CuInGaSe_2$, CIGS, 125
current-transport, 203

die cast, 191
die-casting growth, 197
discotic liquid crystals, 89
dislocation nucleation, 235
dye-sensitized solar cell(s), 3, 17,
        35, 107, 131

electrodeposition, 49
electroless deposition, 155
electroluminescence, 259
electron irradiation, 241
electrospinning, 107
excimer laser, 279

GaAs/Si, 241
glancing angle deposition, 131
grain boundary, 273

heterojunction, 273
hexabenzocoronene, 93
hybrid solar cells, 61
hydrothermal method, 23

lightweight and flexible, 3

MEHPPV, 101
mercury cadmium telluride, 265
metal-induced growth, 203
metamorphic, 223
multi-junction solar cells, 223

nanocrystalline titanium dioxide, 3
nanofiber, 137
nanoparticles, 119, 125
nanophase separation, 143
nanorod, 49, 113

oligomers, 89
organic photovoltaics, 101

PCBM, 69, 75
pH dependence, 161
phase-segregated composite
    interfaces, 69
photoconductivity, 43
photocurrent, 113
photoemission, 185
photoluminescence, 259
photolysis, 29
photosensitizer, 149
photovoltaic(s), 143, 185, 249, 273,
        285
    cells, 149
    diode, 93
    structures, 173
P_3HT, 75
plasticizer, 75
poly(3-hexylthiophene), organic
    photovoltaic, 69
polyelectrolyte multilayer, 17
polymer solar cell, 55
poly-Si, 203
porous
    diode, 285
    silicon, 61

PPV-C_{60}, 81
purity and crystallinity, 23

quantum
  dot, 55
  efficiency, 279

Raman spectroscopy, 179
RF sputtering, 179, 265
ruthenium dye, 11

scanning tunneling luminescence, 249
screen printing, 81
SiGe substrates, 211
silicon(-), 197, 291
  germanium, 191
    alloy, 235
silver antimony selenide, 173
single wall carbon nanotube, 55
sol-gel, 43
  synthesis, 35
solar cell(s), 11, 119, 125, 137, 155, 197, 241
STM, 249
surface
  energy, 93
  photovoltage, 185
switchable windows, 35

tandem solar cell, 265
III-V
  multi-junction photovoltaics, 211
  on Si integration, 235
III-V/Si integration, 211
TiO_2, 11, 17, 101, 137
  nanofiber, 107
titania, 29
titanium dioxide, 43, 131
transition metal complex, 149
tritium, 291

ZnO, 113

Printed in the United States
By Bookmasters